The Golgi Apparatus and the Plant Secretory Pathway

Annual Plant Reviews

A series for researchers and postgraduates in the plant sciences. Each volume in this annual series will focus on a theme of topical importance and emphasis will be placed on rapid publication.

Editorial Board:

Titles in the series:

1. Arabidopsis
Edited by M. Anderson and J. Roberts

2. Biochemistry of Plant Secondary Metabolism
Edited by M. Wink

3. Functions of Plant Secondary Metabolites and their Exploitation in Biotechnology
Edited by M. Wink

4. Molecular Plant Pathology
Edited by M. Dickinson and J. Beynon

5. Vacuolar Compartments
Edited by D.G. Robinson and J.C. Rogers

6. Plant Reproduction
Edited by S.D. O'Neill and J.A. Roberts

7. Protein–Protein Interactions in Plant Biology
Edited by M.T. McManus, W.A. Laing and A.C. Allan

8. The Plant Cell Wall
Edited by J. Rose

9. The Golgi Apparatus and the Plant Secretory Pathway
Edited by D.G. Robinson

The Golgi Apparatus and the Plant Secretory Pathway

Edited by

DAVID G. ROBINSON
Professor of Plant Cell Biology
Director, Heidelberg Institute for Plant Sciences
University of Heidelberg
Germany

Blackwell
Publishing

CRC Press

© 2003 by Blackwell Publishing Ltd

Editorial Offices:
9600 Garsington Road, Oxford OX4 2DQ, UK
    Tel: +44 (0) 1865 776868
108 Cowley Road, Oxford OX4 1JF, UK
    Tel: +44 (0) 1865 791100
Blackwell Munksgaard, 1 Rosenørns Allé,
P.O. Box 227, DK-1502 Copenhagen V,
Denmark
    Tel: +45 77 33 33 33
Blackwell Publishing Asia Pty Ltd,
550 Swanston Street, Carlton South,
Victoria 3053, Australia
    Tel: +61 (0)3 9347 0300
Blackwell Verlag, Kurfürstendamm 57,
10707 Berlin, Germany
    Tel: +49 (0)30 32 79 060
Blackwell Publishing, 10 rue Casimir
Delavigne, 75006 Paris, France
    Tel: +33 1 53 10 33 10

Published in the USA and Canada (only) by
CRC Press LLC
2000 Corporate Blvd., N.W.
Boca Raton, FL 33431, USA
Orders from the USA and Canada (only) to
CRC Press LLC

USA and Canada only:
ISBN: 0-8493-2812-8

First published 2003 by
Blackwell Publishing Ltd

A catalogue record for this title is available
from the British Library

ISBN 1-8412-7329-5
Originated as Sheffield Academic Press

Library of Congress
Cataloging in Publication Data
is available

Set in 10/12 Times
by Integra Software Services Pvt Ltd,
Pondicherry, India
Printed and bound in Great Britain by
MPG Books, Bodmin, Cornwall

For further information on
Blackwell Publishing, visit our website:
www.blackwellpublishing.com

Contents

F. ANIENTO, J. BERND HELMS and ABDUL R. MEMON

CHRIS HAWES, CLAUDE SAINT-JORE
and FEDERICA BRANDIZZI

List of Contributors

F. Aniento
Departament de Bioquimica i Biologia Molecular, Facultat de Farmacia, Universitat de Valencia, 46100 Burjassot (Valencia), Spain

M.R. Blatt
Laboratory of Plant Physiology and Biophysics, Institute of Biomedical and Life Sciences, Bower Building, University of Glasgow, Glasgow G12 8QQ, UK

Jürgen Denecke
Centre for Plant Sciences, Faculty of Biological Sciences, The University of Leeds, Leeds LS2 9JT, UK

Benjamin S. Glick
Department of Molecular Genetics and Cell Biology, The University of Chicago, 920 East 58th Street, Chicago, IL 60637, USA

Chris Hawes, Claude Saint-Jore, and Federica Brandizzi
Research School of Biological & Molecular Sciences, Oxford Brookes University, Headington, Oxford, OX3 0BP, UK

J. Bernd Helms
Biochemie-Zentrum Heidelberg, Univ. Heidelberg, Im Neuenheimer Feld 328, D-69120 Heidelberg, Germany

Eliot M. Herman
USDA-ARS, SGIL Building 006, 13000 Baltimore Avenue, Beltsville, Maryland 20705, USA

Giselbert Hinz
Abteilung für Strukturelle Zellphysiologie, Albrecht-von-Hall Institut für Pflanzenwissenschaften, Georg-August Universität Göttingen, Untere Karspüle 2, 37073 Göttingen, Germany

Liwen Jiang
Department of Biology, The Chinese University of Hong Kong, Shatin, N.T., Hong Kong, China

Gerd Jürgens and Tobias Pacher
ZMBP, Entwicklungsgenetik, Universität Tübingen, Auf der Morgenstelle 3, D-72076 Tübingen, Germany

Ken Matsuoka
RIKEN Plant Science Center, Hirosawa 2-1, Wako, 351-0198 Japan

Abdul R. Memon
TÜBITAK, Institute for Genetic Engineering and Biotechnology, 41470 Gebze, Kocaeli, Turkey

Andreas Nebenführ
Department of Botany, University of Tennessee, Knoxville, TN, 37996-1100, USA

Margit Pavelka
Institute of Histology and Embryology, University of Vienna, Schwarzspanierstr. 17, A-1090 Vienna, Austria

Christophe Ritzenthaler
Institut de Biologie Moléculaire des Plantes, 12 rue du Général Zimmer, F-67084 Strasbourg Cedex, France

David G. Robinson
Heidelberg Institute for Plant Sciences, University of Heidelberg, Im Neuenheimer Feld 230, D-69120 Heidelberg, Germany

John C. Rogers
Institute of Biological Chemistry, Washington State University, Pullman, WA 99164-6340, USA

Stephen Rutherford and Ian Moore
Department of Plant Sciences, University of Oxford, South Parks Rd., Oxford OX1 3RB, UK

Herta Steinkellner and Richard Strasser
Zentrum für Angewandte Genetik, Universität für Bodenkultur Vienna, Muthgasse 18, A-1190, Vienna, Austria

G. Thiel
Membrane Biophysics, Institute of Botany, Darmstadt University of Technology, Schnittspahnstrasse 3, D64287 Darmstadt, Germany

Preface

The Golgi apparatus is at the fulcrum of secretory traffic beginning at the endoplasmic reticulum (ER) and culminating at the cell surface or lytic/vacuolar compartments. Its major functions are protein glycosylation and polysaccharide synthesis, on one hand, and protein sorting on the other. Correct functioning of the Golgi apparatus is therefore essential for cell viability. Morphologically, it can assume diverse forms, ranging from a single fenestrated cisterna in baker's yeast to a complex interwoven network of cisternae and tubules in mammalian cells. It is also a highly dynamic organelle, whose existence is dependent on the unimpeded bidirectional movement of membrane and cargo between the ER and the Golgi.

The discovery of the Golgi apparatus is a little over 100 years old. To celebrate this, an international conference was convened in 1998 in Pavia where Camillo Golgi lived and worked. Of over 30 oral presentations, only one was focused on plants. A similar bias is seen in the two previous books devoted to the Golgi apparatus: the plant organelle receives 14 out of 190 pages in Whaley's 1975 monograph 'The Golgi Apparatus', while only one out of 10 chapters deals with plants in Berger and Roth's 1997 book of the same title. It is therefore not surprising that the description of the Golgi apparatus and its key role in intracellular protein trafficking in today's major cell biology textbooks is essentially a portrayal of the organelle in mammalian cells. Partly responsible for this is the fact that, up to 1995, plant Golgi research was mainly morphological, with relatively little biochemistry and virtually no molecular biology. The last few years have, however, witnessed dramatic changes in this situation, due to the availability of many new tools and technologies – such as generation of antibodies against recombinant Arabidopsis homolog proteins, preparation of Golgi- and endoplasmic reticulum-directed GFP constructs, and the screening of dominant-negative mutants. While some of this research indicates that the plant Golgi apparatus does indeed follow the general blueprint for this organelle, as established in other eukaryotic cells, it is also becoming increasingly apparent that there are many interesting variations and some features that are unique to plants.

This volume focuses on these major developments, drawing attention to the distinct differences between the plant and non-plant Golgi apparatus, while at the same time highlighting unsolved problems. The stage is set with two introductory chapters in which the Golgi apparatus in yeast and mammalian cells is compared to that in plants. This is followed by a chapter devoted to the unique interaction between the actin cytoskeleton and the plant Golgi apparatus. Two chapters are devoted to each of glycosylation and protein (lytic and storage) sorting.

There are chapters on vesicle formation and intra-Golgi transport – the latter, a highly controversial theme, even in animal cells. Additional chapters on Rab proteins and SNAREs reflect the growing importance of these factors in controlling vesicle-mediated trafficking events. The book concludes with a chapter focused on a distinctive feature of the plant Golgi apparatus: its continued presence – perhaps even increased activity – during mitosis and cytokinesis.

With advances in our understanding of how the Golgi apparatus operates in plants, the possibility of manipulating the timing, the type and site of delivery of secretory macromolecules is therefore becoming more and more feasible. This knowledge will be not only important in terms of improving crop production, but will be essential to the increasing usage of plants as bioreactors for vaccines and other therapeutic agents.

David G. Robinson

1 To stack or not to stack: the yeast Golgi apparatus

Benjamin S. Glick

1.1 Introduction

My high school biology textbook stated that yeasts have cell walls and are therefore plants. But today's biologists do not classify yeasts as plants, so why should a book on the plant Golgi include a chapter on yeasts? The general reason is that yeasts are excellent model systems for understanding eukaryotic cell biology. More specifically, key organizational features of the plant Golgi are shared by certain yeasts. Recent reviews focusing on the yeast *Saccharomyces cerevisiae* offer excellent coverage of Golgi-related topics, including carbohydrate processing (Munro, 2001) and the molecular basis of membrane traffic through the secretory pathway (Kaiser *et al.*, 1997; Duden & Schekman, 1997). Rather than duplicating those efforts, this chapter will concentrate on the relationship between membrane dynamics and Golgi structure.

1.2 Golgi organization in budding yeasts

The budding yeast *S. cerevisiae* has an unusual Golgi in that the individual cisternae are not found in ordered stacks, but instead are scattered throughout the cytoplasm (Preuss *et al.*, 1992). Despite this challenging morphology, researchers have generated a wealth of data about the compartmentation, dynamics, and inheritance of the *S. cerevisiae* Golgi. A second budding yeast called *Pichia pastoris* is attracting interest because it contains ordered Golgi stacks (Rossanese *et al.*, 1999). Comparative analysis of these two yeasts can shed light on the mechanisms that define Golgi organization.

1.2.1 The Golgi apparatus in S. cerevisiae

1.2.1.1 Golgi structure and compartmentation

When *S. cerevisiae* cells are examined by electron microscopy, organelles resembling conventional Golgi stacks are rarely observed. A detailed morphological study (Preuss *et al.*, 1992) indicated that a typical *S. cerevisiae* cell contains about 30 disk- or cup-shaped Golgi cisternae. Most of these cisternae seem to be physically isolated, although stacks of two or more cisternae are occasionally seen. The morphology of the *S. cerevisiae* Golgi is essentially unaffected by growth conditions or the stage of the cell cycle.

Fluorescence microscopy of the *S. cerevisiae* Golgi typically reveals about 10–15 spots, which presumably represent individual cisternae (Segev *et al.*, 1988; Franzusoff *et al.*, 1991; Antebi & Fink, 1992; Lussier *et al.*, 1995). Early and late Golgi markers reside in separate cisternae that do not visibly associate with one another (Fig. 1.1A). Different Golgi markers show varying degrees of colocalization, demonstrating that the *S. cerevisiae* Golgi is biochemically polarized (Lussier *et al.*, 1995; Jungmann & Munro, 1998). Biochemical and genetic experiments have defined four Golgi compartments (Brigance *et al.*, 2000) (Fig. 1.2). Resident Golgi enzymes involved in N- and O-linked glycosylation can be assigned to *cis*, medial and *trans* compartments based on the order in which these enzymes act (Munro, 2001). The fourth compartment, which is analogous to the vertebrate *trans*-Golgi

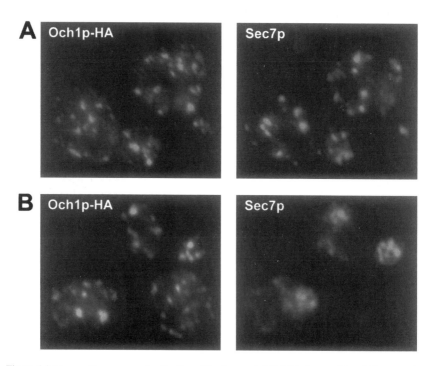

Figure 1.1 Immunofluorescence visualization of the fragmented Golgi in *S. cerevisiae*. Cells expressing a hemagglutinin (HA) epitope-tagged version of Och1p were fixed and incubated with a mouse monoclonal anti-HA antibody and a rabbit polyclonal anti-Sec7p antibody. Detection was performed with Oregon Green-conjugated anti-mouse antibody and Texas Red-conjugated anti-rabbit antibody. Spots represent individual cisternae. Panel A shows two wild-type *S. cerevisiae* cells. Panel B shows three *sec14* mutant cells that had been incubated for 1 h at 37°C to promote the formation of exaggerated late Golgi structures. In both cases, the *cis* cisternae marked with Och1p-HA are not associated with the TGN cisternae marked with Sec7p. Adapted with permission from Rossanese *et al.*, from *The Journal of Cell Biology*, 1999, **145**, 69–81, by copyright permission of The Rockefeller University Press.

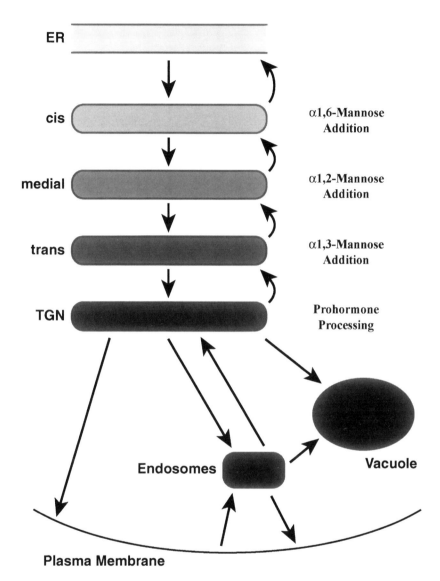

Figure 1.2 Simplified view of Golgi compartmentation and membrane traffic in *S. cerevisiae*. Arrows represent known or inferred transport pathways that involve either vesicular carriers or compartmental maturation. The processes listed on the right side of the diagram begin in the indicated Golgi compartments, but may continue in later compartments. Adapted with permission from Brigance *et al.*, from *Molecular Biology of the Cell*, 2000, **11**, 171–182, with permission by the American Society for Cell Biology, with additional data from Pelham (2002).

network (TGN), functions in prohormone processing (Bryant & Boyd, 1993). Newly synthesized proteins are sorted at the TGN into various types of transport carriers for delivery to endosomes, the vacuole, or the plasma membrane (Kaiser *et al.*, 1997; Graham & Nothwehr, 2002; Pelham, 2002).

Is the fragmented appearance of the *S. cerevisiae* Golgi an artifact of the imaging methods? This interpretation was initially appealing. Like the Golgi of higher eukaryotes, the *S. cerevisiae* Golgi is compartmentalized, polarized, and active in carbohydrate processing and protein sorting, so a reasonable assumption was that *S. cerevisiae* would also exhibit Golgi stacking. In support of this idea, when proteins involved in exit from the TGN are inactivated by mutation, the cells accumulate exaggerated Golgi organelles that often have a stack-like appearance (Duden & Schekman, 1997). However, these abnormal organelles are not ordered stacks of *cis*, medial, *trans* and TGN cisternae. For example, Fig. 1.1 shows double-label immunofluorescence of the *cis*-Golgi marker Och1p and the TGN marker Sec7p, in wild-type cells (Fig. 1.1A) and in *sec14* mutant cells that are blocked in exit from the TGN (Fig. 1.1B). The Och1p and Sec7p markers do not colocalize in either strain, but the *sec14* mutant contains large structures that label with Sec7p. A likely explanation is that in mutants such as *sec14*, the TGN expands but the Golgi as a whole does not form ordered stacks. The physical separation of early and late Golgi markers is seen even in living *S. cerevisiae* cells (Wooding & Pelham, 1998), indicating that the fragmented nature of the *S. cerevisiae* Golgi is a real phenomenon.

Despite these findings, it would be premature to rule out some type of physical links between Golgi cisternae in *S. cerevisiae*. Electron microscopy occasionally reveals two or more cisternae in close apposition (Preuss *et al.*, 1992). The stack-like appearance of the Golgi in mutants such as *sec14* could reflect an exaggeration of a normally weak association between cisternae. Resolving this issue will require more detailed morphological analyses, as well as an improved molecular understanding of cisternal stacking.

1.2.1.2 *Golgi membrane dynamics*

The mechanism of protein transport through the Golgi is still a matter of debate (Glick, 2000; Pelham & Rothman, 2000). According to the stable compartments model, Golgi cisternae are long-lived structures. In this view, newly synthesized cargo molecules move from the ER to the *cis*-Golgi in COPII coated vesicles (Barlowe, 2002), and then move from one cisterna to the next in anterograde COPI coated vesicles (Rothman & Wieland, 1996). Resident Golgi proteins would remain within the stable cisternae. By contrast, the cisternal maturation model postulates that Golgi cisternae are transitory structures. In this view, COPII vesicles fuse homotypically to generate a new *cis* cisterna, and the newly synthesized cargo molecules remain within this cisterna as it matures (Glick & Malhotra, 1998). Maturation would be accomplished by the recycling of resident Golgi proteins in retrograde COPI vesicles (Glick *et al.*, 1997). While many uncertainties remain, current data suggest that cisternal maturation is the primary mode of intra-Golgi transport.

The evidence for cisternal maturation in *S. cerevisiae* is still indirect, and is based largely on mutations that block various steps of membrane traffic. An electron microscopic study showed that when ER export is inhibited, recognizable Golgi cisternae disappear in the order *cis* to *trans* (Morin-Ganet *et al.*, 2000). Reactivating ER export regenerates Golgi cisternae in the same order. Fluorescence microscopy revealed that when COPI vesicle fusion is blocked, resident Golgi proteins accumulate in a dispersed compartment, suggesting that COPI vesicles contain Golgi proteins (Wooding & Pelham, 1998). Moreover, glycosyltransferases of the *cis*-Golgi have been shown to cycle through both the TGN and the ER (Harris & Waters, 1996; Todorow *et al.*, 2000), as expected if resident Golgi proteins are carried forward by cisternal maturation and stochastically recycled in COPI vesicles (Glick *et al.*, 1997). A prediction of the maturation model is that secretory carriers destined for the plasma membrane do not form by a classical vesicle budding pathway, but instead are terminally mature portions of TGN cisternae. This idea may explain why yeast genetics has not implicated a vesicle coat in TGN-to-plasma membrane transport (Kaiser *et al.*, 1997).

Although the cisternal maturation model accounts for many observations, it also raises many questions. For example, if Golgi organization is maintained by a balance of anterograde and retrograde membrane flow (Glick, 2000), how is this balance achieved? Assuming that all transmembrane Golgi proteins are recycled in COPI vesicles (Lanoix *et al.*, 2001; Martínez-Menárguez *et al.*, 2001), what mechanisms regulate this recycling? Many peripherally membrane-associated Golgi proteins bind reversibly from the cytosol (Munro, 2002), but how are these proteins recruited to the correct compartment? Some questions are particularly relevant to *S. cerevisiae*. For example, if Golgi cisternae in *S. cerevisiae* are not physically connected, how do COPI vesicles travel from one cisterna to another? The implication is that COPI vesicles diffuse freely in the cytosol and carry specific targeting determinants. By contrast, it has been proposed that in the mammalian Golgi, each newly formed COPI vesicle immediately becomes tethered to an adjacent cisterna (Orci *et al.*, 1998). A plausible hypothesis is that different eukaryotes all use the same basic mechanism for COPI vesicle targeting, but this mechanism can be modified to accomodate either stacked or non-stacked Golgi cisternae. Studies of *S. cerevisiae* also highlight the close relationship between the TGN and endosomes, because most of the known TGN components are also found in endosomes (Lewis *et al.*, 2000). A speculative interpretation is that part of the TGN matures into an endosomal compartment (Pelham, 2002).

1.2.1.3 *Golgi inheritance and the actin cytoskeleton*
In *S. cerevisiae*, cytoplasmic organelles typically interact with actin but not with microtubules (Botstein *et al.*, 1997). Cells that are growing rapidly or producing mating projections have clusters of late Golgi cisternae near sites of polarized growth, and this clustering requires actin cables and the type V myosin Myo2p (Baba *et al.*, 1989; Preuss *et al.*, 1992; Rossanese *et al.*, 2001). However, Golgi cisternae in *S. cerevisiae* show little directed motion (Wooding & Pelham, 1998),

indicating that actin-dependent Golgi transport is a sporadic event. Actin and Myo2p are thought to play a major role in the transport of secretory carriers to the plasma membrane (Karpova *et al.*, 2000; Pruyne & Bretscher, 2000). Thus, a possible explanation for the clustering of the late Golgi is that as TGN cisternae mature into secretory carriers, these cisternae are sometimes transported toward sites of polarized growth before their maturation is complete.

In addition to undergoing actin-dependent transport, a subset of the late Golgi cisternae in *S. cerevisiae* are retained in the growing bud (Rossanese *et al.*, 2001). This retention mechanism requires actin, Myo2p, and a protein of unknown function called Cdc1p. Myo2p may act as a tether that links Golgi cisternae to actin-containing structures at the bud tip. Growing buds almost always contain early Golgi cisternae, but this inheritance process is independent of actin, Myo2p and Cdc1p (Rossanese *et al.*, 2001). Moreover, early Golgi cisternae rarely cluster at sites of polarized growth. These data suggest that the early Golgi in *S. cerevisiae* has little or no interaction with the actin cytoskeleton. Early Golgi cisternae may form in the bud by emerging from cortical ER membranes, which are present even in very small buds (Preuss *et al.*, 1991; Du *et al.*, 2001; Fehrenbacher *et al.*, 2002). The inheritance mechanisms for the early and late Golgi seem to operate in parallel, and as a result, small buds of *S. cerevisiae* usually contain a full complement of Golgi cisternae (Rossanese *et al.*, 2001).

1.2.2.   The Golgi apparatus in P. pastoris

1.2.2.1   Golgi structure and compartmentation

P. pastoris is unusual among budding yeasts in that it has stacked Golgi cisternae (Gould *et al.*, 1992; Glick, 1996) (Fig. 1.3A). A typical cell contains 2–5 discrete stacks of approximately four cisternae each (Rossanese *et al.*, 1999; Mogelsvang *et al.*, 2003). The *cis*-most cisterna is normally a smooth disc, but the remaining cisternae have multiple fenestrations (Mogelsvang *et al.*, 2003). Immunoelectron microscopy reveals that Och1p is concentrated in *cis* cisternae whereas Sec7p is concentrated in *trans* cisternae (Mogelsvang *et al.*, 2003). When visualized by immunofluorescence, Och1p and Sec7p are closely juxtaposed (Fig. 1.3B). The combined data indicate that *P. pastoris* contains ordered, polarized Golgi stacks.

A comprehensive electron microscopic analysis of *P. pastoris* showed that *cis*, medial and *trans* cisternae are tightly associated with one another, but that TGN cisternae separate from the stack in an apparent *peeling off* process (Mogelsvang *et al.*, 2003). This phenomenon is also seen by time-lapse three-dimensional (4D) fluorescence microscopy, which indicates that Sec7p-containing TGN elements are transported toward the growing bud (Bevis *et al.*, 2002). Thus, even though Golgi architecture is quite different in *S. cerevisiae* and *P. pastoris*, both yeasts probably transport maturing TGN cisternae along actin cables to sites of polarized growth.

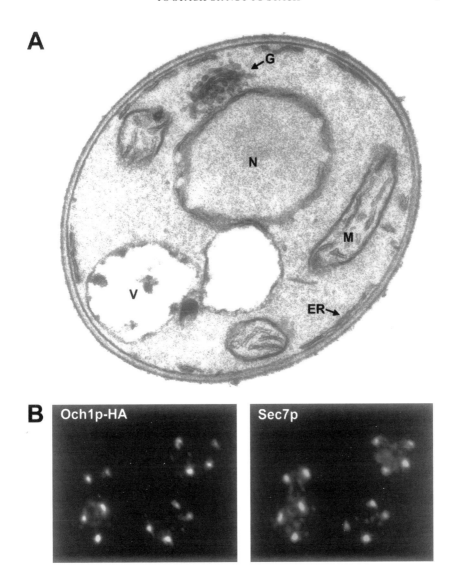

Figure 1.3 Visualization of the ordered Golgi in *P. pastoris*. Part A shows a thin-section electron micrograph of a *P. pastoris* cell. A Golgi stack is located near a vesiculating region of the nuclear envelope. G, Golgi stack; N, nucleus; M, mitochondrion; ER, cortical endoplasmic reticulum; V, vacuole. Reproduced with permission from Glick (1996). Part B shows immunofluorescence images of three cells from a *P. pastoris* strain expressing Och1p-HA. Experimental procedures were as in Fig. 1.1. The *cis* marker Och1p-HA is closely associated with the TGN marker Sec7p, confirming that *P. pastoris* has ordered Golgi stacks. Adapted with permission from Rossanese *et al.*, from *The Journal of Cell Biology*, 1999, **145**, 69–81, by copyright permission of The Rockfeller University Press.

1.2.2.2   Golgi dynamics and the ER–Golgi relationship

How do *S. cerevisiae* and *P. pastoris* assemble such different Golgi structures? The cisternal maturation model offers a possible answer. This model views the Golgi as a dynamic outgrowth of the transitional ER (tER), which is an ER sub-domain specialized for the production of COPII vesicles (Palade, 1975; Bannykh & Balch, 1997). A typical *P. pastoris* cell contains 2–5 discrete tER sites (Rossanese *et al.*, 1999). The Golgi stacks in *P. pastoris* are immediately adjacent to these tER sites, suggesting that discrete tER sites give rise to ordered Golgi stacks. *S. cerevisiae* lacks discrete tER sites (Rossanese *et al.*, 1999). Instead, the entire ER in *S. cerevisiae* seems to function as tER, suggesting that new Golgi cisternae are generated throughout the cytoplasm (Fig. 1.4). Therefore, Golgi structure in budding yeasts may be a consequence of tER organization.

In principle, a simple kinetic mechanism could generate the stacked Golgi in *P. pastoris*. If new cisternae are assembled repeatedly next to a tER site, and if the diffusion of these cisternae in the cytoplasm is relatively slow, the cisternae will pile up to form a stack (Glick, 2000). This model is consistent with the finding that tER sites are long-lived (Bevis *et al.*, 2002), but it is difficult to reconcile with recent data indicating that each *P. pastoris* Golgi stack is enclosed in a ribosome-free *zone of exclusion* (Mogelsvang *et al.*, 2003). The ribosome-free zone spans from the tER site to the *trans* cisterna. TGN cisternae are often partially or entirely outside the ribosome-free zone. These observations suggest that Golgi cisternae in *P. pastoris* are physically linked to one another and to tER sites by a matrix, which is dynamic enough to allow the escape of TGN cisternae.

As a *P. pastoris* cell grows, it must produce new tER-Golgi units. Four-dimensional microscopy has provided strong evidence that tER sites and Golgi stacks in *P. pastoris* form *de novo* (Bevis *et al.*, 2002). When the tER is marked with a fluorescently tagged COPII protein, a new tER site becomes visible as a small fluorescent spot that progressively grows to a steady-state size. tER sites rarely if ever disappear, but when two tER sites collide, they fuse. A fused tER site that is unusually large will often shrink to a more typical size. As a result of these various processes, the size and number of tER sites in a given *P. pastoris* cell can change substantially over time. Similar dynamics are observed with fluorescently tagged Golgi stacks. These results support the idea that tER sites arise *de novo* and then give birth to adjacent Golgi stacks (Bevis *et al.*, 2002).

1.2.2.3   Functional relevance of Golgi stacking

P. pastoris resembles most other eukaryotes in having discrete tER sites and stacked Golgi cisternae, but this feature was evidently lost during the evolution of *S. cerevisiae*. What are the advantages and disadvantages of an organized tER-Golgi system? No experimental data are available on this point. Presumably, the efficiency of vesicle formation and targeting is enhanced by confining the relevant components to compact tER-Golgi units. When the ancestor of *S. cerevisiae* abandoned this compact organization, some adaptive benefit must have offset the

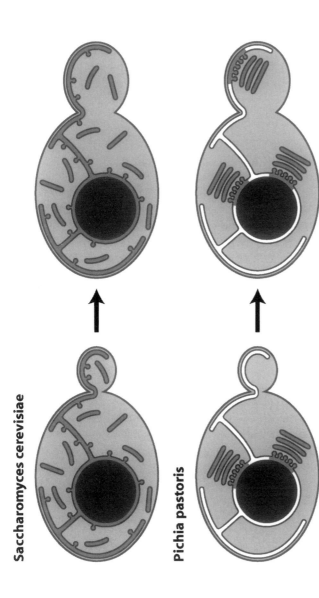

Figure 1.4 Model for the organization and inheritance of the tER and Golgi in budding yeasts. General ER membranes are shown in white, and tER and Golgi compartments are shaded. In *S. cerevisiae*, the entire ER functions as tER, so COPII vesicles bud throughout the ER network. This delocalized tER generates Golgi cisternae throughout the cytoplasm. Because cortical ER membranes are present even in very small buds, Golgi inheritance occurs early in the cell cycle. In *P. pastoris*, the ER contains discrete tER sites, so COPII vesicles bud repeatedly from the same locations. The tER sites generate ordered Golgi stacks. Because tER sites are not always present in the small buds, Golgi inheritance may not occur until the bud has grown relatively large.

reduced transport efficiency. One potential benefit of a delocalized tER is that Golgi inheritance becomes more robust, as defined by the percentage of small buds that contain Golgi cisternae (Fig. 1.4). Newly formed buds of *S. cerevisiae* almost always contain Golgi cisternae, because these buds also contain cortical ER membranes that function as tER (Rossanese *et al.*, 2001). The Golgi cisternae in the bud may facilitate the localized processing and secretion of certain proteins for which the mRNA is delivered to the bud (Takizawa *et al.*, 2000). By contrast, the buds of *P. pastoris* often grow relatively large before acquiring tER sites and Golgi structures (Bevis *et al.*, 2002), so bud-localized processing and secretion cannot occur. Future work may reveal other potential advantages of the dispersed tER-Golgi system in *S. cerevisiae*.

1.3   Golgi apparatus organization in fission yeasts

Golgi stacks are readily observed by electron microscopy in the fission yeast *Schizosaccharomyces pombe* (Kopecka, 1972; Smith & Svoboda, 1972; Johnson *et al.*, 1973). Each cell contains up to ten or more discrete stacks of about 3–7 cisternae each (Ayscough *et al.*, 1993; Ayscough & Warren, 1994). One face of the stack often has a single distended cisterna that may derive from the TGN (Chappell & Warren, 1989) (Fig. 1.5). Unlike budding yeasts, *S. pombe* adds galactose to its *N*-linked oligosaccharides, and the corresponding galactosyltransferase serves as a convenient Golgi marker (Chappell *et al.*, 1994). When this marker is used for fluorescence microscopy, the *S. pombe* Golgi appears as a set of spots that are

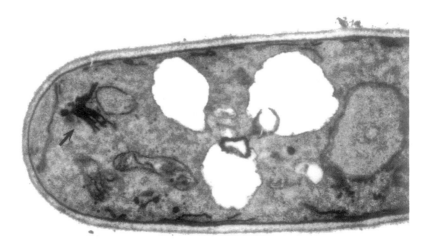

Figure 1.5 Golgi morphology in *S. pombe*. This thin-section electron micrograph shows a Golgi stack (arrow) composed of about six flattened cisternae plus a single distended cisterna. The Golgi stack is not closely associated with ER membranes. Reproduced with permission from Chappell and Warren, from *The Journal of Cell Biology*, 1989, **109**, 2693–2702, by copyright permission of The Rockfeller University Press.

distributed throughout the cytoplasm (Chappell & Warren, 1989; Ayscough *et al.*, 1993; Brazer *et al.*, 2000).

Microtubules are important for maintaining the stacked morphology of the *S. pombe* Golgi. When microtubule function is disrupted in a cold-sensitive β-tubulin mutant strain or in a wild-type strain treated with thiabendazole, the cisternal structure of the Golgi is disrupted as judged by electron microscopy (Ayscough *et al.*, 1993). However, the number of Golgi spots that are detected by fluorescence microscopy does not increase, suggesting that the basic organizational units of the Golgi remain intact (Ayscough *et al.*, 1993; Brazer *et al.*, 2000). A microtubule-dependent kinesin motor is involved in Golgi-to-ER membrane recycling (Brazer *et al.*, 2000), but otherwise the interaction between microtubules and the *S. pombe* Golgi has not been further characterized.

Interestingly, when *S. pombe* cells are treated with cycloheximide to block protein synthesis, Golgi stacks shrink and ultimately disappear (Ayscough *et al.*, 1993). A similar cycloheximide-induced loss of Golgi cisternae has been seen with the budding yeast *S. cerevisiae* (Morin-Ganet *et al.*, 2000). This effect of blocking protein synthesis may be due to a slowing of ER export, with a consequent disruption of the balance between anterograde and retrograde membrane flow (Glick, 2000).

By contrast to the situation with *P. pastoris*, the Golgi stacks in *S. pombe* show no obvious association with the endoplasmic reticulum (Fig. 1.5). Moreover, tER sites have not been observed in *S. pombe* by electron microscopy. It is unclear whether this fission yeast has discrete tER sites or a delocalized tER.

1.4   Comparison of yeast and plant Golgi structures

S. cerevisiae has been a powerful system for identifying and characterizing molecules involved in secretion, and many of the resulting insights are directly applicable to plants. However, *S. cerevisiae* is not as useful for analyzing the supramolecular process of Golgi organization. *P. pastoris* and *S. pombe* seem to be better models for understanding the architecture of the plant Golgi.

P. pastoris resembles algae in displaying a close association between Golgi stacks and tER sites (Whaley, 1975; Becker & Melkonian, 1996). Thus, an analysis of secretory pathway organization in *P. pastoris* should be relevant to algae. One interesting difference between the two cell types is that in algae, the location and number of tER-Golgi units seem to be tightly controlled (Becker & Melkonian, 1996), but in *P. pastoris*, tER-Golgi units form and fuse in an apparently stochastic fashion (Bevis *et al.*, 2002). tER sites and Golgi structures in *P. pastoris* arise *de novo*, whereas the corresponding compartments in algae undergo an ordered mitotic fission (Di Orio & Millington, 1978; Noguchi, 1978; Ueda, 1997). A likely interpretation is that both cell types use the same conserved mechanism to generate tER sites and Golgi stacks, but algae impose additional regulation on this process. In particular, the *de novo* formation of tER sites in *P. pastoris* has been proposed to result in the *de novo* formation of Golgi stacks (Bevis *et al.*, 2002), and the fission of tER sites in algae has been proposed to result in fission of the associated Golgi stacks (Di Orio & Millington, 1978; Bracker *et al.*, 1996; Rossanese & Glick, 2001).

Like Golgi stacks in higher plants, Golgi stacks in *P. pastoris* are embedded in a ribosome-excluding matrix (Staehelin & Moore, 1995; Mogelsvang *et al.*, 2003). The composition and function of the Golgi matrix have long been enigmatic (Mollenhauer & Morré, 1978), but this structure may be amenable to molecular analysis in *P. pastoris*.

S. pombe resembles higher plants in that Golgi stacks are not closely apposed to tER sites (Staehelin & Moore, 1995). Indeed, discrete tER sites have yet to be identified in either cell type, suggesting that COPII vesicles might bud throughout the ER. This model does not explain how ER-derived vesicles would be delivered to the Golgi stacks. Plant Golgi stacks translocate along actin filaments, and it has been speculated that these mobile stacks capture vesicles from nearby regions of the ER (Boevink *et al.*, 1998; Nebenfuhr *et al.*, 1999). A similar process might occur in *S. pombe*, except that the Golgi stacks are likely to move along microtubules. Thus, a central question for both higher plants and *S. pombe* is whether ER-derived vesicles are delivered to Golgi stacks in a directed manner, or whether these vesicles diffuse in the cytosol and collide randomly with the Golgi stacks.

In conclusion, several of the most intriguing properties of the plant Golgi are mirrored in *P. pastoris* and *S. pombe*. These experimentally tractable yeasts offer plant researchers an opportunity to solve some old puzzles.

Acknowledgements

This work was supported by grant GM61156 from the National Institutes of Health.

References

Antebi, A. & Fink, G.R. (1992) The yeast Ca^{2+}-ATPase homologue, PMR1, is required for normal Golgi function and localizes in a novel Golgi-like distribution. *Mol. Biol. Cell*, **3**, 633–654.

Ayscough, K., Hajibagheri, N.M.A., Watson, R. & Warren, G. (1993) Stacking of Golgi cisternae in *Schizosaccharomyces pombe* requires intact microtubules. *J. Cell Sci.*, **106**, 1227–1237.

Ayscough, K. & Warren, G. (1994) Inhibition of protein synthesis disrupts the Golgi apparatus in the fission yeast, *Schizosaccharomyces pombe*. *Yeast*, **10**, 1–11.

Baba, M., Baba, N., Ohsumi, Y., Kanaya, K. & Osumi, M. (1989) Three-dimensional analysis of morphogenesis induced by mating pheromone α factor in *Saccharomyces cerevisiae*. *J. Cell Sci.*, **94**, 207–216.

Bannykh, S.I. & Balch, W.E. (1997) Membrane dynamics at the endoplasmic reticulum-Golgi interface. *J. Cell Biol.*, **138**, 1–4.

Barlowe, C. (2002) COPII-dependent transport from the endoplasmic reticulum. *Curr. Opin. Cell Biol.*, **14**, 417–422.

Becker, B. & Melkonian, M. (1996) The secretory pathway of protists: spatial and functional organization and evolution. *Microbiol. Rev.*, **60**, 697–721.

Bevis, B.J., Hammond, A.T., Reinke, C.A. & Glick, B.S. (2002) Apparent *de novo* formation of transitional ER sites and Golgi structures in *Pichia pastoris*. *Nat. Cell Biol.*, **4**, 750–756.

Boevink, P., Oparka, K., Santa Cruz, S., Martin, B., Betteridge, A. & Hawes, C. (1998) Stacks on tracks: the plant Golgi apparatus traffics on an actin/ER network. *Plant J.*, **15**, 441–447.

Botstein, D., Amberg, D., Mulholland, J., Huffaker, T., Adams, A., Drubin, D. & Stearns, T. (1997) The yeast cytoskeleton, in *The Molecular and Cellular Biology of the Yeast Saccharomyces*, Vol. 3 (eds J.R. Pringle, J.R. Broach & E.W. Jones), Cold Spring Harbor Laboratory Press, pp. 1–90.

Bracker, C.E., Morré, D.J. & Grove, S.N. (1996) Structure, differentiation and multiplication of Golgi apparatus in fungal hyphae. *Protoplasma*, **194**, 250–274.

Brazer, S.W., Williams, H.P., Chappell, T.G. & Cande, W.Z. (2000) A fission yeast kinesin affects Golgi membrane recycling. *Yeast*, **16**, 149–166.

Brigance, W.T., Barlowe, C. & Graham, T.R. (2000) Organization of the yeast Golgi complex into at least four functionally distinct compartments. *Mol. Biol. Cell*, **11**, 171–182.

Bryant, N.J. & Boyd, A. (1993) Immunoisolation of Kex2p-containing organelles from yeast demonstrates colocalisation of three processing proteinases to a single Golgi compartment. *J. Cell Sci.*, **106**, 815–822.

Chappell, T.G., Hajibagheri, N.M.A., Ayscough, K., Pierce, M. & Warren, G. (1994) Localization of an α 1,2 galactosyltransferase activity to the Golgi apparatus of *Schizosaccharomyces pombe*. *Mol. Biol. Cell*, **5**, 519–528.

Chappell, T.G. & Warren, G. (1989) A galactosyltransferase from the fission yeast *Schizosaccharomyces pombe*. *J. Cell Biol.*, **109**, 2693–2702.

Di Orio, J. & Millington, W.F. (1978) Dictyosome formation during reproduction in colchicine-treated *Pediastrum boryanum* (Hydrodictyaceae). *Protoplasma*, **97**, 329–336.

Du, Y., Pypaert, M., Novick, P. & Ferro-Novick, S. (2001) Aux1p/Swa2p is required for cortical endoplasmic reticulum inheritance in *Saccharomyces cerevisiae*. *Mol. Biol. Cell*, **12**, 2614–2628.

Duden, R. & Schekman, R. (1997) Insights into Golgi function through mutants in yeast and animal cells, in *The Golgi Apparatus* (eds E.G. Berger & J. Roth), Birkhäuser Verlag, Basel, pp. 219–246.

Fehrenbacher, K.L., Davis, D., Wu, M., Boldogh, I. & Pon, L.A. (2002) Endoplasmic reticulum dynamics, inheritance, and cytoskeletal interactions in budding yeast. *Mol. Biol. Cell*, **13**, 854–865.

Franzusoff, A., Redding, K., Crosby, J., Fuller, R.S. & Schekman, R. (1991) Localization of components involved in protein transport and processing through the yeast Golgi apparatus. *J. Cell Biol.*, **112**, 27–37.

Glick, B.S. (1996) Cell biology: alternatives to baker's yeast. *Curr. Biol.*, **6**, 1570–1572.

Glick, B.S. (2000) Organization of the Golgi apparatus. *Curr. Opin. Cell Biol.*, **12**, 450–456.

Glick, B.S., Elston, T. & Oster, G. (1997) A cisternal maturation mechanism can explain the asymmetry of the Golgi stack. *FEBS Lett.*, **414**, 177–181.

Glick, B.S. & Malhotra, V. (1998) The curious status of the Golgi apparatus. *Cell*, **95**, 883–889.

Gould, S.J., McCollum, D., Spong, A.P., Heyman, J.A. & Subramani, S. (1992) Development of the yeast *Pichia pastoris* as a model organism for a genetic and molecular analysis of peroxisome assembly. *Yeast*, **8**, 613–628.

Graham, T.R. & Nothwehr, S.F. (2002) Protein transport to the yeast vacuole, in *Protein Targeting, Transport & Translocation* (eds R.E. Dalbey & G. von Heijne), Academic Press, Amsterdam, pp. 322–357.

Harris, S.L. & Waters, M.G. (1996) Localization of a yeast early Golgi mannosyltransferase, Och1p, involves retrograde transport. *J. Cell Biol.*, **132**, 985–998.

Johnson, B.F., Yoo, B.Y. & Calleja, G.B. (1973) Cell division in yeasts: movement of organelles associated with cell plate growth of *Schizosaccharomyces pombe*. *J. Bacteriol.*, **115**, 358–366.

Jungmann, J. & Munro, S. (1998) Multi-protein complexes in the *cis* Golgi of *Saccharomyces cerevisiae* with α-1,6-mannosyltransferase activity. *EMBO J.*, **17**, 423–434.

Kaiser, C.A., Gimeno, R.E. & Shaywitz, D.A. (1997) Protein secretion, membrane biogenesis, and endocytosis, in *The Molecular and Cellular Biology of the Yeast Saccharomyces*, Vol. 3 (eds J.R. Pringle, J.R. Broach & E.W. Jones), Cold Spring Harbor Laboratory Press, pp. 91–227.

Karpova, T.S., Reck-Peterson, S.L., Elkind, N.B., Mooseker, M.S., Novick, P.J. & Cooper, J.A. (2000) Role of actin and Myo2p in polarized secretion and growth of *Saccharomyces cerevisiae*. *Mol. Biol. Cell*, **11**, 1727–1737.

Kopecka, M. (1972) Dictyosomes in the yeast *Schizosaccharomyces pombe*. *Antonie van Leeuwenhoek*, **38**, 27–31.

Lanoix, J., Ouwendijk, J., Stark, A., Szafer, E., Cassel, D., Dejgaard, K., Weiss, M. & Nilsson, T. (2001) Sorting of Golgi resident proteins into different subpopulations of COPI vesicles: a role for ArfGAP1. *J. Cell Biol.*, **155**, 1199–1212.

Lewis, M.J., Nichols, B.J., Prescianotto-Baschong, C., Riezman, H. & Pelham, H.R.B. (2000) Specific retrieval of the exocytic SNARE Snc1p from early yeast endosomes. *Mol. Biol. Cell*, **11**, 23–38.

Lussier, M., Sdicu, A.M., Ketela, T. & Bussey, H. (1995) Localization and targeting of the *Saccharomyces cerevisiae* Kre2p/Mnt1p α1,2-mannosyltransferase to a *medial*-Golgi compartment. *J. Cell Biol.*, **131**, 913–927.

Martínez-Menárguez, J.A., Prekeris, R., Oorschot, V.M.J., Scheller, R., Slot, J.W., Geuze, H.J. & Klumperman, J. (2001) Peri-Golgi vesicles contain retrograde but not anterograde proteins consistent with the cisternal progression model of intra-Golgi transport. *J. Cell Biol.*, **155**, 1213–1224.

Mogelsvang, S., Gomez-Ospina, N., Soderholm, J., Glick, B.S. & Staehelin, L.A. (2003) Tomographic evidence for continuous turnover of Golgi cisternae in *Pichia pastoris*. *Mol. Biol. Cell*, **14**, in press.

Mollenhauer, H.H. & Morré, D.J. (1978) Structural compartmentation of the cytosol: zones of exclusion, zones of adhesion, cytoskeletal and intercisternal elements. *Subcell. Biochem.*, **5**, 327–359.

Morin-Ganet, M.-N., Rambourg, A., Deitz, S.B., Franzusoff, A. & Képès, F. (2000) Morphogenesis and dynamics of the yeast Golgi apparatus. *Traffic*, **1**, 56–68.

Munro, S. (2001) What can yeast tell us about *N*-linked glycosylation in the Golgi apparatus? *FEBS Lett.*, **498**, 223–227.

Munro, S. (2002) Organelle identity and the targeting of peripheral membrane proteins. *Curr. Opin. Cell Biol.*, **14**, 506–514.

Nebenfuhr, A., Gallagher, L.A., Dunahay, T.G., Frohlick, J.A., Mazurkiewicz, A.M., Meehl, J.B. & Staehelin, L.A. (1999) Stop-and-go movements of plant Golgi stacks are mediated by the actomyosin system. *Plant Physiol.*, **121**, 1127–1142.

Noguchi, T. (1978) Transformation of the Golgi apparatus in the cell cycle, especially at the resting and earliest developmental stages of a green alga, *Micrasterias americana*. *Protoplasma*, **95**, 73–88.

Orci, L., Perrelet, A. & Rothman, J.E. (1998) Vesicles on strings: morphological evidence for processive transport within the Golgi stack. *Proc. Natl. Acad. Sci. USA*, **95**, 2279–2283.

Palade, G. (1975) Intracellular aspects of the process of protein synthesis. *Science*, **189**, 347–358.

Pelham, H.R. & Rothman, J.E. (2000) The debate about transport in the Golgi – two sides of the same coin? *Cell*, **102**, 713–719.

Pelham, H.R.B. (2002) Insights from yeast endosomes. *Curr. Opin. Cell Biol.*, **14**, 454–462.

Preuss, D., Mulholland, J., Franzusoff, A., Segev, N. & Botstein, D. (1992) Characterization of the *Saccharomyces* Golgi complex through the cell cycle by immunoelectron microscopy. *Mol. Biol. Cell*, **3**, 789–803.

Preuss, D., Mulholland, J., Kaiser, C.A., Orlean, P., Albright, C., Rose, M.D., Robbins, P.W. & Botstein, D. (1991) Structure of the yeast endoplasmic reticulum: localization of ER proteins using immunofluorescence and immunoelectron microscopy. *Yeast*, **7**, 891–911.

Pruyne, D. & Bretscher, A. (2000) Polarization of cell growth in yeast. II. The role of the cortical actin cytoskeleton. *J. Cell Sci.*, **113**, 571–585.

Rossanese, O.W. & Glick, B.S. (2001) Deconstructing Golgi inheritance. *Traffic*, **2**, 589–596.

Rossanese, O.W., Reinke, C.A., Bevis, B.J., Hammond, A.T., Sears, I.B., O'Connor, J. & Glick, B.S. (2001) A role for actin, Cdc1p and Myo2p in the inheritance of late Golgi elements in *Saccharomyces cerevisiae*. *J. Cell Biol.*, **153**, 47–61.

Rossanese, O.W., Soderholm, J., Bevis, B.J., Sears, I.B., O'Connor, J., Williamson, E.K. & Glick, B.S. (1999) Golgi structure correlates with transitional endoplasmic reticulum organization in *Pichia pastoris* and *Saccharomyces cerevisiae*. *J. Cell Biol.*, **145**, 69–81.

Rothman, J.E. & Wieland, F.T. (1996) Protein sorting by transport vesicles. *Science*, **272**, 227–234.

Segev, N., Mulholland, J. & Botstein, D. (1988) The yeast GTP-binding YPT1 protein and a mammalian counterpart are associated with the secretion machinery. *Cell*, **52**, 915–924.

Smith, D.G. & Svoboda, A. (1972) Golgi apparatus in normal cells and protoplasts of *Schizosaccharo-myces pombe*. *Microbios*, **5**, 177–182.

Staehelin, L.A. & Moore, I. (1995) The plant Golgi apparatus: structure, functional organization and trafficking mechanisms. *Annu. Rev. Plant Physiol. Plant Mol. Biol.*, **46**, 261–288.

Takizawa, P.A., DeRisi, J.L., Wilhelm, J.E. & Vale, R.D. (2000) Plasma membrane compartmentalization in yeast by messenger RNA transport and a septin diffusion barrier. *Science*, **290**, 341–344.

Todorow, Z., Spang, A., Carmack, E., Yates, J. & Schekman, R. (2000) Active recycling of yeast Golgi mannosyltransferase complexes through the endoplasmic reticulum. *Proc. Natl. Acad. Sci. USA*, **97**, 13643–13648.

Ueda, K. (1997) The synchronous division of dictyosomes at the premitotic stage. *Ann. Bot.*, **80**, 29–33.

Whaley, W.G. (1975) *The Golgi Apparatus*, Springer-Verlag, Vienna.

Wooding, S. & Pelham, H.R.B. (1998) The dynamics of Golgi protein traffic visualized in living yeast cells. *Mol. Biol. Cell*, **9**, 2667–2680.

2 The Golgi apparatus in mammalian and higher plant cells: a comparison

Margit Pavelka and David G. Robinson

2.1 Introduction

As indicated by numerous reviews, the Golgi apparatus occupies a pivotal position in intracellular trafficking in both animal and plant cells (Barr, 2002; Farquhar & Hauri, 1997; Glick, 2000; Klumperman, 2000; Lippincott-Schwartz, 2001; Mellman & Warren, 2000; Nebenführ & Staehelin, 2001; Pavelka, 1987; Pelham, 2001; Staehelin & Moore, 1995; Warren & Malhotra, 1998). However, despite some similarities (e.g. stacked cisternae, similar locations for COP1- and clathrin-coated vesicle production), the architecture of the Golgi apparatus and its relationship to other endomembrane compartments in these two major groups of organisms is otherwise quite different. Some of these differences may reflect basic characteristics of the two cell types. For example, plant cells are more involved in the secretion of polysaccharides than proteins, a feature which will certainly affect the manifestation of ER-Golgi transport. Secondly, plant cells as a rule are less differentiated and less polarly organized than animal cells and this may be the reason underlying the polydisperse nature of the plant Golgi apparatus. Thirdly, the lack of fragmentation of the plant Golgi apparatus during mitosis is certainly connected with the increased demand of cell wall polysaccharide production during cytokinesis.

The purpose of this review chapter is therefore to highlight the differences in Golgi structure between mammalian and higher plant cells. However, in order to begin to understand the organization of the Golgi apparatus, it is first of all necessary to be aware of the extreme dynamics of this organelle in terms of bi-directional membrane flow. In addition, the Golgi architecture must always be seen in the light of specific cellular functions, since the Golgi apparatus is an organelle at the crossroads of biosynthetic and endocytic trafficking pathways (for review Farquhar & Hauri, 1997; Glick, 2000; Pavelka, 1987; Pelham, 2001). Golgi organization also alters during cell differentiation and rapidly responds to changes in cellular function. Therefore, the amalgamation of microscopical and biochemical data into a unified model for Golgi structure and function represents a major challenge for those in the field of Golgi research. It should be mentioned that currently two major concepts exist for the organization of intra-Golgi transport: the *stationary cisternae/vesicular traffic* and *cisternal progression* models (see Chapter 5) which represent extreme, but not necessarily mutually exclusive, possibilities. Unfortunately, the pictures of the Golgi apparatus obtained in the electron microscope are at best snapshots of a continuously changing structure, so that whichever organizational model is favoured, one may only get a momentary glimpse of the total dynamic potential of this very important organelle.

2.2   Localization in the cell

2.2.1   Mammalian cells

In mammalian cells, the Golgi apparatus frequently holds a position in close proximity to the cell nucleus. This area corresponds to the cytocentre of cells, where the centrioles are localized and which is the central site of microtubule organization. Recent observations suggest that the centrosome is involved in nucleation of the Golgi apparatus through interactions with Golgi-resident proteins (Takatsuki *et al.*, 2002). In many cell types, a compact Golgi apparatus consisting of multiple stacks of cisternae is organized around a pair of centrioles. Sometimes, characteristic shapes are formed, such as are seen in goblet cells which exhibit huge calyx-like Golgi complexes. By comparison, in undifferentiated cells, Golgi stacks are small and inconspicuous, and may be dispersed in the cytoplasm. Upon stimulation of cells and increased activities of biosynthetic and endocytic systems, the Golgi apparatus becomes reorganized.

Extended Golgi regions in a paranuclear position are characteristically present in cells which are highly active in secretion and membrane trafficking. Such structures are discernible even with the light microscope: in classic histological preparations appearing as *juxtanuclear vacuoles* and recognized as paranuclear bodies in cytochemical preparations and immunofluorescence. They are particularly visible in nerve cells, as discovered by Camillo Golgi more than 100 years ago (Dröscher, 1998; Golgi, 1898), in plasma cells, fibroblasts and hemopoetic cells, and reflect the extremely high activities of these cells in secretion and membrane synthesis. In polarized epithelial cells, such as intestinal absorptive cells, kidney tubular cells and pancreatic acinar cells, the Golgi apparatus typically resides in the supranuclear cytoplasm.

Maintenance of the Golgi apparatus in a position close to the nucleus requires an intact cytoskeleton (for review see Kreis *et al.*, 1997). Microtubules are necessary for the transport of pre-Golgi intermediates from the sites where they are formed close to the export regions of the endoplasmic reticulum (see Section 2.2.2). Upon disruption of the microtubules by treatment with colchicine or nocodazole, the Golgi apparatus disappears from its typical position and a *mini-Golgi apparatus* is formed with small Golgi stacks distributed throughout the cytoplasm (for review see Dinter & Berger, 1998). This condition somewhat resembles the Golgi apparatus in plant cells.

2.2.2   Plant cells

In marked contrast to the Golgi complex of animal cells, but similar to some yeasts (see Chapter 1), the Golgi apparatus of higher plants is polydisperse. By this, we mean that it is distributed throughout the cytoplasm in the form of small units which are called dictyosomes (here called Golgi stacks) – a term seldom used by animal cell biologists. They can be visualized as punctate light sources by immuno-fluorescence microscopy

(i)   Using antibodies against (a) content markers e.g. JIM84 – a monoclonal antibody directed against a complex carbohydrate epitope that is synthesized in *trans* cisternae (Horsley *et al.*, 1993); (b) transmembrane proteins e.g. p58 – a *cis*-localized protein involved in microtubule attachment present in both mammalian (Lahtinen *et al.*, 1992) and plant (Li & Yen, 2001) cells; (c) *trans*-localized pyrophosphatase (Mitsuda *et al.*, 2001); (d) the *trans*-localized, but surface-associated reversibly glycosylated polypeptide (RGP, Dhugga *et al.*, 1997; Mitsuda *et al.*, 2001).

(ii)   Using GFP-constructs targeted to *cis*-based (mannosidase I, Nebenführ *et al.*, 1999), or *trans*-located (sialyl transferase, Saint-Jore *et al.*, 2002; Wee *et al.*, 1998) processing enzymes, or to proteins apparently distributed throughout the Golgi stack (Boevink *et al.*, 1998).

(iii) Using antibodies directed against coat components of COPI-coated vesicles, plant Golgi stacks appear as discs (with γ-COP antibodies) or as discrete doughnuts (with Arf1p antibodies), thus reflecting the peripheral nature of this vesiculating activity (Ritzenthaler *et al.*, 2002; see Fig. 2.1).

The number of individual Golgi stacks in a higher plant cell can vary enormously, from several hundred (Garcia-Herdugo *et al.*, 1988), down to 25 or less (Anton-Lamprecht, 1967; see also Fig. 2.1). Accurate data on how these numbers change with respect to the different demands placed on secretion during differentiation are not available. Moreover, it is generally thought that each Golgi stack in a plant cell, at any one time, fulfils the same function, although the evidence for this assumption is lacking.

2.3   Ultrastructure and organization

2.3.1   *Mammalian cells*

2.3.1.1   *Golgi stacks – an overview*
In mammalian cells, the Golgi apparatus is made up of stacked cisternae, which are interconnected by tubular elements to form a single continuous body (for reviews see Farquhar & Hauri, 1997; Hermo & Smith, 1998; Pavelka, 1987; Rambourg & Clermont, 1997). Although multiple variations exist, the construction of the Golgi apparatus in mammalian cells follows a general plan which always must be seen in connection with Golgi function. In many types of cells each Golgi stack exhibits a clear polarity in terms of cisternae. As an example, the Golgi apparatus in a HepG2 human hepatoma cell is presented in Fig. 2.2. The *cis* side of the stacks

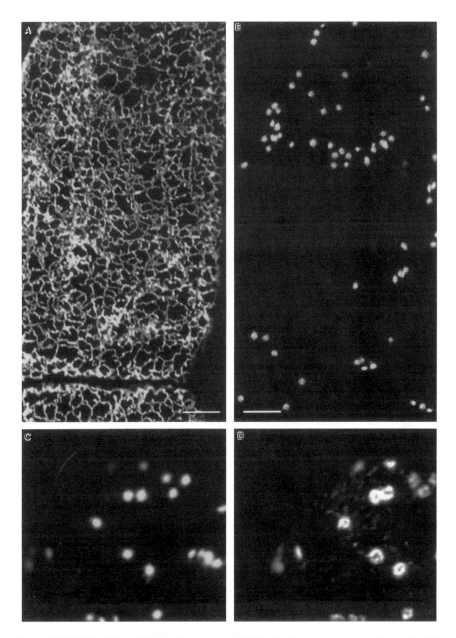

Figure 2.1 CLSM of ER and Golgi stacks in tobacco BY-2 cells. (A) network pattern of the cortical ER in a cell expressing KDEL::GFP. (B) punctate appearance of individual Golgi stacks in a cell expressing mannosidase-1::GFP. (C, D) paired micrographs of several Golgi stacks visualized with mannosidase-1::GFP (c) and Arf1 antibodies (D); the latter form a doughnut around the central portion of the stack. Unpublished micrographs courtesy of Dr. C. Ritzenthaler (Strasbourg). Bar = 2 μm throughout.

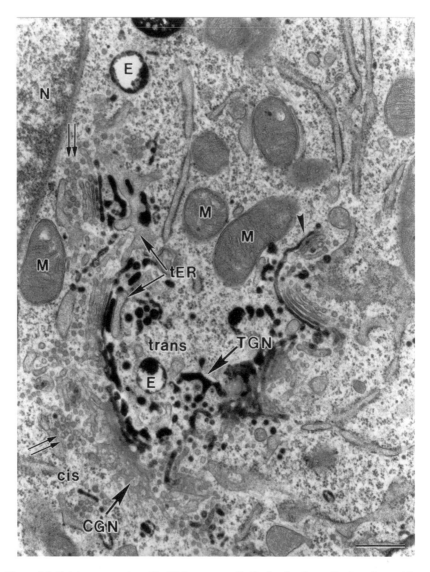

Figure 2.2 Golgi apparatus in a HepG2 hepatoma cell 60 min after internalisation of peroxidase-labelled *Limax flavus* lectin (LFA). At endoplasmic reticulum exit sites, small vesicles known to be COPII coated, bud from transitional ER-elements (double arrows). Pleomorphic structures nearby represent pre-Golgi intermediates (pGI/ERGIC – ER-Golgi intermediate compartment). Networks characterize the *cis* as well as *trans* Golgi faces, the CGN (thick arrow, *cis*-Golgi network) and the TGN (thick arrow, *trans*-Golgi network). CGN and TGN enwrap the central body of stacked Golgi cisternae. *Trans* Golgi networks, showing clathrin-coated buds and multiple neighbouring clathrin-coated vesicles and large endosomes (E), are densely filled with internalized LFA. LFA is also present in cisternae positioned closely opposed to or integrated into the Golgi stacks at the *trans* side (arrowhead). *Trans* Golgi endoplasmic reticulum (thin arrows, tER) is in a close spatial relationship to the TGN and *trans* Golgi cisternae. Bar = 0.5 μm.

is characterized by a fenestrated cisterna and shows neighbouring polymorphous pre-Golgi intermediates (pGI or ERGIC – endoplasmic reticulum Golgi intermediate compartment). Compact, non-fenestrated medial cisternae are interposed between the *cis* cisternae and cisternae of the *trans* side, here prominently labelled with internalized *Limax flavus* lectin. Membranous networks may be apparent at the *cis* side (*cis* Golgi network – CGN), as well as at the *trans* side (*trans* Golgi network – TGN). An increasing bulk of evidence suggests that the networks at both sides represent import as well as export sites.

2.3.1.2   Pre-Golgi intermediates (ERGIC)

After having passed quality controls within the endoplasmic reticulum, newly synthesized lumenal as well as membrane glycoproteins leave this compartment at specialized export sites (the ER transitional elements) where they are packed into COPII-coated vesicles for further transport (Bannykh & Balch, 1998; Bannykh *et al.*, 1996; Barlowe, 2002). Assembly of the COPII-coat and vesicle formation are well characterized (see Chapter 3). After export out of the ER, polymorphous tubular-vesicular structures with variable morphology and different membrane domains are formed, which constitute the pre-Golgi intermediates or the ER-Golgi intermediate compartment (ERGIC). ER to Golgi transport of newly synthesized glycoproteins may be facilitated by lectin binding to cargo receptors localized in the membranes of pre-Golgi intermediates (for review see Hauri *et al.*, 2001). The mannose-specific lectin ERGIC-53 is a marker for the ER-Golgi intermediate compartment and is well characterized as a transport receptor for glycoproteins (Hauri *et al.*, 2000). Lack of functional ERGIC-53 results in a selective defect in the secretion of glycoproteins. In humans, it is necessary for efficient secretion of coagulation factors V and VIII (for review see Hauri *et al.*, 2001). It is clear that the ER export sites and formation of the pre-Golgi intermediates are not confined to areas close to the *cis* Golgi region, such as is shown in Fig. 2.2 but may be distributed throughout the entire cytoplasm, possibly located many micrometers away from the Golgi region.

2.3.1.3   The cis Golgi

The network-like formations at the *cis* Golgi side (open thick arrows in Fig. 2.2) are seen as early stages in the formation of the *cis* Golgi subcompartment. Of particular interest for the initial formation of the *cis* Golgi cisterna is the *cis* Golgi matrix comprised of GM130 and the Golgi reassembly and stacking protein GRASP65, which have been found to cycle via membranous tubules between the Golgi apparatus and a late intermediate compartment (Marra *et al.*, 2001). GM130 complexed with GRASP65 and other proteins can be recognized by the vesicle docking and tethering machinery (Nakamura *et al.*, 1997). It is thought that complexed matrix proteins may represent blueprints for the assembly of the first *cis* cisterna, which are recognized by incoming vesicles containing newly synthesized cargoes. A *homing in-hypothesis* has been proposed to program heterotypic membrane fusion between transport intermediates and their Golgi target compartments (Moyer *et al.*, 2001; for review Barr, 2002; Pfeffer, 2001).

The *cis* Golgi is not only the main input site for the newly synthesized molecules coming from the ER, but also plays a central role in the retrograde Golgi-to-ER traffic, which serves the recycling of both the ER and Golgi molecules (see Chapter 6).

2.3.1.4   The medial Golgi

Cis and *trans* Golgi import and export sites peripherally enwrap the central body of the Golgi apparatus, where the main Golgi functions are the modification of proteins and lipids. The medial matrix is defined by golgin 45 and GRASP55 (for reviews see Barr, 2002; Lowe, 2002; Pfeffer, 2001) and the stacks of cisternae are composed of compact and non-compact regions being accompanied by multiple COPI-type vesicles (Hermo & Smith, 1998; Ladinsky *et al.*, 1999; Rambourg & Clermont, 1997). In this central Golgi region, newly synthesized molecules coming from the ER and entering the Golgi apparatus at the *cis* side and internalized molecules coming from the plasma membrane and entering the Golgi apparatus at the *trans* side, are modified before being targeted to their final destinations. Bearing in mind this continuous bi-directional traffic across this organelle and its dynamics, it is clear that neither a morphological nor a functional clear-cut boundary can exist between the *cis*, medial and *trans* Golgi.

Medial stacked Golgi cisternae represent a central station in the secretory pathways to be passed by the majority of newly synthesized secretory and membrane proteins, as well as lysosomal enzymes. Within the stacked cisternae, *O*-linked oligosaccharides are synthesized and *N*-linked glycans which have been preformed in the ER are trimmed and elongated, thus high-mannose oligosaccharide chains are converted into complex ones (for review see Roth, 1997, 2002). Furthermore, the synthesis of hybrid oligosaccharides, characteristic for lysosomal enzymes, is finished in the Golgi stacks, and the Golgi apparatus holds a pivotal function in the synthesis and transport of lipids (Bretscher & Munro, 1993; De Matteis *et al.*, 2002; Holthuis *et al.*, 2001; Simons & Ikonen, 2000; Van Meer, 2001). It is well established that molecules passing through the Golgi apparatus are modified by a sequential action of enzymes and a clear *cis*-to-*trans* orientation is evident. Trimming of the oligosaccharide chains proceeds through the sequential actions of endomannosidase, Golgi mannosidase IA, IB, and II. Depending on these activities, sugar chains are elongated by the sequential insertion of sugars through the actions of families of *N*-acetyl-glucosaminyl-, fucosyl-, galactosyl- and sialyltransferases. Most of the glycosidases and glycosyltransferases involved have been studied in detail, and the enzymes as well as their products have been localized to either more *cis* or medial or more *trans*-oriented Golgi cisternae (for review see Farquhar & Hauri, 1997; Roth, 1997). However, it has become increasingly clear that there are not distinct cisternae reserved for distinct enzymes, as has been suggested from the results of earlier studies, but enzymes are localized in a broader distribution throughout the stacks (for a recent review see Roth, 2002). A clear overlap in the localization and activities exists, and glycan-modifying enzymes recognizing each other act in unison (Giraudo *et al.*, 2001; Nilsson *et al.*, 1994, kin recognition model). Furthermore, glycosyltransferases show cell type-specific distributions and the enzyme patterns change during differentiation and with altered cell functions. In summary, the data available indicate that glycosylation in the Golgi apparatus is compartmentalized, but there do not exist clear-cut sub-compartment boundaries.

2.3.1.5 The trans Golgi

The *trans* Golgi is a central junction between biosynthetic and endocytic systems, and represents both the stacks' export site in the biosynthetic pathways and import site in connection with endocytosis. *Trans* Golgi sorting events not only involve proteins, but are closely connected with the synthesis and organization of sphingo-lipids and cholesterol. Lipid subdomains can be built up within TGN-membranes, and lipid-based sorting mechanisms involving the formation of lipid rafts play an important role in the apical delivery of proteins in epithelial cells and mediate axonal delivery of proteins in neurons (for review see Holthuis *et al.*, 2001; Simons & Ikonen, 2000).

Newly synthesized secretory, lysosomal and membrane glycoproteins, which have been modified within the stacks of cisternae, leave the Golgi apparatus at the *trans* side. Prominent membranous TGN may be formed, characterized as main compartments for the export out of the Golgi. Here, sorting to final destinations takes place, being closely connected with the packaging of cargo into transport carriers (Arvan *et al.*, 2002; Griffiths & Simons, 1986; Roth & Taatjes, 1998). Some of the *trans* Golgi sorting mechanisms and subsequent transport routes are well characterized. These include the mannose-6-phosphate system active in the targeting of lysosomal enzymes to the lysosome (for a recent review see Mullins & Bonifacino, 2001), which involves the formation of clathrin-coated vesicles for export out of *trans* Golgi, and well-defined recycling pathways for the transport of mannose-6-phosphate receptors back from late endosomes to the TGN. Other carriers function in the traffic between the Golgi apparatus and the cell surface and have been shown to be large, measuring more than $1\,\mu m$, pleomorphic structures that fuse directly with the plasma membrane (Polishchuk *et al.*, 2000). In some cases, an entire Golgi cisterna appears to form the carrier.

Export out of the Golgi is not necessarily connected with the formation of a prominent secretory TGN and sorting may already take place in the stacked *trans* Golgi cisternae, as revealed by electron tomography studies of the Golgi apparatus in rat kidney cells (Ladinsky *et al.*, 1999). In this case, a prominent TGN was not apparent, but instead two types of *trans* cisternae were present, each of them producing exclusively vesicles with only one type of coat, either a clathrin- or a non-clathrin-type. It has been clearly shown that more than one *trans* cisterna may be involved in the exit from the Golgi apparatus, and molecules to be transported to the plasma membrane are sorted from a different cisterna than those destined for the lysosomal system (Ladinsky *et al.*, 1999).

As mentioned above, the *trans* Golgi is not only an export site, but is also a focus for import, as exemplified by the retrograde traffic of several toxins which need to reach the ER and cytosol for exerting their inhibitory effects (for reviews, see Johannes, 2002; Pavelka *et al.*, 1998; Sandvig & van Deurs, 2000). Endocytosis is obviously connected with extensive *trans* Golgi dynamics: an endocytic TGN is newly formed, rapidly increases in size, is consumed again and parts of it are integrated into the Golgi stacks (Vetterlein *et al.*, 2002; Fig. 2.2).

On the other hand, TGN and *trans* Golgi cisternae are closely associated with parts of the ER (Fig. 2.2). Extended contact regions may be formed, which are particularly prominent in differentiating cells, such as in the crypt cells of the small intestine and in embryonic pancreatic cells (Pavelka & Ellinger, 1983, 1986). Three-dimensional reconstructions of the Golgi apparatus based on electron tomography studies (Ladinsky *et al.*, 1999) show that this domain of ER is *in contact* with different *trans* Golgi cisternae, and forms a complex and extended compartment. The contact regions have been hypothesized as sites where a direct exchange of lipids between ER and *trans* Golgi (Kok *et al.*, 1988) may take place (Ladinsky *et al.*, 1999; Vetterlein *et al.*, 2002).

One of the great current challenges in the field of Golgi research is to bring together the clear structural identity of the various Golgi subcompartments with the continuous antero- and retrograde membrane flow which occurs across this organelle. In this respect, we draw attention to the fact that two main models for Golgi organization have been discussed in the literature: the stationary cisternae/ vesicular transport model and the cisternae progression-maturation model, together with multiple variations (e.g. Becker *et al.*, 1995; Bonfanti *et al.*, 1998; Martínez-Menárguez *et al.*, 2001; Mironov *et al.*, 2001; Orci *et al.*, 2000; Volchuk *et al.*, 2000; Weidman, 1995). These models are discussed in detail by Nebenführ in this volume (Chapter 5). However, it is important to point out that electron tomography studies suggest that not all cisternae have to be visited by traffic across the stack (Ladinsky *et al.*, 1999), and retrograde endocytic flow across the Golgi apparatus (Sandvig & van Deurs, 2000; Vetterlein *et al.*, 2002; see also Section 3.1.5) further impedes attempts to work out a model of general validity.

2.3.2   Plant cells

2.3.2.1   Parameters of stack polarity

As a rule, Golgi stacks in higher plants are very polar structures. This has been apparent for over 20 years from observations on chemically fixed material (e.g. Mollenhauer & Morré, 1978; Robinson & Kristen, 1982). The introduction of cryo-methods (high-pressure freezing, freeze substitution) has certainly improved the general quality of structural preservation (Kiss *et al.*, 1990; Staehelin *et al.*, 1990), but has not revealed any new parameter of stack polarity. In well-fixed (chemical or cryo-) samples, clear gradients in

width of the cisternal lumina (*cis* >> *trans*)
intercisternal spacing (*cis* << *trans*)
stainability of the cisternal membranes (*cis* << *trans*)

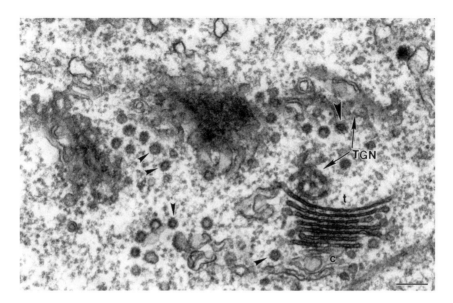

Figure 2.3 Golgi stacks in tobacco BY-2 cells. Typical parameters of stack polarity are visible in the Golgi stack at the bottom right. c = *cis*, t = *trans*, TGN = *trans* Golgi network, large arrowhead = clathrin-coated vesicle, small arrowheads = COPI-coated vesicles. Bar = 250 nm.

are usually easy to discern (see for example Figs 2.3 and 2.4). These purely morphological parameters of polarity go hand-in-hand with a *cis*-medial-*trans* functional compartmentation of the plant Golgi stack in terms of the immunogold detection of certain cell wall polymers (Zhang & Staehelin, 1992), and the presence of particular glycosyl transferases (see Chapter 11).

Two other features of polarity, which are typical of plant Golgi stacks are a relatively small (in comparison to the other cisternae in the stack) *cis*-most cisterna, and the presence of intercisternal filaments, which are usually found between the two or three cisternae at the *trans*-face (Mollenhauer & Morré, 1975; see also Fig. 2.3). These 2–4 nm diameter filaments, whose nature is still not known, lie parallel to one another (spacing around 10 nm), even in those exceptional cases e.g. the outer root cap cells of maize, where they are situated between hypertrophied slime-containing vesicles. Although there has been much speculation as to their possible function as stack-stabilizing elements, in conjunction with trans-membrane proteins (Staehelin *et al.*, 1990), they do not seem to be universally required for Golgi stack production e.g. Golgi stacks in meristematic cells do not have intercisternal filaments (Staehelin *et al.*, 1990), as are sycamore cells (Driouich *et al.*, 1994).

Plant Golgi stacks usually have between four to ten cisternae, whose diameter varies from 0.5–1.5 μm (Zhang & Staehelin, 1992; Ritzenthaler *et al.*, 2002).

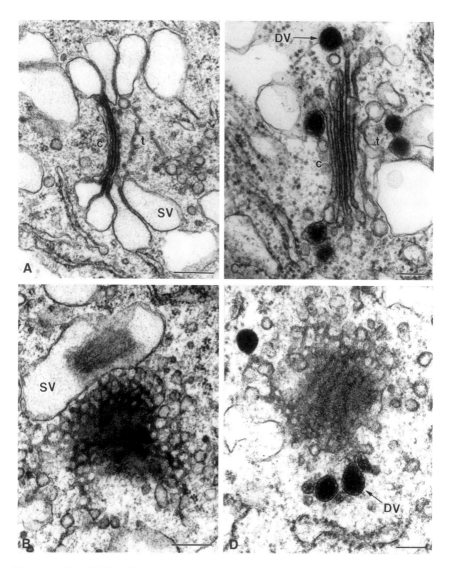

Figure 2.4 Plant Golgi stacks with unusual vesicles. (A, B) slime vesicles (SV) produced by Golgi stacks in the mantle cells of maize root tips. (C, D) storage protein-containing dense vesicles (DV) produced by parenchyma cells in developing pea cotyledons. (A, C) cross sections; (B, D) flat sections through the stack. Bars = 250 nm (A, B), 200 nm (C, D).

Within a single cell, cisternal number per stack and diameter are constant. It is normally considered that the cisternae in a particular stack are individual, separate entities. However, there have been reports, based on 3D reconstructions of serial sections (Kristen *et al.*, 1984, 1989; Lockhausen *et al.*, 1990), of the occasional peripheral tubular connection.

2.3.2.2   Vesicle types

At least four different classes of vesicles have been reported to bud off the cisternae of plant Golgi stacks:

(i)   Small, 60 nm vesicles with nap-like coats (see small arrowheads Fig. 2.3). These are most likely to be COPI-coated vesicles, since they stain positively in the cryosections with antibodies against the COPI-coat proteins Arf1p and γ-COP (Pimpl *et al.*, 2000). Statistically, they are seen more frequently at the *cis*-pole of the stack (Pimpl *et al.*, 2000).

(ii)   80–100 nm spiny vesicles (see large arrowheads Fig. 2.3). These are clathrin-coated vesicles, and their presence is restricted to the *trans*-face of the stack (Hillmer *et al.*, 1988; Staehelin *et al.*, 1990).

(iii) Smooth surfaced secretory vesicles. These tend to have electron-denser contents in cryo-fixed samples as compared to the chemically fixed cells (Kiss *et al.*, 1990). In slime-secreting root-cap cells, these vesicles are extremely large and kidney-shaped, and are each attached to an individual cisternae through tubules at several positions (Robinson, 1980; see also Fig. 2.4A,B). The secretory vesicles seem to bud off the stack from *trans*-cisternae.

(iv)   Vesicles roughly 200 nm in diameter, which have osmiophilic contents. These are the so-called *dense vesicles*, typical of many cell types, especially legume cotyledons, engaged in storage protein deposition (Robinson & Hinz, 1999; see also Fig. 2.4C,D). They are filled with storage protein aggregates, a condensation process which starts at the periphery of the very first *cis*-cisterna and gradually proceeds towards the *trans*-pole, where they are released from the stack (Hillmer *et al.*, 2001). Clathrin caps and budding clathrin-coated vesicles can often be seen on dense vesicles which are still attached at the TGN (Robinson & Hinz, 1997).

2.3.2.3   Structural relationships to the ER

The structural basis for ER-Golgi transport in plants has to take into account two characteristic features of the plant secretory pathway. Firstly, plants secrete polysaccharides rather than proteins (Robinson, 1985). Therefore, protein transport between the ER and the Golgi apparatus in plant cells does not occur to the same extent that it does in mammalian cells. Secondly, as we now know from live cell observations with Golgi-targeted GFP-constructs, plant Golgi stacks show stop-go tumbling movements, often along *tracks* defined by tubular ER and actin microfilaments (Boevink *et al.*, 1998; Nebenführ *et al.*, 1999; see also Chapter 4). These three features (less cargo, Golgi mobility, and shorter ER-Golgi transport distances) will automatically reduce the chances of observing anterograde and retrograde transport events between the two organelles in the electron microscope. If transport is squeezed into a narrow time frame (*tanking up and moving on*), the percentage of Golgi stacks which might be caught at any one time in the act of *tanking* will not be large. On the other hand, much might be gained by concentrating on those plants/organs where *sui generis* protein transport between the ER and the Golgi apparatus is particularly intense e.g. developing cotyledons or endosperm. Certainly such objects have been investigated in the past (see Robinson & Hinz, 1999; Chapter 9 for references), but the actuality of this problem recommends a reinvestigation with state-of-the-art cryo-preparations.

Whereas vesicle-budding at the *cis*-face of plant Golgi stacks has been demonstrated on innumerable occasions, the corresponding event at the level of the ER has proved extremely difficult to visualize in higher plants (for an isolated example see Ritzenthaler *et al.*, 2002). For this reason it has been questioned whether ER-to-Golgi transport in plants is actually achieved by vesicles (Saint-Jore *et al.*, 2002; Brandizzi *et al.*, 2002; see also Chapter 4). The only feasible alternative is a direct physical continuity between ER and Golgi membranes, possibly through a tubule, albeit temporary. Given the small number of papers where such contacts have been reported (e.g. Juniper *et al.*, 1982; Harris & Oparka, 1983; compare however Robinson, 1980; Parker & Hawes, 1982, using similar methods and tissues), and considering the multitude of published electron micrographs where an obvious spatial, or *intimate* relationship between sectioned Golgi stacks and ER is lacking, one must conclude that both possible transfer events are most ephemeral, and therefore difficult *to catch*, i.e. to visualize in the electron microscope. Clearly, the jury is still out on the decision of vesicles versus tubules!

Of great help will be the identification of ER-exit sites, as has already been done in yeast (see Chapter 1) and in mammalian cells (e.g. Rust *et al.*, 2001). One wants to know their size, their number (are there more than there are Golgi stacks?) and their distribution (tubular versus cisternal type ER). With modern CLSM technology, and the appropriate combinations of antibodies it should be possible to obtain single images revealing Golgi stacks and ER-export sites displayed against the total ER network. With COPII-GFP constructs it should be possible to reveal the dynamics of ER-export sites with respect to Golgi stacks in living cells. Thus, it can be confidently expected that a lot of this information will become available in the next few years.

2.3.2.4   TGN: cisternal release and the partially coated reticulum

The TGN of plant cells is a ramifying, tubular membrane system usually showing profiles of budding clathrin-coated vesicles which emanate from the last *trans*-cisterna (Staehelin *et al.*, 1990). Sometimes it can extend a considerable distance from the stack (Hillmer *et al.*, 1988). At other times Golgi stacks can (sometimes) be visualized without a clear TGN. The reason for this difference is the sloughing or displacement of the TGN/*trans*-most cisterna, a particular feature of the *trans*-face of the plant Golgi stack, which has been recorded on numerous occasions (Hillmer *et al.*, 1988; Mollenhauer, 1971; Robinson & Kristen, 1982). It is virtually impossible to distinguish the TGN released from the Golgi stack from the so-called *partially coated reticulum* (PCR) (Pesacreta & Lucas, 1985) on the basis of morphology. Studies involving the internalization of electron dense markers by protoplasts have indicated that the PCR lies on the endocytic pathway of plant cells (Tanchak *et al.*, 1988; Galway *et al.*, 1993). Hillmer *et al.* (1988) already pointed out 15 years ago that it is difficult to reconcile the sorting demands of the biosynthetic pathway at the TGN with those of the endocytic pathway of the PCR, if in fact TGN and PCR are the same structure. This problem still exists.

2.3.2.5   Golgi matrix

The current interest in matrix proteins, which are thought to constitute a scaffold necessary for the reassembly of the Golgi apparatus after completion of mitosis in animal cells (Seeman *et al.*, 2000; Ward *et al.*, 2001), provokes the question about the existence of plant homologs even though plant Golgi stacks do not disassemble during mitosis (see below). Pertinent data are not yet available, but reports concerning a *zone of exclusion* surrounding each Golgi stack have circulated in the plant literature for over 20 years, without further experimental substantiation (Mollenhauer & Morré, 1980; Staehelin & Moore, 1995).

2.4   The Golgi apparatus and cell division

2.4.1   Mammalian cells

In mammalian cells, the characteristic stacked Golgi architecture disassembles at the onset of mitosis, the cisternae fragment and Golgi membranes are absorbed into the endoplasmic reticulum (Rabonille & Warren, 1997; Zaal *et al.*, 1999). In fact, recent findings suggest that Golgi fragmentation and dispersal is required for entry into mitosis in mammalian cells (Sütterlin *et al.*, 2002). Golgi disassembly is connected to phosphorylation of a *cis*-Golgi matrix protein GM130 and subsequent inhibition of binding to the vesicle tethering protein p115 (Lowe *et al.*, 2000). GM130 is dephosphorylated again in telophase coinciding with the onset of Golgi reconstitution in the daughter cells. Golgi fragmentation during mitosis seems to occur in several steps, the first leading to a conversion of the stacks into membranous *blobs* that further reorganize into tubular-vesicular compartments, which finally become integrated into the ER. MEK1 (mitogen-activated protein kinase kinase) and Cdc2 are involved and appear to work sequentially in this process (Colanzi *et al.*, 2000; Kano *et al.*, 2000; for review see Nelson, 2000).

The question whether the Golgi apparatus completely disappears during mitosis and is *in toto* redistributed and absorbed into the ER, or whether some Golgi remnants remain in the cytoplasm is highly controversial. It has been suggested that the Golgi apparatus contains stable components that serve as a template neces-sary for its biogenesis and reconstitution after completion of mitosis (Jesch & Linstedt, 1998; Jokitalo *et al.*, 2001; Pelletier *et al.*, 2000; Seemann *et al.*, 2000).

Results of studies with cells expressing a constitutively inactive form of Sar1, in which ER-exit sites are completely disrupted, contradict this model and suggest that all classes of Golgi components including glycosylation enzymes, matrix and itinerant proteins, are dynamically associated with the organelle. It has therefore been proposed that the Golgi apparatus is a dynamic steady-state system and that most, if not all, proteins of the Golgi region cycle through the ER or are in a dynamic equilibrium with soluble cytoplasmic pools (Miles *et al.*, 2001; Ward *et al.*, 2001).

2.4.2   Plant cells

Higher plant cells do not fragment or disassemble their Golgi apparatus during mitosis. In fact the opposite appears to be the case. Working with onion roots, Garcia-Herdugo *et al.* (1988) have reported that the number of Golgi stacks in interphase is around 420 per cell, but doubles by metaphase/anaphase. Nebenführ *et al.* (2000) have indirectly verified this observation by showing that roughly equal numbers of Golgi stacks are distributed into the cytoplasm of daughter cells of dividing tobacco BY-2 cells. It has been claimed that Golgi stack replication occurs by cisternal fission, beginning at the *cis*-face (Hirose & Komamine, 1989), but in the absence of other studies a *de novo* biogenesis cannot be ruled out.

Nebenführ *et al.* (2000) have carefully followed the behaviour of Golgi stacks in BY-2 cells during mitosis and cell division. As cells prepare for division, the distribution of Golgi stacks changes so that over 60% are now found in the perinuclear cytoplasm, and movement of Golgi stacks ceases. During mitosis, approximately 20% of the Golgi stacks remain in the immediate vicinity of the spindle, whereas another 20% relocate to the equatorial region near the plasma membrane, roughly where the preprophase band had been. These regions are devoid of mitochondria and plastids. During cell division Golgi stacks surround the expanding phragomoplast releasing secretory vesicles required for cell plate formation. Interestingly, and in contrast to the situation in interphase, while cytoskeletal elements may be necessary for the initial segregation of Golgi stacks and other large organelles at the start of mitosis and cell division, they are not required for their positional maintenance during these stages in the cell cycle.

Acknowledgements

The authors gratefully acknowledge the excellent technical assistance of Mrs. Manuela Cazalla-Martinez, Mrs. Stefanie Gold, Mrs. Beatrix Mallinger, Mrs. Elfriede Scherzer, and Mr. Ulrich Kaindl and are particularly indebted to Monika Vetterlein for critically reading the manuscript. Parts of the work were supported by the Österreichische National Bank (Jubiläumsfondsprojekt No.6582), and the Deutsche Forschungsgemeinschaft.

References

Anton-Lamprecht, I. (1967) Anzahl und Vermehrung der Zellorganellen im Scheitelmeristem von *Epilobium. Ber. Dtsch. Bot. Ges.*, **80**, 747–754.

Arvan, P., Zhang, B., Feng, L., Liu, M. & Kuliawat, R. (2002) Luminal protein multimerization in the distal secretory pathway/secretory granules. *Curr. Opin. Cell Biol.*, **14**, 448–453.

Bannykh, S.I. & Balch, W.E. (1998) Selective transport of cargo between endoplasmic reticulum and Golgi compartments. *Histochem. Cell Biol.*, **109**, 463–475.

Bannykh, S.I., Rowe, T. & Balch, W.E. (1996) The organization of endoplasmic-reticulum export complexes. *J. Cell Biol.*, **135**, 19–35.

Barlowe, C. (2002) COPII-dependent transport from the endoplasmic reticulum. *Curr. Opin. Cell Biol.*, **14**, 417–422.

Barr, F.A. (2002) The Golgi apparatus: going round in circles. *Trends Cell Biol.*, **12**, 101–153.

Becker, B., Bölinger, B. & Melkonian, M. (1995) Anterograde transport of algal scales through the Golgi complex is not mediated by vesicles. *Trends Cell Biol.*, **5**, 305–307.

Boevink, P., Oparka, K., Santa Cruz, S., Martin, B., Betteridge, A. & Hawes, C. (1998) Stacks on tracks: the plant Golgi apparatus traffics on an actin/ER network. *Plant J.*, **15**, 441–447.

Bonfanti, L., Mironov, A.A. Jr., Martínez-Menárguez, J.A., Martella, O. & Fusella, A. (1998) Procollagen traverses the Golgi stack without leaving the lumen of cisternae, evidence for cisternal maturation. *Cell*, **95**, 993–1003.

Brandizzi, F., Snapp, E., Roberts, A., Lippicott-Schwartz, J. & Hawes, C. (2002) Membrane protein transport between the ER and Golgi in tobacco leaves is energy dependent but cytoskeleton independent: evidence from selective photobleaching. *Plant Cell*, **14**, 1293–1309.

Bretscher, M.S. & Munro, S. (1993) Cholesterol and the Golgi apparatus. *Science*, **261**, 1280–1281.

Colanzi, A., Deerinck, T.J., Ellisman, M.H. & Malhotra, V. (2000) A specific activation of the mitogen activated protein kinase kinase 1 (MEK1) is required for Golgi fragmentation during mitosis. *J. Cell Biol.*, **149**, 331–339.

De Matteis, M.A., Godi, A. & Corda, D. (2002) Phosphoinositides and the Golgi complex. *Curr. Opin. Cell Biol.*, **14**, 434–447.

Dhugga, K.S., Tiwari, S.C. & Ray, P.M. (1997) A reversibly glycosylated polypeptide (RGP1) possibly involved in plant cell wall synthesis: purification, gene cloning and trans Golgi localization. *Proc. Natl. Acad. Sci. USA*, **94**, 7679–7684.

Dinter, A. & Berger, E.G. (1998) Golgi-disturbing agents. *Histochem. Cell Biol.*, **109**, 571–590.

Driouich, A., Levy, S., Staehelin, L.A. & Faye, L. (1994) Structural and functional organization of the Golgi apparatus in plant cells. *Plant Physiol. Biochem.*, **32**, 731–749.

Dröscher, A. (1998) Camillo Golgi and the discovery of the Golgi apparatus. *Histochem. Cell Biol.*, **109**, 425–430.

Farquhar, M.G. & Hauri, H.P. (1997) Protein sorting and vesicular traffic in the Golgi apparatus, in *The Golgi apparatus* (eds E.G. Berger & J. Roth), pp. 63–129.

Galway, M.E., Rennie, P.J. & Fowke, L.C. (1993) Ultrastructure of the endocytic pathway in glutar-aldehyde-fixed and high-pressure frozen/freeze-substituted protoplasts of white spruce (*Picea glauca*). *J. Cell Sci.*, **106**, 847–858.

Garcia-Herdugo, G., Gonzáles-Reyes, J.A., Garcia-Navarro, F. & Navas, P. (1988) Growth kinetics of the Golgi apparatus during the cell cycle in onion root meristems. *Planta*, **175**, 305–312.

Giraudo, C.G., Daniotti, J.L. & Maccioni, J.F. (2001) Physical and functional association of glycolipid N-acetyl-galactosaminyl and galactosyl transferases in the Golgi apparatus. *Proc. Natl. Acad. Sci.*, **98**, 1625–1630.

Glick, B. (2000) Organization of the Golgi apparatus. *Curr. Opin. Cell Biol.*, **12**, 450–456.

Golgi, C. (1898) Sur la structure des cellules nerveuses. *Arch. Ital. Biol.*, **30**, 60–71.

Griffiths, G. & Simons, K. (1986) The trans Golgi network: sorting at the exit site of the Golgi complex. *Science*, **234**, 438–443.

Harris, N. & Oparka, K. (1983) Connections between dictyosomes, ER and GERL in cotyledons of mung bean (*Vigna radiata* L.). *Protoplasma*, **114**, 93–102.

Hauri, H.-P., Appenzeller, Ch., Kuhn, F. & Nufer, O. (2001) Lectins and traffic in the secretory pathway. *FEBS Lett.*, **476**, 32–37.

Hauri., H.-P., Kappeler, F., Andersson, H. & Appenzeller, Ch. (2000) ERGIC-53 and traffic in the secretory pathway. *J. Cell Sci.*, **113**, 587–596.

Hermo, L. & Smith, C.E. (1998) The structure of the Golgi apparatus: a sperm's eyeview in principal epithelial cells of the epididymis. *Histochem. Cell Biol.*, **109**, 431–447.

Hillmer, S., Freundt, H. & Robinson, D.G. (1988) The partially coated reticulum and its relationship to the Golgi apparatus in higher plant cells. *Eur. J. Cell Biol.*, **47**, 206–212.

Hillmer, S., Movafeghi, A., Robinson, D.G. & Hinz, G. (2001) Vacuolar storage proteins are sorted in the *cis*-cisternae of the pea cotyledon Golgi apparatus. *J. Cell Biol.*, **152**, 41–50.

Hirose, S. & Komamine, A. (1989) Changes in ultrastructure of Golgi apparatus during the cell cycle in a synchronous culture of *Catharanthus roseus*. *New Phytol.*, **111**, 599–605.

Holthuis, J.C.M., Pomorski, Th., Raggers R.J., Sprong, H. & van Meer, G. (2001) The organizing potential of sphingolipids in intracellular membrane transport. *Physiol. Rev.*, **81**, 1689–1723.

Horsley, D., Coleman, J., Evans, D., Crooks, K., Peart, J., Satiat-Jeunemaitre, B. & Hawes, C. (1993) A monoclonal antibody, JIM 84, recognizes the Golgi apparatus and plasma membrane in plant cells. *J. Exp. Bot.*, **44**, 223–229.

Jesch, S.A. & Linstedt, A.D. (1998) The Golgi and endoplasmic reticulum remain independent during mitosis in HeLa cells. *Mol. Biol. Cell*, **9**, 623–635.

Johannes, L. (2002) The epithelial cell cytoskeleton and intracellular trafficking. I. Shiga toxin B-subunit system: retrograde transport, intracellular vectorization, and more. *Amer. J. Physiol.*, **283**, G1–G7.

Jokitalo, E., Cabrera-Poch, N., Warren, G. & Shima, D.T. (2001) Golgi clusters and vesicles mediate mitotic inheritance independently of the endoplasmic reticulum. *J. Cell Biol.*, **154**, 317–330.

Juniper, B., Hawes, C.R. & Horne, J.C. (1982) The relationship between the dictyosomes and the form of endoplasmic reticulum in plant cells with different export programmes. *Bot. Gaz.*, **143**, 1545–1552.

Kano, F.K., Takenada, A., Yamamoto, K., Nagayama, K., Nishida, E. & Murata, M. (2000) MEK and Cdc2 kinase are sequentially required for Golgi disassembly in MDCK cells by the mitotic *Xenopus* extracts. *J. Cell Biol.*, **149**, 357–368.

Kiss, J.Z., Giddings, T.H., Staehelin, L.A. & Sack, F.D. (1990) Comparison of the ultrastructure of conventionally fixed and high pressure frozen/freeze substituted root tips of *Nicotiana* and *Arabidopsis*. *Protoplasma*, **157**, 64–74.

Klumperman, J. (2000) The growing Golgi: in search for its independence. *Nature Cell Biol.*, **2**, E217–E219.

Kok, J.W., Babia, T., Klappe, K., Egea, G. & Hoekstra, D. (1988) Ceramide transport from endoplasmic reticulum to Golgi apparatus is not vesicle mediated. *Biochem. J.*, **333**, 779–786.

Kreis, T.E., Goodson, H.V., Perez, F. & Rönnholm, R. (1997) Golgi apparatus-cytoskeleton interactions in *The Golgi apparatus* (eds E.G. Berger & J. Roth), pp. 179–193.

Kristen, U., Lockhausen, J. & Robinson, D.G. (1984) Three-dimensional reconstruction of a dictyosome using serial sections. *J. Exp. Bot.*, **35**, 1113–1118.

Kristen, U., Lockhausen, J., Menhardt, W. & Dallas, W.J. (1989) Computer-generated three-dimensional representation of dictyosomes reconstructed from electron micrographs of serial ultrathin sections. *J. Cell Sci.*, **93**, 385–389.

Ladinsky, M.S., Mastronarde, D.N., McIntosh, J.R., Howell, K.E. & Staehelin, L.A. (1999) Golgi structure in three dimensions: functional insights from the normal rat kidney cell. *J. Cell Biol.*, **144**, 1135–1149.

Lahtinen, U., Dahllöf, B. & Saraste, J. (1992) Characterization of a 58 kDa *cis*-Golgi protein in pancreatic exocrine cells. *J. Cell Sci.*, **103**, 321–333.

Li, Y. & Yen, L.-F. (2001) Plant Golgi-associated vesicles contain a novel α-actinin-like protein. *Eur. J. Cell Biol.*, **80**, 703–710.

Lippincott-Schwartz, J. (2001) The secretory membrane system studied in real-time. *Histochem. Cell Biol.*, **116**, 67–107.

Lockhausen, J., Kristen, U., Menhardt, W. & Dallas, W.J. (1990) Three-dimensional reconstruction of a plant dictyosome from series of ultra-thin sections using computer image processing. *J. Microscopy*, **158**, 197–205.

Lowe, M. (2002) Golgi complex biogenesis de novo? *Current Biol.*, **12**, R166–R167.

Lowe, M., Gonatas, M.K. & Warren, G. (2000) The mitotic phosphorylation cycle of the cis-Golgi matrix protein GM130. *J. Cell Biol.*, **149**, 341–356.

Marra, P., Maffucci, T., Daniele, G., di Tullio, G., Ikehara, Y., Chan, E.K.L., Luini, A., Beznoussenko, G., Mironov, A. & de Matteis, M.A. (2001) The Golgi matrix proteins GM130 and GRASP65 cycle through a novel subdomain of the intermediate compartment. *Nature Cell Biol.*, **3**, 1101–1113.

Martínez-Menárguez, J.A., Prekeris, R., Oorschot, V., Scheller, R., Slot, J.W., Geuze, H.J. & Klumperman, J. (2001) Peri-Golgi vesicles contain retrograde but not anterograde proteins favouring the cisternal-progression model of intra-Golgi transport. *J. Cell Biol.*, **155**, 1213–1224.

Mellman, I. & Warren, G. (2000) The road taken: past and future foundations of membrane traffic. *Cell*, **100**, 99–112.

Miles, S., McManus, H., Forsten, K.E. & Storrie, B. (2001) Evidence that the entire Golgi apparatus cycles in interphase HeLa cells: sensitivity of Golgi matrix proteins to an ER exit block. *J. Cell Biol.*, **155**, 543–556.

Mironov, A.A., Beznoussenko, G.V., Nicoziani, P., Martella, O., Trucco, A., Kweon, H.S., Di Giandomenico, D., Polishchuk, R.S., Fusella, A., Lupetti, P., Berger, E.G., Geerts, W.J.C., Koster, A.J., Burger, K.N.J. & Luini, A. (2001) Small cargo proteins and large aggregates can traverse the Golgi by a common mechanism without leaving the lumen of cisternae. *J. Cell Biol.*, **155**, 1225–1238.

Mitsuda, N., Enami, K., Nakata, M., Takeyasu, K. & Sato, M.H. (2001) Novel type *Arabidopsis thaliana* H⁺-PPase is localized to the Golgi apparatus. *FEBS Lett.*, **488**, 29–33.

Mollenhauer, H.H. (1971) Fragmentation of mature dictyosome cisternae. *J. Cell Biol.*, **49**, 212–214.

Mollenhauer, H.H. & Morré, D.J. (1975) A possible role for intercisternal elements in the formation of secretory vesicles in plant Golgi apparatus. *J. Cell Sci.*, **19**, 231–237.

Mollenhauer, H.H. & Morré, D.J. (1978) Structural differences contrast higher plant and animal Golgi apparatus. *J. Cell Sci.*, **32**, 357–362.

Mollenhauer, H.H. & Morré, D.J. (1980) The Golgi apparatus, in *The Biochemistry of Plants* (ed. P. Stumpf & E. Conn), Academic Press, New York, pp. 437–488.

Moyer, B.D., Allan, B.B & Balch, W.E. (2001) Rab1 interaction with GM130 effector complex regulates COP II vesicle *cis*-Golgi tethering. *Traffic*, **2**, 268–276.

Mullins, C. & Bonifacino, J.S. (2001) The molecular machinery for lysosome biogenesis. *BioEssays*, **23**, 333–343.

Nakamura, N., Lowe, M., Levine, T.P., Rabouille, C. & Warren, G. (1997) The vesicle docking protein p115 binds GM130, a cis-Golgi matrix protein, in a mitotically regulated manner. *Cell*, **89**, 445–455.

Nebenführ, A., Gallagher, L.A., Dunahay, T.G., Frohlick, J.A., Mazurkiewicz, A.M., Meehl, J.B. & Staehelin, L.A. (1999) Stop-and-go-movements of plant Golgi stacks are mediated by the actomyosin system. *Plant Physiol.*, **121**, 1127–1141.

Nebenführ, A., Frohlick, J.A. & Staehelin, L.A. (2000) Redistribution of Golgi stacks and other organelles during mitosis and cytokinesis in plant cells. *Plant Physiol.*, **124**, 135–161.

Nebenführ, A. & Staehelin, L.A. (2001) Mobile factories: Golgi dynamics in plant cells. *Trends in Plant Sci.*, **6**, 160–167.

Nelson, W.J. (2000) W(h)ither the Golgi during mitosis? *J. Cell Biol.*, **149**, 243–248.

Nilsson, T., Hoe, M.H., Slusarewicz, P., Rabouille, C., Watson, R., Hunte, F., Watzele, G., Berger, E.G. & Warren, G. (1994) Kin recognition between medial Golgi enzymes in HeLa cells. *EMBO J.*, **13**, 562–574.

Orci, L., Ravazzola, M., Volchuk, A., Engel, T., Gmachl, M., Amherst, M., Perrelet, A., Sollner, Th. & Rothman, J.E. (2000) Anterograde flow of cargo across the Golgi stack potentially mediated by bidirectional "percolating" COP I vesicles. *Proc. Natl. Acad. Sci. USA*, **97**, 10400–10405.

Parker, M.L. & Hawes, C.R. (1982) The Golgi apparatus in developing endosperm of wheat (*Triticum aestivum* L.). *Planta*, **154**, 277–283.

Pavelka, M. (1987) Functional morphology of the Golgi apparatus. *Adv. Anat. Embryol. Cell Biol.*, **106**, 1–94.

Pavelka, M. & Ellinger, A. (1983) The trans Golgi face in small intestinal absorptive cells. *Eur. J. Cell Biol.*, **29**, 253–261.

Pavelka, M. & Ellinger, A. (1986) The Golgi apparatus in the acinar cells of the developing embryonic pancreas. I. Morpholgy and enzyme cytochemistry. *Am. J. Anat.*, **178**, 215–223.

Pavelka, M., Ellinger, A., Debbage, P., Loewe, C., Vetterlein, M. & Roth, J. (1998) Endocytic routes to the Golgi apparatus. *Histochem. Cell Biol.*, **109**, 555–570.

Pelletier, L., Jokitalo, E. & Warren, G. (2000) The effect of Golgi depletion on exocytic transport. *Nature Cell Biol.*, **2**, 840–846.

Pelham, H.R. (2001) Traffic through the Golgi apparatus. *J. Cell Biol.*, **155**, 1099–1101.

Pesacreta, T. & Lucas, W.J. (1985) Presence of a partially-coated reticulum in angiosperms. *Protoplasma*, **125**, 173–184.

Pfeffer, S. (2001) Constructing a Golgi complex. *J. Cell Biol.*, **155**, 873–875.

Pimpl, P., Movafeghi, A., Coughlan, S., Denecke, J., Hillmer, S. & Robinson, D.G. (2000) In situ localization and in vitro induction of plant COPI-coated vesicles. *Plant Cell*, **12**, 2219–2236.

Polishchuk, R.S., Polishchuk, E.V., Marra, P., Alberti, S., Buccione, R., Luini, A. & Mironov, A.A. (2000) Correlative light-electron microscopy reveals the tubular-saccular ultrastructure of carriers operating between Golgi apparatus and plasma membrane. *J. Cell Biol.*, **148**, 45–58.

Rabonille, C. & Warren, G. (1997) Changes of the architecture of the Golgi apparatus during mitosis, in *The Golgi apparatus* (eds E.G. Berger & J. Roth), pp. 195–217.

Rambourg, A. & Clermont, Y. (1997) Three dimensional structure of the Golgi apparatus in mammalian cells, in *The Golgi apparatus* (eds E.G. Berger & J. Roth), pp. 37–61.

Ritzenthaler, C., Nebenführ, A., Movafeghi, A., Stussi-Garaud, C., Behnia, L., Pimpl, P., Staehelin, L.A. & Robinson, D.G. (2002) Reevaluation of the effects of brefeldin A on plant cells using tobacco bright yellow 2 cells expressing Golgi-targeted green fluorescent protein and COPI antisera. *Plant Cell*, **14**, 237–261.

Robinson, D.G. (1980) Dictyosome–endoplasmic reticulum associations in higher plant cells? A serial section analysis. *Eur. J. Cell Biol.*, **23**, 22–36.

Robinson, D.G. (1985) Plant Membranes. Endo- and plant membranes of plant cells. John Wiley, New York.

Robinson, D.G. & Hinz, G. (1997) Vacuole biogenesis and protein transport to the plant vacuole: a comparison with the yeast vacuole and the mammalian lysosome. *Protoplasma*, **197**, 1–25.

Robinson, D.G. & Hinz, G. (1999) Golgi-mediated transport of seed storage proteins. *Seeds Sci. Res.*, **9**, 267–283.

Robinson, D.G. & Kristen, U. (1982) Membrane flow via the Golgi apparatus of higher plant cells. *In. Rev. Cytol.*, **77**, 89–127.

Roth, J. (1997) Topology of glycosylation in the Golgi apparatus, in *The Golgi apparatus* (eds E.G. Berger & J. Roth), pp. 131–161.

Roth, J. (2002) Protein N-glycosylation along the secretory pathway: relationship to organelle topography and function, protein quality control and cell interaction. *Chem. Rev.*, **102**, 285–303.

Roth, J. & Taatjes D.J. (1998) Tubules of the trans Golgi apparatus visualized by immunoelectron microscopy. *Histochem. Cell Biol.*, **109**, 545–553.

Rust, R., Landmann, L., Gosert, R., Tang, B.L., Hong, W., Hauri, H.P., Egger, D. & Bienz, K. (2001) Cellular COPII proteins are involved in production of the vesicles that form the poliovirus replication complex. *J. Virol.*, **75**, 9808–9818.

Saint-Jore, C.M., Evins, J., Batoko, H., Brandizzi, F., Moore, I. & Hawes, C. (2002) Redistribution of membrane proteins between the Golgi apparatus and endoplasmic reticulum in plants is reversible and not dependent on cytoskeletal networks. *Plant Cell*, **29**, 1–20.

Sandvig, K. & van Deurs, B. (2000) Entry of ricin and Shiga toxin into cells: molecular mechanisms and medical perspectives. *EMBO J.*, **19**, 1–8.

Seeman, J., Jokitalo, E., Pypaert, M. & Warren, G. (2000) Matrix proteins can generate the higher order architecture of the Golgi apparatus. *Nature*, **407**, 1022–1026.

Simons, K. & Ikonen, E. (2000) How cells handle cholesterol. *Science*, **290**, 1721–1726.

Staehelin, L.A. & Moore, I. (1995) The plant Golgi apparatus. *Annu. Rev. Plant Physiol. Plant Mol. Biol.*, **46**, 261–288.

Staehelin, L.A., Giddings, T.H., Kiss, J.Z. & Sack, F.D. (1990) Macromolecular differentiation of Golgi stacks in root tips of *Arabidopsis* and *Nicotiana* seedlings as visualized in high pressure frozen and freeze-substituted samples. *Protoplasma*, **157**, 75–91.

Sütterlin, Ch., Hsu, P., Mallabiabarrena, A. & Malhotra, V. (2002) Fragmentation and dispersal of the pericentriolar Golgi complex is required for entry into mitosis in mammalian cells. *Cell*, **109**, 359–369.

Takatsuki, A., Nakamura, M. & Kono, Y. (2002) Possible implication of Golgi-nucleating function for the centrosome. *Biochem. Biophys. Res.Comm.*, **291**, 494–500.

Tanchak, M.A., Rennie, P.J. & Fowke, L.C. (1988) Ultrastructure of the partially coated reticulum and dictyosomes during endocytosis by soybean protoplasts. *Planta*, **175**, 433–441.

Van Meer, G. (2001) What sugar next? Dimerization of sphingolipid glycosyltransferases. *Proc. Natl. Acad. Sci. USA*, **98**, 1321–1323.

Vetterlein, M., Ellinger, A., Neumüller, J. & Pavelka, M. (2002) Golgi apparatus and TGN during endocytosis. *Histochem. Cell Biol.*, **117**, 143–150.

Volchuk, A., Amherdt, M., Ravazzola, M., Brugger, B., Rivera, V.M., Clackson, T., Perrelet, A., Sollner, T.H., Rothman, J.E. & Orci, L. (2000) Vegavesicles implicated in the rapid transport of intracisternal aggregates across the Golgi stack. *Cell*, **102**, 335–348.

Ward, T.H., Polishschuk, R.S., Caplan, S., Hirschberg, K. & Lippincott-Schwartz, J. (2001) Maintenance of Golgi structure and function depends on the integrity of ER export. *J. Cell Biol.*, **155**, 557–570.

Warren, G. & Malhotra, V. (1998) The organisation of the Golgi apparatus. *Curr. Opin. Cell Biol.*, **10**, 493–498.

Wee, E.-G., Sherrier, D.J., Prime, T. & Dupree, P. (1998) Targeting of active sialyltransferase to the plant Golgi apparatus. *Plant Cell*, **10**, 1759–1768.

Weidman, P. (1995) Anterograde transport through the Golgi complex: do Golgi tubules hold the key? *Trends Cell Biol.*, **5**, 302–305.

Zaal, K.J.M., Smith, C.L., Polishchuk, R.S., Altan, N., Cole, N.B., Ellenberg, J., Hirschberg, K., Presley, J.F., Roberts, T.H., Siggia, E., Phair, R.D. & Lippincott-Schwartz, J. (1999) Golgi membranes are reabsorbed into and reemerge from the ER during mitosis. *Cell*, **99**, 589–601.

Zhang, G.F. & Staehelin, G.F. (1992) Functional compartmentalization of the Golgi apparatus of plant cells. An immunocytochemical analysis of high pressure frozen and freeze-substituted sycamore maple suspension culture cells. *Plant Physiol.*, **99**, 1070–1083.

3 How to make a vesicle: coat protein–membrane interactions

F. Aniento, J. Bernd Helms and Abdul R. Memon

3.1 Introduction

The secretory apparatus within all eukaryotic cells comprises a dynamic membrane system with bi-directional transport. The movement of newly synthesized proteins through the cell's secretory system involves several specific cycles of membrane vesicle budding and fusion (Rothman & Wieland, 1996; Schekman & Orci, 1996; Jahn & Südhof, 1999). Proteins destined for secretion are translocated in the endoplasmic reticulum, packaged into transport vesicles and delivered to the cell surface via the Golgi apparatus. At the same time, endocytosed proteins from the plasma membrane or escaped ER proteins to the Golgi complex during normal ER to Golgi transport can travel back to the ER. Thus, anterograde and retrograde transport of biosynthetic proteins occurs simultaneously in order to achieve a balance that retains the individual character of each participating membrane (Rothman & Wieland, 1996; Nickel & Wieland, 1997; Mellman & Warren, 2000). The major protein components of a number of vesicular trafficking pathways have been identified and characterized. A combination of biochemical, molecular, genetic and morphological methods has resulted in a widely accepted transport model in which small coated vesicles (around 60–80 nm in diameter) act as carriers that mediate uni- or bi-directional transport between two adjacent membranes in the secretory pathway (Rothman & Wieland, 1996; Nickel & Wieland, 1997; Lippincott-Schwartz *et al.*, 1998; Pelham & Rothman, 2000).

The formation of three basic types of coat complexes has been studied extensively in yeast and mammalian secretory systems: COPII vesicles, which mediate ER to Golgi traffic; COPI vesicles, which are responsible for retrograde traffic from Golgi to ER and for traffic between the cisternae of the Golgi; and clathrin-coated vesicles, which mediate certain endocytic and post-Golgi vesicular trafficking steps. Despite recent advances in understanding the mechanism of vesicular transport in the secretory system in animal and yeast cells, our knowledge about the mechanism underlying the equivalent processes in plant cells is largely fragmentary. Several plant homologue proteins of secretory pathway have been identified and it appears that plants essentially use the same protein machinery for trafficking in the secretory pathway as described for animals and yeasts (Robinson *et al.*, 1998a; Blatt *et al.*, 1999; Nebenführ & Staehelin, 2001; Phillipson *et al.*, 2001; Ritzenthaler *et al.*, 2002). However, there are a number of features that are unique to plant pathways. For example, *de novo* assembly of the cell plate during plant cell division has no counterpart in other systems (Verma, 2001). In addition, plant cells appear to possess a unique and complex vacuolar targeting machinery (Paris *et al.*, 1996; Sanderfoot *et al.*, 1998). Recent studies on vacuolar protein targeting have identified two critical molecular components of sorting machinery, which partially overlap in their subcellular locations (Sanderfoot *et al.*, 1998). This multiple targeting to vacuolar compartments in plant cells requires sorting and targeting mechanisms that differ from their counterparts in yeast or animal cells (Marty, 1999; Hadlington & Denecke, 2000; Sanderfoot *et al.*, 2000). These studies indicate that, although plant cells essentially use the same machinery for vesicular trafficking as yeast or mammalian cells, they have adapted several of these proteins to the unique requirements of their secretory system by using them for transport between membrane compartments which are not found in other organisms. In this chapter, we describe the recent progress in understanding the mechanisms underlying vesicular trafficking along the secretory pathway in yeast and mammalian cells, and compare this with the plant cells.

3.2 Clathrin-coated vesicles

3.2.1 Clathrin coat components

3.2.1.1 Clathrin

The major protein component of clathrin-coated vesicles is clathrin, a name given by Pearse (1975) in reference to the cage-like structure it forms. Clathrin consists of three 190-kDa heavy chains (HCs) each with an associated ~25 kDa light chain (LCs). It is a spider-like molecule, with three legs radiating from a central hub, named a triskelion. The central hub of a triskelion contains three regions: a small globular domain at the extreme C-terminus; a trimerization domain that constitutes the vertex; and a proximal leg, to which the LCs are bound. The distal leg segment and the globular ~50-kDa terminal domain located at the N-terminus of each HC are connected to the hub through a protease-sensitive bend, called a knee. An anti-parallel interaction of the legs from triskelions centred on adjacent vertices of the lattice allows assembly of the cage (for reviews see Schmidt, 1997; Robinson *et al.*, 1998a; Kirchhausen, 1999; Kirchhausen, 2000a,b). A novel clathrin heavy chain homologue has been described which displays distinct biochemistry, distribution and function compared with conventional clathrin heavy chain. It is expressed at high levels in muscle, associates with the AP-1 and AP-3 adaptor complexes (see below) but not with AP-2, codistributes with cytoskeletal elements and functions at the *trans*-Golgi network (Liu *et al.*, 2001).

3.2.1.2 AP complexes

The most abundant proteins in a clathrin-coated vesicle, after clathrin, are those of the heterotetrameric adaptor protein (AP) complexes, which promote assembly of clathrin coats. Present in brain coated vesicles at a ratio of about one per two triskelions, the AP complexes contain one each of an α-type large chain (α, γ, δ or ϵ), a β-type large chain (β), a μ-chain, and a σ chain. There are at least four such complexes in mammalian cells: AP-1 (γ, $\beta1$, $\mu1$, $\sigma1$), found in coated vesicles associated with the TGN; AP-2 (α, $\beta2$, $\mu2$, $\sigma2$), found in coated vesicles associated with the plasma membrane; AP-3 ($\delta3$, $\beta3$, $\mu3$, $\sigma3$), found at the TGN and endosomes and AP-4 (ϵ, $\beta4$, $\mu4$, $\sigma4$) also found at the TGN. The AP complexes link clathrin assembly to the incorporation of certain classes of cargo (Robinson & Bonifacino, 2001). Recognition of the cytoplasmic tail of membrane proteins is often the function of the μ-chain, through the binding of the Yppϕ sorting signal (see below), but the β-chain appears to have recognition sites for other sorting signals, such as the LL-motif (see below). The heterotetrameric adaptors associate with clathrin through a relatively short β-chain segment, the so-called hinge, which links a large (60 kDa) N-terminal *trunk* domain to a projecting ~30 kDa C-terminal *ear*. AP β-chains interact with the same site on clathrin and the consensus sequence LLpL(−) within their interacting segments has been called a clathrin box (see Kirchhausen, 2000a, for a review).

3.2.2   Clathrin-coated vesicle formation

Clathrin-coated vesicle (CCV) formation involves the following steps:

3.2.2.1   Activation of coat assembly
At the TGN, this step involves activation of the GTPase ARF1 and subsequent recruitment of ARF1-GTP to the *trans*-Golgi membrane. The mechanism of ARF1 recruitment is unknown. However, recent observations indicate that BIG2 (a guanine nucleotide exchange factor for ARF) is implicated in membrane association of AP-1, but not in that of COPI, through activating ARF1 (Shinotsuka *et al.*, 2002).

3.2.2.2   Binding of adaptor complexes and cargo recruitment
The AP-1 adaptor complex is recruited from the cytosol onto the *trans*-Golgi network membrane, where it co-assembles with clathrin into a coat that drives vesicle budding. AP-1 (but not AP-2) recruitment requires ARF1 activation and is sensitive to brefeldin A (Stamnes & Rothman, 1993). A differential direct interaction of class 1, 2 and 3 ARFs with AP-1 and AP-3 (but not AP-2) adaptor protein complexes has been recently demonstrated (Austin *et al.*, 2002). In addition, a number of membrane proteins in the TGN have been proposed to act as AP-1 docking sites (Mallet & Brodsky, 1996; Seaman *et al.*, 1996).

Two main types of motifs in the cytoplasmic domain of membrane receptors have been shown to interact with AP-1 and/or AP-3 adaptor complexes: Yppϕ and LL motifs (Kirchhausen, 1999). The *Yppϕ motif* (where Y is tyrosine, p is a polar residue and ϕ is a residue with a bulky hydrophobic side chain) is found in a large number of membrane proteins, including the synaptic vesicle Zn (ZnT-3) transporter and the lysosomal membrane proteins CD63, lamp-1 and lamp-2. It is specifically recognized by the μ-chain of APs. It is not only used as an endocytic motif but also for direct traffic within the endosomal and secretory pathways (Kirchhausen *et al.*, 1997; Marks *et al.*, 1997). Some Yppφ motifs bind with higher affinity AP-2, while others are more readily recognized by AP-1, such as the YQTI sorting motif found in lamp-1. The dileucine, or LL, motif takes the form (−)(2–4) xLL, where x is usually a polar residue and (−) is often a negatively charged residue (aspartic or glutamic acid or phosphoserine). The acidic residues containing LL motifs are constitutively active, whereas those with serine are regulated by phosphorylation (Geisler *et al.*, 1998). The LL motif is found in a large number of trafficking proteins, including the mannose 6-P receptor, Limp-II, tyrosinase or GLUT4 (glucose) transporter, and it is recognized by the β-chain of APs (Kirchhausen, 2000a).

The predominant cargo of lysosomally directed TGN-derived coated vesicles is the cation-independent or the cation-dependent mannose 6-phosphate receptors (MPRs). Both MPRs mediate recruitment of the lysosomal hydrolases (containing mannose 6-phosphate groups) to clathrin-coated vesicles which deliver the MPR–hydrolase complexes to endosomes. The acidic pH of endosomes induces the release of the hydrolases from the MPRs, after which the hydrolases are transported to lysosomes while the MPRs return, again via CCV, to the TGN for additional rounds of sorting (see Rohn *et al.*, 2000 for a review). Sorting of both MPRs from the TGN to endosomes is mediated by the signals present in the cytosolic tails of the receptors which consist of a cluster of acidic amino acid residues followed by two leucine residues (Johnson & Kornfeld, 1992; Chen *et al.*, 1997). The cytoplasmic tail of MPRs has been implicated as an essential component in the AP-1 docking site, in co-operation with ARF (LeBorgne *et al.*, 1996; LeBorgne & Hoflack, 1997). However, recent data indicate that AP-1 does not bind the acidic cluster-dileucine signals from the MPRs (Zhu *et al.*, 2001). Thus, clathrin-associated proteins other than AP-1 might be responsible for the signal-mediated sorting of MPRs at the TGN. Prime candidates for this role are the Golgi localized γ-adaptin-ear-containing, ARF-binding proteins (GGAs) (Boman *et al.*, 2000; Dell'Angelica *et al.*, 2000; Hirst *et al.*, 2000; Poussu *et al.*, 2000). These proteins are monomeric and display a modular structure consisting of a VHS (VPS27, Hrs, and STAM) domain which binds the acidic cluster-dileucine motif in the cytoplasmic tail of MPRs (Zhu *et al.*, 2001; Puertollano *et al.*, 2001b; Takatsu *et al.*, 2001), a GAT domain that interacts with the GTP-bound form of ARF (Boman *et al.*, 2000; Dell'Angelica *et al.*, 2000; Puertollano *et al.*, 2001a), a hinge domain that interacts with clathrin (Puertollano *et al.*, 2001b), and a GAE domain that interacts with γ-synergin and other potential regulators of coat assembly (Hirst *et al.*, 2000). AP-1 adaptor, instead, may be involved in retrograde transport of the MPRs from endosomes to the TGN (Meyer *et al.*, 2000). Another protein involved in MPR sorting is TIP47 (Tail-interacting 47-kDa protein), a soluble protein that recognizes a phenylalanine/tryptophane-based sorting signal in the cytoplasmic tail of the mannose 6-phosphate receptor, and that is involved in its traffic from the endosome to the TGN (Diaz & Pfeffer, 1998; Orsel *et al.*, 2000). Sorting of other proteins at the TGN may use the AP-1 or AP-3 adaptor complexes (Cowles *et al.*, 1997;

Stepp *et al.*, 1997; Kantheti *et al.*, 1998; Vowels & Payne, 1998; Dell'Angelica *et al.*, 1999; Simmen *et al.*, 1999; Teuchert *et al.*, 1999).

3.2.2.3   Coat polymerization and vesicle budding

Once adaptor proteins are bound to the TGN membrane, cytosolic clathrin is recruited and the assembly of the coat starts. A model for the assembly and disassembly of the clathrin coat has been described (Kirchhausen, 2000a) and an animation of this process can be downloaded from the website of Harvard University (http://www.hms.harvard.edu/news/clathrin/). Propagation of the coat probably includes ARF-GTP hydrolysis, further cargo recruitment and coat assembly, and membrane deformation coupled to the growth of the coat which originates vesicle budding (Kirchhausen *et al.*, 2000a,b).

At the plasma membrane, budding of clathrin-coated vesicles requires binding of amphiphysin to clathrin and AP-2, which acts as a receptor for dynamin. Dynamin-GDP is then recruited to the neck of the budding vesicle and polymerizes into a dynamin ring. Endophylin is then recruited to the ring. The acyltransferase activity of endophylin coupled to the neck constriction imparted by the dynamin ring facilitates membrane deformation and fission. Uncoating of plasma membrane derived CCVs involves the ATPase Hsc70 (See Kirchhausen, 2000a for a review). Neither vesicle budding nor uncoating has been characterized for TGN-derived CCVs.

3.2.3   CCVs in plant cells

Plant clathrin heavy chains have a number of well-conserved regions in common with animal and yeast cells (Blackbourn & Jackson, 1996). A putative plant clathrin light chain ortholog has been recently identified by homology searches of the *Arabidopsis* genome, and was found to interact specifically with mammalian clathrin heavy chains in heterologous binding experiments using recombinant triskelion hubs (Scheele & Holstein, 2002). Plant counterparts of adaptins are not as well characterized as their mammalian counterparts, but current evidence supports a role for γ- and β-type adaptins in plasma membrane recycling (Drucker *et al.*, 1995; Keon *et al.*, 1995). Another feature of animal adaptins is the presence of autophosphorylating and casein kinase activity, which has also been shown in adaptin-like polypeptides from zucchini (Drucker *et al.*, 1998). Two genomic clones that encode homologues of AP19, the smallest polypeptide (σ1 subunit) component of AP-1 and the clathrin associated protein complex found in clathrin-coated vesicles of the Golgi apparatus, have been isolated from *Arabidopsis thaliana* (Maldonado-Mendoza & Nessler, 1997). Amino acid sequence comparisons with mammalian, yeast and plant clathrin associated sequences indicate that the *Arabidopsis* genes encode polypeptides that are more closely related to the AP19 proteins associated with clathrin-coated Golgi vesicles than to AP17, which is part of the AP-2 complex of endocytic clathrin-coated pits (σ2 subunit). The cDNA and genomic sequences coding for AP17 from maize and its corresponding mRNA accumulation have been also characterized (Roca *et al.*, 1998).

A variety of other molecules involved in clathrin-coated vesicle formation have also been reported in plant cells. One plant homologue of dynamin has been isolated from *Arabidopsis* with significant similarity to yeast Vps1p, rat dynamin and murine Mx1 protein (Dombrowski and Raikhel, 1995). Another dynamin homologue, phragmoplastin, has been found at the cell plate in soybean (Gu & Verma, 1996), which probably acts to squeeze vesicles to allow the exocytic release of vesicle contents during the growth of the cell plate (Battey *et al.*, 1999). Finally, preliminary evidence suggests that uncoating ATPases involved in the removal of the clathrin coat from CCVs also exist in plant cells (Beevers, 1996). In plants, the best characterized membrane receptor involved in clathrin-mediated transport along the secretory pathway is the vacuolar protein sorting receptor BP-80, a type I integral membrane protein abundant in clathrin-coated vesicles with high affinity to vacuole-targeting determinants containing asparagine-proline-isoleucine-arginine (Kirsch *et al.*, 1994; Ahmed *et al.*, 1997; Robinson *et al.*, 1998a). Its structure resembles that of EGF in the lumenal extracytoplasmic domain with several cysteine-rich domains, while its cytoplasmic domain contains two tyrosine-based sorting signals (residues 589–594 and 606–609), with only the latter one fitting the consensus motif Yppϕ (Ahmed *et al.*, 1997; Paris *et al.*, 1997). Due to its homology with EGF, its *Arabidopsis* homologue has been named AtELP (*A. thaliana* epidermal growth factor receptor-like protein) (Ahmed *et al.*, 1997). Its sequence, and the sequences of homologues from pea (*Pisum sativum*), rice (*Oryza sativa*), and maize (*Zea mays*) define a novel family of proteins unique to plants that is highly conserved in both monocotyledons and dicotyledons. The BP-80 protein is present in the dilated ends of Golgi cisternae and in *prevacuoles*, which are small vacuoles separated from but capable of fusing with lytic vacuoles. When transiently expressed in tobacco (*Nicotiana tabacum*) suspension-culture protoplasts, a truncated form lacking transmembrane and cytoplasmic domains was secreted (Paris *et al.*, 1997). These results, coupled with the previous studies of ligand-binding specificity and pH dependence, strongly support the hypothesis that BP-80 is a vacuolar sorting receptor that traffics in CCVs between the *trans*-Golgi and a prevacuolar compartment (Paris *et al.*, 1997). In contrast, vacuolar storage proteins are sorted in the *cis*-cisternae of the Golgi apparatus as dense vesicles containing α-TIP but not BP-80, which is concentrated in *trans*-cisternae (Robinson *et al.*, 1998a; Hinz *et al.*, 1999; Hillmer *et al.*, 2001). Recently, BP-80 has been shown to interact specifically with the proaleurain vacuolar-sorting determinant *in vivo*, an interaction leading to the transport of the protein through the yeast secretory pathway to the vacuole. This finding confirms the function of BP-80 as a plant vacuolar sorting receptor (Humair *et al.*, 2001).

BP-80 from pea cotyledons has been shown to bind the adaptors from bovine brain, wheat germ and pea cotyledons (Butler *et al.*, 1997). Competition experiments using the cytoplasmic tails from lamp-1 and TGN38 have demonstrated a specific interaction of the AtELP cytoplasmic tail with the mammalian AP-1 adaptor complex (Sanderfoot *et al.*, 1998). Similarly, the cytoplasmic domain of the CI-MPR was shown to bind adaptors from zucchini hypocotyls (Robinson *et al.*,

1998b). These type of experiments point to a similarity in adaptor–receptor binding mechanisms between plants and animals.

3.3   COPI vesicles

COPI vesicles are uniform in size with a diameter of about 75 nm and have an electron dense, fuzzy protein coat, different from clathrin-coated vesicles (Orci *et al.*, 1986, 1989; Malhotra *et al.*, 1989). These vesicles bud preferentially from the rims of Golgi cisternae and were termed COP-coated, and later COPI-coated vesicles (Rothman & Wieland, 1996; Orci *et al.*, 1997). COPI vesicles have been implicated in both anterograde and retrograde transport through the Golgi stack (Nickel & Wieland, 1997; Orci *et al.*, 1997).

3.3.1   Molecular architecture of the COPI coat

The coat of COPI-coated vesicles consists of a multiprotein complex named coatomer, ADP-ribosylation factor 1 (ARF1), and p24 family of integral membrane proteins (Serafini *et al.*, 1991a,b; Waters *et al.*, 1991; Bremser *et al.*, 1999).

3.3.1.1   COPI proteins

The COPI complex is composed of seven subunits (α-, β-, β'-, γ-, δ-, ϵ-, ζ-COPs) (Waters *et al.*, 1991) which have been characterized at the molecular level (Table 3.1). To date, more than 40 genes encoding the yeast, fungal, plant, nematode, *Drosophila*, rat, hamster, bovine and human homologues of seven COP subunits have been cloned and sequenced (Chow *et al.*, 2001).

Table 3.1 Coat proteins of COPI vesicles

Protein Mammals	Yeast	M_r (kDa)	Interacting proteins	Features	References
α-COP	Ret1p	~140	β'-COP, ϵ-COP	WD-40 repeats	Gerich *et al.*, 1995; Faulstich *et al.*, 1996
β-COP	Sec26p	~107	δ-COP, ARF	homology with β-/β'-adaptin	Duden *et al.*, 1991; Serafini *et al.*, 1991a
β'-COP	Sec27p	~102	α-COP	WD-40 repeats	Harrison-Lavoie *et al.*, 1993
γ-COP	Sec21p	~97	ζ-COP; KKXX and p23; ARF1		Stenbeck *et al.*, 1992; Harter & Wieland, 1998
δ-COP	Ret2p	~57	β-COP	homology with μ-adaptin	Faulstich *et al.*, 1996
ϵ-COP	Sec28p	~35	α-COP		Hara-Kuge *et al.*, 1994
ζ-COP	Ret3p	~20	γ-COP	homology with σ-adaptin	Kuge *et al.*, 1993
ARF1	yARF1/2/3	~21	β-COP; γ-COP	GTPase of Ras superfamily	Kahn & Gilman, 1986; Serafini *et al.*, 1991b

Once assembled, coatomer is a stable complex with a half-life of ~28 h in mammalian cells without exchange of subunits (Lowe & Kreis, 1996). Using the two-hybrid system, four interacting pairs of coatomer subunits have been identified (Faulstich *et al.*, 1996): β/δ-COPs, γ/ζ-COPs, α/ε-COPs and α/β′-COPs. *In vitro*, coatomer dissociates in high salt buffers into subcomplexes that retain partial functions (Cosson & Leutourneur, 1994; Lowe & Kreis, 1995; Fiedler *et al.*, 1996; Pavel *et al.*, 1998). Under these conditions, a stable α-, β′- and ε-COPs subcomplex is generated which interacts with KKXX motifs (Cosson & Leutourneur, 1994; Lowe & Kreis, 1995). In addition, a β-, γ-, δ- and ζ-COP subcomplex of coatomer can be generated (Fiedler *et al.*, 1996), which tends to disassemble further into two stable heterodimers, consisting of β/δ COP and γ/ζ COP (Lowe & Kreis, 1995; Pavel *et al.*, 1998). The β/δ COP heterodimer can bind to Golgi membranes in an ARF- and GTPγS-dependent manner (Pavel *et al.*, 1998). The γ-COP subunit binds the cytoplasmic tail peptide of p23 and this binding can be inhibited by an excess of a typical KKXX peptide, indicating that both p23 and ER-resident membrane proteins share a common binding site within γ-COP.

3.3.1.2 *ADP-ribosylation factor 1 (ARF1)*

ADP-ribosylation factors (ARFs) are small GTP-binding proteins (21 kDa) implicated in the maintenance of organelle structure, the formation of two types of coated vesicles in the secretory and endocytic pathways and the activation of enzymes that modify phospholipids such as phospholipase D or phosphatidylinositol-4-OH kinase (Donaldson & Jackson, 2000). Six ARF proteins have been identified in mammals, three in yeast, and three in *Arabidopsis thaliana*, and several ARF-related proteins have been found in mammals and plants (Bischoff *et al.*, 1999; Donaldson & Jackson, 2000). Specific classes of guanine nucleotide exchange factors (GEFs) and GTPase activating proteins (GAPs) control the GDP-GTP cycle of ARF protein. All GEFs for ARFs share a common catalytic domain: the Sec7 domain (Chardin *et al.*, 1996). Several GEFs have been identified in mammals, in yeast and in plants (Jackson and Casanova, 2000) and are classified into four subfamilies: Gea/Gnom/GBF, Sec7/BIG, ARNO/cytohesin/GRP, EFA6 (Jackson & Casanova, 2000). The *large* GEFs are sensitive to brefeldin A, a fungal metabolite that induces the fusion of the Golgi apparatus with the ER, and that acts by stabilizing an abortive complex between ARF1-GDP and the Sec7 domain of GEFs (Chardin & McCormick, 1999). Several ARF-GAP proteins for mammals, yeast and plant ARFs have been identified and they all show a characteristic zinc finger motif (Bischoff *et al.*, 1999; Donaldson & Jackson, 2000) which is essential to promote hydrolysis of ARF-bound GTP (Goldberg, 1999).

3.3.2 *COPI-coated vesicle formation*

3.3.2.1 *ARF1 activation and COP binding*

Upon activation by a nucleotide exchange factor (Donaldson *et al.*, 1992b; Helms & Rothman, 1992; Chardin *et al.*, 1996; Franco *et al.*, 1996; Peyroche *et al.*, 1996;

Togawa *et al.*, 1999; Yamaji *et al.*, 2000), cytosolic ARF1-GDP is converted into a membrane-bound ARF1-GTP (Regazzi *et al.*, 1991; Serafini *et al.*, 1991a; Haun *et al.*, 1993; Helms *et al.*, 1993; Randazzo *et al.*, 1995; Goldberg, 1998). This, in turn, triggers membrane recruitment of coatomer (Donaldson *et al.*, 1992a; Palmer *et al.*, 1993; Teal *et al.*, 1994). Direct interactions between ARF1 and coatomer have been reported and were shown to be GTP-dependent (Zhao *et al.*, 1997, 1999). While only ARF1-GTP is stably associated with the membrane (Regazzi *et al.*, 1991; Serafini *et al.*, 1991a; Haun *et al.*, 1993; Helms *et al.*, 1993; Randazzo *et al.*, 1995; Goldberg, 1998), ARF1-GDP can associate with membranes as well. Biochemical studies established that the membrane recruitment of ARF1 must take place as a prerequisite for nucleotide exchange to proceed (Beraud-Dufour *et al.*, 1999). In addition, it has recently been shown that before the binding of ARF in GTP form to the membrane, it first interacts directly to the cytoplasmic domain of p23 cargo receptor in GDP form (Gommel *et al.*, 2001). These results establish a specific binding to the Golgi of ARF1-GDP as the first step of ARF1 recruitment and identify the cyoplasmic domain of p23 as a receptor for ARF1-GDP.

Several reports indicate that GTP hydrolysis by ARF1 is required for efficient cargo sorting and uptake into COPI vesicles, suggesting that the ARF1 GTPase cycle may allow for cargo selection and loading during COPI budding (Nickel *et al.*, 1998; Lanoix *et al.*, 1999; Malsam *et al.*, 1999). It has also been suggested that GDP/GTP exchange cycles and GTP hydrolysis open a time window that allows cargo to diffuse into the budding zone. In this context, it is to be noted that coatomer has been reported to potentially stimulate ARF-GAP activity (Goldberg, 1999).

3.3.2.2 p24 family of proteins in vesicle formation

Recently, a family of type I transmembrane proteins of 23–27 kDa, termed the p24 family of proteins, has been implicated in the formation of vesicles and the selection of cargo in both directions between the ER and the Golgi apparatus (Fiedler *et al.*, 1996). These proteins have a lumenal domain of about 180 amino acids and a single membrane-spanning domain with a short (10–15 amino acids) cytosolic tail (Stamnes *et al.*, 1995; Dominguez *et al.*, 1998; Marzioch *et al.*, 1999). The cytosolic sequences of all p24 proteins are similar, and some contain a KKXX motif or closely related sequence for the COPI-dependent Golgi-to-ER retrieval (Fiedler *et al.*, 1996). In the Golgi membranes, p23, p24, and other members of the p24 family form hetero-oligomeric complexes (Dominguez *et al.*, 1998; Blum *et al.*, 1999; Fligge *et al.*, 2000). The coiled-coil structures may promote association between identical or different members of p24 proteins, like Emp24p and Erv25p in yeast (Belden & Barlowe, 1996), or between p23 and p24 in mammals (Gommel *et al.*, 1999; Emery *et al.*, 2000). It has been hypothesized that the p24 proteins may have a coupled function: serving as cargo receptors on the lumenal side, possibly sorting proteins selectively into a certain type of coated vesicles and to interact with both small GTPase (e.g. ARF1) and cytosolic coatomers on the cytoplasmic side.

The disruption of both p23 alleles caused early embryonic lethality in mice and the inactivation of one allele led not only to the reduction of p23 levels but also reduced the levels of other family members (Denzel *et al.*, 2000). These results reveal that p23 plays an essential and non-redundant role in the early stages of mammalian development. In contrast, yeast strains carrying (multiple) mutations in p24 genes grow normally, and overt defects in either COPI or COPII vesicle functions have not been seen (Schimmöller *et al.*, 1995; Belden & Barlowe, 1996). Interestingly, an octuple mutant yeast strain (where all 8 homologues of p24 were knocked out) showed no detectable defect in the rate of ER to Golgi transport (Springer *et al.*, 2000). Similar results were also obtained with p24 mutants in *Caenorhabditis elegans* (Wen & Greenwald, 1999).

3.3.2.3   Minimal machinery for forming a transport vesicle
COPI-coated vesicle formation and budding from defined liposomes requires three protein components: ARF-GTPase, GTP and the cytoplasmic domains of putative cargo receptors (p24 family proteins) or membrane cargo proteins containing the KKXX retrieval signal projecting from the membrane bilayer surface (Bremser *et al.*, 1999). Wieland's group has recently shown that vesicle formation from liposomes is independent of any specific lipid requirement of donor liposomes, provided that the cytoplasmic tail of p23 (or another p24 family member) is present (Bremser *et al.*, 1999). This also implies that the interaction of p24 lumenal domains with cargo proteins does not seem to be necessary for the coat assembly.

3.3.2.4   COPI coatomer polymerization and bud formation
Coatomer, in conjunction with the GTP binding protein ARF1 and p24 membrane receptors, forms an electron-dense coat that, when assembled onto Golgi membranes, is thought to drive membrane deformation, budding and fission events associated with the Golgi membrane traffic (Springer *et al.*, 1999; Wieland & Harter, 1999). How could binding of coatomer complexes to membranes induce membrane curvature? *In vitro* binding studies of coatomer and the cytoplasmic tail peptide of p23 suggest that this peptide binds to coatomer in an oligomeric state. This tetrameric interaction induces a conformational change in coatomer, possibly causing polymerization of coatomer complexes (Reinhard *et al.*, 1999).

3.3.2.5   A model for COPI-coated vesicle formation
The above review of the literature indicates that the minimal requirements for COPI coated vesicle formation are heptameric cytosolic coat proteins, ARF1 and p24 family membrane proteins. A model to illustrate early steps of ARF1 recruitment to membranes is depicted in Fig. 3.1. Our recently published data (Gommel *et al.*, 2001) and earlier findings of Antonny and co-workers (Beraud-Dufour *et al.*, 1999) demonstrate that ARF1-GDP is recruited to the membrane before nucleotide exchange. Binding of ARF1-GDP to the purified Golgi membranes is virtually abolished in the presence of dimeric form of p23, indicating that an interaction between ARF1 and p23 at the membrane is necessary for efficient recruitment of

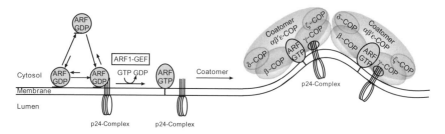

Figure 3.1 A model for the recruitment of ARF1 to Golgi membranes. ARF-GDP binds to membrane phospholipids at low affinity. Upon binding to a p23 oligomer, the membrane interaction is stabilized. Subsequently, a nucleotide exchange factor acts on ARF1-GDP and the resulting ARF-GTP is released from the p23 receptor. Two binding sites of coatomer have now been generated (membrane-bound ARF1-GTP and a p23 oligomer) and this interaction induces the recruitment of coatomer and a subsequent conformational change and polymerization of the complex, which shapes the membrane into a coated bud.

the GTPase (Gommel *et al.*, 2001). This interaction is likely to promote the activation of ARF1 by a nucleotide exchange factor (Franco *et al.*, 1996; Paris *et al.*, 1997). This GDP-GTP exchange results in a change in conformational state of ARF1, which leads to its dissociation from p23 (Amor *et al.*, 1994; Greasley *et al.*, 1995; Randazzo *et al.*, 1995; Goldberg, 1998). Our recent observations suggest that the nucleotide exchange and the ARF1 release from p23 are concerted events. As a result, the membrane becomes primed for coatomer recruitment both by membrane-associated ARF1-GTP, which interacts with β- and γ-COP (Zhao *et al.*, 1997, 1999) and, in close proximity, by p23 cytoplasmic tails, which interact with γ-COP (Harter & Wieland, 1998). It is also possible that the p23-mediated ARF1-GDP binding and the subsequent nucleotide exchange modulate the oligomeric status of the complex of p24 proeins in order to generate the coatomer-binding sites (Wieland & Harter, 1999).

ARF1-mediated GTP hydrolysis has been shown to be required for the efficient uptake by the biosynthetic cargo of COPI vesicles (Malsam *et al.*, 1999; Pepperkok *et al.*, 2000), and therefore, ARF1-GDP is continuously produced during coat assembly. A mechanism to prevent the release of ARF1-GDP from membranes during early stages of coat recruitment would enhance the assembly of a pre-budding complex (Gommel *et al.*, 2001). Recent studies with *ldlF* cell lines expressing ε-COP-GFP or ε-COP-YFP have showed that after the association of ARF and coatomer to the Golgi membranes, both proteins have different residence times on the membrane. The average residence time for coatomer on Golgi membrane is 43 s, in contrast with 26 s for ARF1 (Presley *et al.*, 2002). This indicates that the coatomer is stabilized on the membranes for additional periods after ARF1 undergoes GTP hydrolysis and dissociation. This repeated binding and dissociation cycle of ARF1 to the Golgi membrane could facilitate the cargo uptake and coat polymerization which result in the formation of a coated bud, driven by the curvature of the assembling coat (Huttner & Zimmerberg, 2001).

3.3.3 COP1-coat components in plant cells

3.3.3.1 Plant COPI homologues

Sequence databases have been used to search for potential plant homologues of yeast and mammalian proteins (Andreeva *et al.*, 1998; Sanderfoot *et al.*, 1999, 2000), including components of the coats of COPI vesicles. Although many potential plant homologues of the proteins involved in anterograde and retrograde transport have been found (Andreeva *et al.*, 1998), very few component proteins have been identified experimentally. Several putative COPI subunits have been identified in rice cells by using antibodies raised against several mammalian COPI proteins (Contreras *et al.*, 2000).

Evidence is accumulating for COPI homologues playing similar roles in plant cells, but this is neither well characterized nor well understood (Robinson *et al.*, 1998a; Movafeghi *et al.*, 1999; Pimpl *et al.*, 2000). Pimpl *et al.* (2000) showed *in situ* localization of COPI-coated vesicles and demonstrated *in vitro* the recruitment of coatomer, the protein complex that makes up the COPI vesicle coat, from a cytosolic fraction onto budding vesicles. Antisera prepared against recombinant plant COP-coat protein homologues have allowed for the identification of cytosolic protein complexes similar to coatomer and the Sec23/24 dimer from mammalian and yeast cells (Movafeghi *et al.*, 1999). These and additional antisera against ARF1p/Sar1p homologues have successfully been employed in immunogold electron microscopy to localize plant ARF1p and coatomer to the budding vesicles at the periphery of *cis*-cisternae in *Arabidopsis* and maize root cells (Pimpl *et al.*, 2000). It has also been possible to recruit ARF1p and coatomer from cauliflower cytosol onto the mixed ER/Golgi membranes, as well as to provide evidence for the release of plant COPI-vesicles *in vitro* (Pimpl *et al.*, 2000). Recently it has been shown that plant coatomer complexes as in mammalian and yeast cells, interact with di-lysine motifs of ER retrieval signal and that interaction was lost when two lysines were substituted by serine or arginine (F. Aniento, unpublished data).

In mammalian and yeast cells, the COP1-coat formation is triggered by the conversion of ARF1p in its GDP-form to a GTP-form, a process which can be inhibited by brefeldin A (Peyroche *et al.*, 1999), leading to the release of membrane bound coatomer (Scheel *et al.*, 1997). It has recently been established that the primary effect of brefeldin A on plant cells is the rapid displacement of coatomer into the cytosol, similar to that in other eukaryotes. The subsequent manifestations of this toxin on the endo-membrane system is, however, very different (Ritzenthaler *et al.*, 2002).

In summary, whilst there are basic similarities in the overall secretory pathway in all eukaryotes, several lines of evidence point to numerous variations at the ultrastructural and molecular levels (Dupree & Sherrier, 1998; Nebenführ & Staehelin, 2001) which are unique to plants.

3.3.3.2 Plant ARF1p and p24-family homologues

The cDNAs of *ARF1* with high sequence similarity to mammalian ARF have been isolated from *Arabidopsis* (Regad *et al.*, 1993), the green alga *Chlamydomonas reinhardtii* (Memon *et al.*, 1995), and several crop plants: carrots (Kiyosue & Shinozaki, 1995), maize (Verwoert *et al.*, 1995), potatoes (Szopa & Mueller-Roeber, 1994), rice (Higo *et al.*, 1994). Rice and wheat *ARF1* are completely identical at the DNA level (Kobayashi-Uehara *et al.*, 2001). Genomic Southern hybridization indicates that wheat *ARF* is encoded by at least 2 or 3 copies of *ARF1* genes. Moreover, it seems that *ARF* transcription and translation is greater in roots than in shoots (Kobayashi-Uehara *et al.*, 2001). ARF1 has been identified in pea plumules by western blotting and is recognized by an antibody prepared against mammalian ARF1 antibodies (Memon *et al.*, 1993). In *Chlamydomonas*, there is a biphasic pattern of ARF1 mRNA accumulation during the light/dark growth cycles, which is also reflected at the protein level (Memon *et al.*, 1995). The *Arabidopsis* GNOM mutant is defective in the transport of auxin, through mislocalization of auxin efflux carriers of the PIN1 type (Steinmann *et al.*, 1999). A similar phenotype is caused by brefeldin A treatment. It turns out that the GNOM mutation lies in an allele for ARF-GEF. When BFA is removed, the normal localization is restored rapidly (Grebe *et al.*, 2000), suggesting that this vesicle-dependent targeting is dynamic in character.

The p24 family membrane proteins also exist in plants, but the low number of available sequences precludes any thorough analysis at this stage. Data bank analysis of the *Arabidopsis* genome reveals about 8 p23-like sequences, but nothing is known about their expression, localization or function (Memon & Wieland – unpublished data). These results indicate that the molecular machinery needed for both anterograde and retrograde vesicle transport between ER and Golgi exists in plants and is mediated by a similar set of proteins as in animal and yeast cells.

3.4   COPII vesicles

3.4.1   COPII components

3.4.1.1   Sar1p
Sar1p is a small (24 kDa) GTP-binding protein. It is converted to its GTP-bound form through interaction with the integral membrane glycoprotein Sec12p (Barlowe & Scheckman, 1993), an event catalyzed by Sec23p of the Sec23/24p complex (Yoshihisa *et al.*, 1993). The Sar1 NH_2-terminus contains two regions: an NH_2-terminal extension containing an evolutionary conserved hydrophobic motif, that facilitates membrane recruitment and activation by the mammalian Sec12p guanine nucleotide exchange factor, and an $\alpha1'$ amphipathic helix that contributes to the interaction with the Sec23/24p GAP complex that is responsible for cargo selection during ER export (Huang *et al.*, 2001).

3.4.1.2   Sec13/31p and Sec23/24p complexes
The COPII coat consists of two coat complexes, which exist separately in the cytosol. The Sec23/24p complex comprises the proteins Sec23p (~85 kDa) and

Sec24p (~105 kDa) (Hicke *et al.*, 1992). The three-dimensional reconstruction of Sec23/24p reveals that it has a bone-shaped structure (17 nm in length), composed of two similar globular domains, one corresponding to Sec23p and the other to Sec24p (Lederkremer *et al.*, 2001). The Sec13/31p complex comprises the proteins Sec13p (~33 kDa) and Sec31p (~140 kDa), both containing numerous WD motifs (Salama *et al.*, 1993). Sec13/31p is a heterotetramer composed of two copies of Sec13p and two copies of Sec31p, which elute as a globular protein of ~700 kDa during gel-filtration chromatography. It has an elongated shape, is 28–30 nm in length, and contains five consecutive globular domains linked by relatively flexible joints (Lederkremer *et al.*, 2001).

3.4.1.3 Structure of the COPII coat

The spatial arrangement of COPII coat protein subunits has been recently analyzed by deep-etch rotary shadowing and electron microscopy after crosslinking the isolated proteins to an artificial membrane surface (Matsuoka *et al.*, 2001). Sec23/24p resembles a bow tie, and Sec13/31p contains terminal bilobed globular structures bordering a central rod. The surface structure of COPII vesicles reveals a coat structure built with polygonal units. The length of the side of the hexagonal/pentagonal units is close to the dimension of the central rod-like segment of Sec13/31p. Partially uncoated profiles reveal strands of Sec13/31p stripped from the vesicle surface. It was concluded that the coat subunits form layers that are displaced from the membrane surface in reverse order of addition to the coat. A detailed electron microscopy analysis of negatively stained COPII subunits showed a globular domain similar to the other segments of the chain of flexibly hinged globular domains (Lederkremer *et al.*, 2001), whereas quick-freeze/deep-etch images reveal a more extended cylindrical central domain (Matsuoka *et al.*, 2001).

3.4.2 COPII-coated vesicle formation

COPII-coated vesicles transport proteins from the ER to the Golgi. The small GTPase Sar1p and the heterodimeric protein complexes Sec23/24p and Sec13/31p are necessary and sufficient to produce COPII vesicles from ER microsomes (Barlowe *et al.*, 1994) and from chemically defined liposomes (Matsuoka *et al.*, 1998b). The sequence of steps in COPII-vesicle formation is similar to the one established for COPI (Matsuoka *et al.*, 1998b; Spang *et al.*, 1998; Springer *et al.*, 1999; Antonny & Schekman, 2001; Antonny *et al.*, 2001).

3.4.2.1 Activation of coat assembly

As with COPI, the small GTPase (Sar1p) must be recruited first to the donor membrane (ER). Initially, Sar1p is complexed with GDP and resides in the cytosol. Binding of GTP is stimulated by the NH_2-terminal, cytoplasmically exposed, domain of the guanine nucleotide exchange factor (GEF) Sec12p, which is a type II transmembrane protein of the ER (Barlowe & Schekman, 1993; Weissman *et al.*, 2001). The restriction of Sec12p to the ER membrane causes association of

GTP-bound Sar1p to the ER membranes and thus restricts COPII-vesicle production to the ER. Once bound to the membrane, Sar1p-GTP attracts the remaining coat subunits.

3.4.2.2 Binding of COPII subunits and cargo capture

To generate ER-to-Golgi vesicles, the COPII protein heterodimers Sec23/24p and Sec13/31p bind sequentially to the donor membrane (Barlowe *et al.*, 1994; Matsuoka *et al.*, 1998a). Remarkably, this assembly also leads to the budding of coated vesicles from synthetic liposomes in the absence of membrane proteins (Matsuoka *et al.*, 1998b). Thus, the coat has an intrinsic ability to deform a lipid bilayer. However, it is becoming clear that membrane proteins of the ER membrane must play an active role to facilitate the budding reaction. Various membrane cargo proteins have been shown to interact with the incoming coat components (Table 3.2), in particular with the Sec23/24p complex, which is probably involved in cargo recognition (Springer & Schekman, 1998; Peng *et al.*, 2000). A direct interaction between COPII and a newly synthesized membrane protein, viral glycoprotein VSV-G, has been demonstrated. This protein is very efficiently sorted into the COPII vesicles and interacts with Sar1p and Sec23/24p through an EXD signal located in its cytosolic carboxy-terminus (Nishimura & Balch, 1997; Sevier *et al.*, 2000; Aridor *et al.*, 2001). Ma and co-workers (2001) have identified two alternative motifs within the carboxy terminus of some potassium channels that are necessary and sufficient to promote efficient ER export. Both motifs contain EXD or EXE sequences.

A second type of signal involved in the interaction with the COPII coat is the one based in hydrophobic residues. The intracellular membrane lectin ERGIC-53 (endoplasmic reticulum (ER)-Golgi intermediate compartment-53) carries a cytosolic

Table 3.2 Coat Protein II (COPII)-membrane interactions. Several membrane cargo proteins have been shown to interact with subunits of the COPII coat. Two different types of signals in the cytosolic domain of these proteins, a di-acidic (EXD, EXE) or a dihydrophobic ($\phi\phi$, FF), have been shown to be involved in these interactions

Interacting coat subunit	Membrane protein	Sorting signal	Reference
Sar1p; Sec23/24p	VSV-G	EXD	Nishimura & Balch, 1997; Sevier *et al.*, 2000; Aridor *et al.*, 2001
Not tested	Potassium channel	EXD/EXE	Ma *et al.*, 2001
Sec23/24p	Sys1p	EXD	Vostmeier & Gallwitz, 2001
Sec23/24p	ERGIC-53	FF	Kappeler *et al.*, 1997
Sec23p	p24	FF	Dominguez *et al.*, 1998
Sec13/31p > Sar1p > Sec23/24p	Emp24p	$\phi\phi$	Belden & Barlowe, 2001b
Sec13/31p > Sec23/24 p	Erv25p	$\phi\phi$	Belden & Barlowe, 2001b
Sec24p	Syntaxin Sed5	Not tested	Peng *et al.*, 1999
Sar1p; Sec23/24p	v-SNAREs (Bet1p and Bos1p)	Not tested	Springer & Schekman, 1998

C-terminal with two lysines at positions −3 and −4 which bind COPI coatomer (and are presumably involved in ERGIC53 recycling) and two terminal phenyl-alanines that bind Sec23p but not Sec13p (Kappeler *et al.*, 1997). Members of the p24 family of putative cargo receptors (in both yeast and humans) also bind to Sec23p through a cytosolic diphenylalanine motif (Dominguez *et al.*, 1998). As these proteins are required for efficient ER-to-Golgi traffic of some cargo proteins (Schimmöller *et al.*, 1995), it has been proposed that these proteins might serve as cargo adaptors (Kaiser, 2000). The cytoplasmically exposed tail sequences of two p24 proteins, Emp24p and Erv25p, bind efficiently to the Sec13/31p subunit of the COPII coat (but also bind to Sar1p and the Sec23/24p complex with lower affinity) in a process dependent on a pair of aromatic residues found at positions −7 and −8 in both tail sequences. COPI subunits also bind to these tails; however, the Erv25p tail sequence, which contains a dilysine motif, binds COPI more efficiently. These results suggest that both the Emp24p and Erv25p cytoplasmic sequences contain a di-aromatic motif that binds subunits of the COPII coat and promotes export from the ER. The Erv25p tail sequence also binds COPI and is responsible for returning this complex to the ER (Belden & Barlowe, 2001a). Other membrane proteins which have been shown to interact with the COPII coat components are summarized in Table 3.2.

Some lumenal proteins are very efficiently exported from the ER, suggesting a specific interaction with the coat through membrane receptors. One example of a receptor for soluble cargo is ERGIC-53, whose lumenal domain interacts with some glycoproteins (Appenzeller *et al.*, 1999). Erv29p, a conserved transmembrane protein of 29 kDa, also has been shown to have a role in collecting soluble secretory proteins (Belden & Barlowe, 2001b). Members of the p24 family may also be involved in the sorting of glycosyl phosphatidyl inositol-anchored proteins (Muñiz *et al.*, 2000). The p24 group does not appear to be essential for the secretion of soluble proteins such as invertase (Springer *et al.*, 2000).

Direct interactions between Sec23/24p and membrane proteins are most likely Sar1p-independent, but as Sar1p is required for Sec23/24p binding to microsomal and liposomal membranes, Sec23/24p must be able to recognize Sar1p as well as the cargo protein domains. This situation could be similar to the COPI system, where ARF1 and the cytosolic domain of cargo proteins seem to form a dual binding site for coatomer. Such a dual mode of Sec23/24p binding has indeed been observed *in vitro* in a complex of Sar1p, Sec23/24p and the ER-to-Golgi v-SNARE Bet1p. In this complex, Sec23/24p simultaneously contacts Sar1p-GTP and Bet1p in a cooperative manner (Springer & Scheckman, 1998). Thus, as in COPI, an initial complex is formed between the incoming coat component (Sec23/24p), the GTPase Sar1p in its GTP-bound form and a membrane protein which will be sub-sequently included into the vesicles. These membrane proteins have been proposed to work as *primers* that would be able to nucleate the COPII coat (Springer & Scheckman, 1998; Springer *et al.*, 1999). This would guarantee their packaging into COPII vesicles, linking the budding of a vesicle to the incorporation of proteins that are essential for its travel or fusion.

3.4.2.3   Polymerization of the coat and vesicle budding

Once priming complexes (Sar1-GTP/membrane protein/Sec23/24p) are established, they associate to form a larger polymeric coat on the ER membrane. Sec23/24p has a slightly curved shape and the concave side likely faces the membrane, with both subunits making contact with the cytosolic portion of membrane proteins selected for transport by COPII vesicles. The relatively fixed concave shape of Sec23/24p can impart local curvature to the lattice. Sec23p must also contact the membrane-bound Sar1p-GTP. Sec23/24p then interacts with different regions of Sec31p, located near its carboxy terminus.

As coat protein complexes on the ER membrane grow, other cargo proteins may diffuse into them and become captured by interaction with the coat. At this point, membrane or lumenal secretory proteins that cannot themselves interact with the vesicular coat may also become included in the vesicle budding site, if they are recruited by a protein that binds to the coat. Finally, in the last stage, the polymerized coat deforms the membrane, and a vesicle buds off. No additional cytosolic or membrane proteins seem to be required for this step to occur *in vitro* from isolated donor membranes or defined liposomes (Matsuoka *et al.*, 1998a, 1998b; Spang *et al.*, 1998), but whether and how this step is regulated *in vivo* remains unknown.

At present, it is unclear at which time point during the budding process Sar1p-GTP is hydrolyzed. Although GTP hydrolysis is required for the release of Sar1p from the vesicle membranes and the subsequent uncoating of the vesicles (to allow fusion with the target membrane), it could occur much earlier, possibly even before the vesicle separates from the donor membrane. This need not result in an early disassembly of the vesicle coat: COPII vesicles that are produced from microsomes in the presence of GTP do not contain Sar1p, but do have a COPII coat that is visible under the electron microscope (Barlowe *et al.*, 1994). The fact that Sar1p is required to assemble, but not to maintain, the COPII coat may be explained by the lateral association of the COPII subunits. Once bound in a poly-meric lattice, Sec23/24p and Sec13/31p may no longer need Sar1p to adhere to the vesicle; instead, their interaction with primer proteins may be sufficient. Using Sar1p-GTP bound to liposomes, it has been shown that a single round of assembly and disassembly of the COPII coat lasts a few seconds (Antonny *et al.*, 2001). The two large COPII complexes Sec23/24p and Sec13/31p bind almost instantaneously (in less than 1 s) to Sar1pGTP-doped liposomes. This binding is followed by a fast (less than 10 s) disassembly due to a 10-fold acceleration of the GTPase-activating protein activity of Sec23/24p by the Sec13/31p complex. This suggests that GTP hydrolysis is concomitant or subsequent to the polymerization of the coat induced by Sec13/31p.

3.4.3   COPII coat components in plant cells

COPII vesicles have not yet been observed in plants using electron microscopy, but there is convincing evidence supporting their presence. Functional plant genes with homology to the *SAR1, SEC12, SEC13* and *SEC23* genes have been isolated (D'Enfert *et al.*, 1992; Bar-Peled *et al.*, 1995; Bar-Peled & Raikhel, 1997). In addition, the identification and characterization of some COPII and COPI coat components in plants (Bar-Peled & Raikhel, 1997; Movafeghi *et al.*, 1999; Phillipson *et al.*, 2001) suggest that the principle of non-clathrin-coated vesicle formation applies to the plant kingdom as well (Pimpl *et al.*, 2000).

Using antibodies against AtSec23p, the *Arabidopsis* homologue to Sec23p, Movafeghi *et al.* (1999) characterized the AtSec23p antigen. In gel filtration experiments, AtSec23p (~85 kDa) eluted with a molecular mass of around 200 kDa, consistent with the molecular weight of the COPII Sec23/24p dimer. In addition, they found that AtSec23p is not an integral membrane protein, but a tightly associated peripheral membrane protein. Finally, they showed its preferential association with ER membranes, in contrast with the distribution of AtSec21p (the *Arabidopsis* homologue to Sec21p, a component of the COPI coat), which appears to be bound to both ER and Golgi membranes. This would also suggest that in plant cells, the COPII vesicles form only at ER membranes while the COPI vesicles can be formed from both ER and Golgi membranes. *In vitro* assays showing recruitment of ARF1 and coatomer to Golgi membranes, also showed recruitment of Sar1p, but not of AtSec23/24p (Pimpl *et al.*, 2000).

AtSar1p and AtSec12p are both associated with the ER. However, about one-half of the cellular AtSar1p is present in the cytosol (Bar-Peled & Raikhel, 1997). When overexpressed, Sar1p is mostly cytosolic (Bar-Peled & Raikhel, 1997; Phillipson *et al.*, 2001). In yeast, overexpression of Sec12p reduces ER export, presumably via the titration of Sar1p (D'Enfert *et al.*, 1991). Stable overexpression of Sec12p in transgenic plants neither affected cell viability nor caused a redistribution of Sar1p (Bar-Peled & Raikhel, 1997). In contrast, the transient overexpression of Sec12p in tobacco protoplasts resulted in the recruitment of the GTPase to the ER membrane (Phillipson *et al.*, 2001). One likely explanation would be that plant cells contain regulatory mechanisms to respond to an imbalance in the Sec12p/Sar1p ratio (Phillipson *et al.*, 2001). Under these conditions, Sec12p overexpression inhibits COPII transport through the depletion of Sar1p, therefore preventing the formation of COPII vesicles. Coexpression of mutant GTP-trapped Sar1p, which is less sensitive to the GTPase-activating activity of Sec23p, inhibits COPII transport at a later stage: it prevents vesicle uncoating and fusion with the target membrane.

Acknowledgements

F.A. was a recipient of grants from the Ministerio de Ciencia y Tecnologia (Grant no. PB98-1425) and Generalitat Valenciana (Grant no. GV99-86-1-05). A.R.M. is a recipient of grants from The Scientific and Technical Research Council of Turkey (TUBITAK-Agriculture and Forestry Section, Grant no. TOGTAG-3022), Turkey and Jülich Forschung Center (TBAG-5-101T121), Germany.

References

Ahmed, S.U., Bar-Peled, M. & Raikhel, N.V. (1997) Cloning and subcellular location of an *Arabidopsis* receptor-like protein that shares common features with protein-sorting receptors of eukaryotic cells. *Plant Physiol.*, **114**, 325–336.

Amor, J.C., Harrison, D.H., Kahn, R.A. & Ringe, D. (1994) Structure of the human ADP-ribosylation factor 1 complexed with GDP. *Nature*, **372**, 704–708.

Andreeva, A.V., Kutuzov, M.A., Evans, D.E. & Hawes, C.R. (1998) Proteins involved in membrane transport between the ER and the Golgi apparatus: 21 putative plant homologues revealed by dbEST searching. *Cell Biol. Int.*, **22**, 145–160.

Antonny, B., Madden, D., Hamamoto, S., Orci, L. & Schekman, R. (2001) Dynamics of the COPII coat with GTP and stable analogues. *Nat. Cell Biol.*, **3**, 531–537.

Antonny, B. & Schekman, R. (2001) ER export: public transportation by the COPII coach. *Curr. Opin. Cell Biol.*, **13**, 438–443.

Appenzeller, C., Andersson, H., Kappeler, F. & Hauri, H.P. (1999) The lectin ERGIC-53 is a cargo transport receptor for glycoproteins. *Nat. Cell Biol.*, **1**, 330–334.

Aridor, M., Fish, K.N., Bannykh, S., Weissman, J., Roberts, T.H., Lippincott Schwartz, J. & Balch, W.E. (2001) The Sar1p GTPase coordinates biosynthetic cargo selection with endoplasmic reticulum export site assembly. *J. Cell Biol.*, **152**, 213–230.

Austin, C., Boehm, M. & Tooze, S.A. (2002) Site-specific cross-linking reveals a differential direct interaction of class 1, 2, and 3 ADP-ribosylation factors with adaptor protein complexes 1 and 3. *Biochemistry*, **41**, 4669–4677.

Barlowe, C. & Schekman, R. (1993) SEC12 encodes a guanine-nucleotide-exchange factor essential for transport vesicle budding from the ER. *Nature*, **365**, 347–349.

Barlowe, C., Orci, L., Yeung, T., Hosobuchi, M., Hamamoto, S., Salama, N., Rexach, M.F., Ravazzola, M., Amherdt, M. & Schekman, R. (1994) COPII: a membrane coat formed by SEC proteins that drive vesicle budding from the endoplasmic reticulum. *Cell*, **77**, 895–907.

Bar-Peled, M., Conceiçao, A.S., Frigerio, L. & Raikhel, N.V. (1995) Expression and regulation of aERD2, a gene encoding the KDEL receptor homolog in plants and other proteins involved in ER-Golgi vesicular trafficking. *Plant Cell*, **7**, 667–676.

Bar-Peled, M. & Raikhel, N.V. (1997) Characterization of AtSec12 and AtSar1, proteins likely involved in endoplasmic reticulum and Golgi transport. *Plant Physiol.*, **114**, 315–324.

Battey, N.H., James, N.C., Greenland, A.J. & Brownlee, C. (1999) Exocytosis and endocytosis. *Plant Cell*, **11**, 643–659.

Beevers, L. (1996) Clathrin-coated vesicles in plants. *Int. Rev. Cytol.*, **167**, 1–35.

Belden, W.J. & Barlowe, C. (1996) Erv 25p, a component of COP II-coated vesicles, forms a complex with Emp24 that is required for efficient endoplasmic reticulum to Golgi transport. *J. Biol. Chem.*, **271**, 26939–26946.

Belden, W.J. & Barlowe, C. (2001a) Distinct roles for the cytoplasmic tail sequences of Emp24p and Erv25p in transport between the endoplasmic reticulum and Golgi complex. *J. Biol. Chem.*, **276**, 43040–43048.

Belden, W.J. & Barlowe, C. (2001b) Role of Erv29p in collecting soluble secretory proteins into ER-derived transport vesicles. *Science*, **294**, 1528–1531.

Beraud-Dufour, S., Paris, S., Charbe, M. & Antonny, B. (1999) Dual interaction of ADP-ribosylation factor1 with Sec7 domain and with lipid membranes during catalysis of guanine nucleotide exchange. *J. Biol. Chem.*, **274**, 37629–37636.

Bischoff, F., Molendijk, A., Rajendrakumar, C.S.V. & Palme, K. (1999) GTP-binding proteins in plants. *Cell. Mol. Life Sci.*, **55**, 233–256.

Blackbourn, H.D. & Jackson, A.P. (1996) Plant clathrin heavy chain: sequence analysis and restricted localisation in growing pollen tubes. *J. Cell Sci.*, **109**, 777–786.

Blatt, M.R., Leyman, B. & Gleen, D. (1999) Molecular events of vesicle trafficking and control by SNARE proteins in plants. *New Phytol.*, **144**, 389–418.

Blum, R., Pfeiffer, F., Feick, P., Nastainczyk, W., Bärbel, K., Schäfer, K.-H. & Schulz, I. (1999) Intracellular localization and *in vivo* trafficking of p24A and p23. *J. Cell Sci.*, **112**, 537–548.

Boman, A.L., Zhang, C., Zhu, X. & Kahn, R.A. (2000) A family of ADP-ribosylation factor effectors that can alter membrane transport through the trans Golgi. *Mol. Biol. Cell*, **11**, 1241–1255.

Bremser, M., Nickel, W., Schweikert, M., Ravazzola, M., Amherdt, M., Hughes, C.A., Söllner, T.H., Rothman, J.E. & Wieland, F.T. (1999) Coupling of coat assembly and vesicle budding to packaging of putative cargo receptors. *Cell*, **96**, 495–506.

Butler, J.M., Kirsch, T., Watson, B., Paris, N., Rogers, J.C. & Beevers, L. (1997) Interaction of the vacuolar targeting receptor BP-80 with clathrin adaptors. *Plant Physiol.*, **114** (Suppl.) Abstract 1210.

Chardin, P., Paris, Antonny, B., Robineau, S., Beraud-Dufour, S., Jackson, C.L. & Charbe, M. (1996) A human exchange factor for ARF contains Sec7 and pleckstrin-homology domains. *Nature*, **384**, 481–484.

Chardin, P. & McCormik, F. (1999) Brefeldin A: the advantage of being uncompetitive. *Cell*, **97**, 153.

Chen, H.J., Yuan, J. & Lobel, P. (1997) Systematic mutational analysis of the cation-independent mannose 6-phosphate/insulin-like growth factor-II receptor cytoplasmic domain. An acidic cluster containing a key aspartate is important for function in lysosomal enzyme sorting. *J. Biol. Chem.*, **272**, 7003–7012.

Chow, W.T.K., Sakharkar, M.K., Lim, D.P.P. & Yeo, W.M. (2001) Phylogenetic relationships of the seven coat protein subunit of the coatomer complex, nad comperative sequence analysis of murine xenin and proxenin. *Biochem. Genet.*, **39**, 201–211.

Contreras, I., Ortiz-Zapater, E., Castilho, L.M. & Aniento, F. (2000) Characterization of COPI coat proteins in plant cells. *Biochem. Biophys. Res. Commun.*, **273**, 176–182.

Cosson, P. & Letourneur, F. (1994) Coatmer interaction with di-lysine endoplasmic reticulum retention. *Science*, **263**, 1629–1631.

Cowles, C.R., Odorizzi, G., Payne, G.S. & Emr, S.D. (1997) The AP-3 adaptor complex is essential for cargo-selective transport to the yeast vacuole. *Cell*, **91**, 109–118.

Dell'Angelica, E.C., Shotelersuk, V., Aguilar, R.C., Gahl, W.A. & Bonifacino, J.S. (1999) Altered trafficking of lysosomal proteins in Hermansky-Pudlak syndrome due to mutations in the β3A subunit of the AP-3 adaptor. *Mol. Cell*, **3**, 11–21.

Dell'Angelica, E.C., Puertollano, R., Mullins, C., Aguilar, R.C., Vargas, J.D., Hartnell, L.M. & Bonifacino, J.S. (2000) GGAs: a family of ADP ribosylation factor-binding proteins related to adaptors and associated with the Golgi complex. *J. Cell Biol.*, **149**, 81–94.

D'Enfert, C., Wuestehube, L.J., Lila, T. & Schekman, R. (1991) Sec12p-dependent membrane binding of the small GTP-binding protein Sar1p promotes formation of transport vesicles from the ER. *J. Cell Biol.*, **114**, 663–670.

D'Enfert, C., Geusse, M. & Gaillardin, C. (1992) Fission yeast and a plant have functional homologues of the Sar1 and Sec12 proteins involved in ER to Golgi traffic in budding yeast. *EMBO J.*, **11**, 4205–4210.

Denzel, A., Otto, F., Girod, A., Pepperkok, R., Watson, R., Rosewell, I., Bergerson, J., Solari, R.C. & Owen, M.J. (2000) The p24 family member p23 is required for early embryonic development. *Curr. Biol.*, **10**, 55–58.

Diaz, E. & Pfeffer, S. (1998) Tip47 – a cargo selection device for mannose 6-phosphate receptor trafficking. *Cell*, **93**, 433–443.

Dombrowski, J.E. & Raikhel, N.V. (1995) Isolation of a cDNA encoding a novel GTP-binding protein of *Arabidopsis thaliana. Plant Mol. Biol.*, **28**, 1121–1126.

Dominguez, M., Dejgaard, K., Fullekrug, J., Dahan, S., Fazel, A., Paccaud, J.P., Thomas, D.Y., Bergeron, J.J. & Nilsson, T. (1998) gp25L/emp24/p24 protein family members of the *cis*-Golgi network bind both COPI and COPII coatomer. *J. Cell Biol.*, **140**, 751–765.

Donaldson, J.G., Cassel, D., Kahn, R.A. & Klausner, R.D (1992a) ADP-ribosylation factor, a small GTP-binding protein, is required for binding the coatomer protein β-COP to Golgi membranes. *Proc. Natl. Acad. Sci. USA*, **89**, 6408–6412.

Donaldson, J.G., Finazzi, D. & Klausner, R.D. (1992b) Brefeldin A inhibits Golgi membrane-catalysed exchange of guanine nucleotide onto AARF protein. *Nature*, **360**, 6408–6412.

Donaldson, J.G. & Jackson, C.L. (2000) Regulators and effectors of the ARF GTPases. *Curr. Opin. Cell Biol.*, **12**, 475–482.

Drucker, M., Herkt, B. & Robinson, D.G. (1995) Demonstration of a β-type adaptin at the plasma membrane. *Cell Biol. Int.*, **19**, 191–201.

Drucker, M., Happel, N. & Robinson, D.G. (1998) Localisation and properties of kinase activities in clathrin coated vesicles from zucchini hypocotyls. *Eur. J. Biochem.*, **240**, 570–575.

Duden, R., Griffiths, G., Frank, R., Argos, P. & Kreis, T.E. (1991) β-COP, a 110 kDa protein associated with non-clathrin coated vesicles and the Golgi complex, shows homology to β-adaptin. *Cell*, **64**, 649–665.

Dupree, P. & Sherrier, D.J. (1998) The plant Golgi apparatus. *Biochim. Biophys. Acta*, **1404**, 259–270.

Emery, G., Rojo, M. & Gruenberg, J. (2000) Coupled transport of p24 family members. *J. Cell Sci.*, **113**, 2507–2516.

Faulstich, D., Auerbach, S., Orci, L., Ravazzola, M., Weghingel, S., Lottspeich, F., Stenbeck, G., Harter, C., Wieland, F.T. & Tschochner, H. (1996) Architecture of coatomer: molecular characterization of δ-COP and protein interactions within the complex. *J. Cell Biol.*, **135**, 53–61.

Fiedler, K., Veit, M., Stamnes, M.A. & Rothman, J.E. (1996) Bimodal interaction of coatomer with the p24 family of putative cargo receptors. *Science*, **273**, 1396–1399.

Fligge, T.A., Reinhard, C., Harter, C., Wieland, F.T. & Przybylski, M. (2000) Oligomerization of peptides analogous to the cytoplasmic domains of coatomer receptors revealed by mass spectrometry. *Biochemistry*, **39**, 8491–8496.

Franco, M., Chardin, P., Charbe, M. & Paris, S. (1996) Myristoylation-facilitated binding of the G protein ARF1 GDP to membrane phospholipids is required for its activation by a soluble nucleotide exchange factor. *J. Biol. Chem.*, **271**, 1573–1578.

Geisler, C., Dietrich, J., Nielsen, B.L., Kastrup, J., Lauritsen, J.P.H., Odum, N. & Christensen, M.D. (1998) Leucine-based receptor sorting motifs are dependent on the spacing relative to the plasma membrane. *J. Biol. Chem.*, **273**, 21316–21323.

Gerich, B., Orci, L., Tschochner, H., Lottspeich, F., Rvazzola, M., Amherdt, M., Wieland, F.T. & Harter, C. (1995) Non-clathrin-coat protein alpha is a conserved subunit of coatomer and in *Saccharomyces cerevisiae* is essential for growth. *Proc. Acad. Sci. USA*, **92**, 3229–3233.

Goldberg, J. (1998) Structural basis for activation of ARF GTPase: mechanisms of guanine nucleotide exchange and GTP-myristoyl switching. *Cell*, **95**, 237–248.

Goldberg, J. (1999) Structural and functional analysis of the ARFI–ARFGAP complex reveals a role for coatomer in GTP hydrolysis. *Cell*, **96**, 893–902.

Gommel, D., Orci, L., Eming, E.M., Hannah, M.J., Ravazzola, M., Nickel, W., Helms, J.B., Wielan, F.T. & Shon, K. (1999) p24 and p23, the major transmembrane proteins of COP-coated transport vesicles, from hetero-oligomeric complexes and cycle between the organelles of the early secretory pathway. *FEBS Lett.*, **447**, 179–185.

Gommel, D.U., Memon, A.R., Heiss, A., Lottspeich, F. Pfannstiel, J., Lechner, J., Reinhard, C., Helms, B.J., Nickel, W. & Wieland, F.T. (2001) Recruitment to Golgi membranes of ADP-ribosylation factor1 is mediated by the cytoplasmic domain of p23. *EMBO J.*, **20**, 6751–6760.

Greasley, S.E., Jhoti, H., Teahan, C., Solari, R., Fensome, A., Thomas, G.M., Cockroft, S. & Bax, B. (1995) The structure of rat ADP-ribosylation factor 1 (ARF-1) complexed to GDP determined from two different crystal forms. *Nature Struct. Biol.*, **2**, 797–806.

Grebe, M., Gadea, J., Steinmann, T., Keintz, M., Rahfeld, J.U., Salchert, K., Koncz, C. & Jürgens, G. (2000) A conserved domain of the *Arabidopsis* GNOM protein mediates subunit interaction and cyclophilin 5 binding. *The Plant Cell*, **12**, 343–356.

Gu, X. & Verma, D.P.S. (1996) Phragmoplastin, a dynamin-like protein associated with cell plate formation in plants. *EMBO J.*, **15**, 695–704.

Hadlington, J.L. & Denecke, J. (2000) Sorting of soluble proteins in the secretory pathway of plants. *Curr. Opin. Plant Biol.*, **3**, 461–468.

Hara-Kuge, S., Kuge, O., Orci, L., Amherdt, M., Rvazzola, M., Wieland, F.T. & Rothman, J.E. (1994) En bloc incorporation of coatomer subunits during the assembly of COP-coated vesicles. *J. Cell Biol.*, **124**, 883–892.

Harrison-Lavoie, K.J., Lewis, V.A., Hynes, G.M., Collison, K.S., Nutland, E. and Willison, K.R. (1993) A 102 kDa subunit of a Golgi-associated particle has homology to β-subunits of trimeric G proteins. *EMBO J.*, **12**, 2847–2853.

Harter, C. & Wieland, F.T. (1998) A single binding site for di-lysine retrieval motifs and p23 within the γ-subunits of coatomer. *Proc. Natl. Acad. Sci. USA*, **95**, 11649–11654.

Haun, R.S., Tsai, S.C., Adamik, R., Moss, J. & Vaughan, M. (1993) Effect of myristoylation on GTP-dependent binding of ADP-ribosylation factor to Golgi. *J. Biol. Chem.*, **268**, 7064–7068.

Helms, J.B. & Rothman, J.E. (1992) Inhibition by Brefeldin A of a Golgi membrane enzyme that catalyses exchange of a guanine nucleotide bound to ARF. *Nature*, **360**, 352–354.

Helms, J.B., Palmer, D.J. & Rothman, J.E. (1993) Two distinct population of ARF bound to Golgi membranes. *J. Cell. Biol.*, **121**, 751–760.

Hicke, L., Yoshiga, T. & Schekman, R. (1992) Sec23p and a novel 105 kDa protein function as a multimeric complex to promote vesicle budding and protein transport from the ER. *Mol. Biol. Cell*, **3**, 667–676.

Higo, H., Kishimoto, N., Saito, A. & Higo, K.I. (1994) Molecular cloning and characterization of a cDNA encoding a small GTP-binding protein related to mammalian ADP ribosylation factor from rice. *Plant Sci.*, **100**, 41–49.

Hillmer, S., Movafeghi, A., Robinson, D.G. & Hinz, G. (2001) Vacuolar storage proteins are sorted in the *cis*-cisternae of the pea cotyledon Golgi apparatus. *J. Cell Biol.*, **152**, 41–50.

Hinz, G., Hillmer, S., Baumer, M. & Hohl, I.I. (1999) Vacuolar storage proteins and the putative vacuolar sorting receptor BP-80 exit the Golgi apparatus of developing pea cotyledons in different transport vesicles. *Plant Cell*, **11**, 1509–1524.

Hirst, J., Lui, W.W., Bright, N.A., Totty, N., Seaman, M.N. & Robinson, M.S. (2000) A family of proteins with γ-adaptin and VHS domains that facilitate trafficking between the trans-Golgi network and the vacuole/lysosome. *J. Cell Biol.*, **149**, 67–80.

Huang, M., Weissman, J.T., Beraud-Dufour, S., Luan, P., Wang, C., Chen, W., Aridor, M., Wilson, I.A. & Balch, W.E. (2001) Crystal structure of Sar1-GDP at 1.7 A resolution and the role of the NH$_2$ terminus in ER export. *J. Cell Biol.*, **155**, 937–948.

Humair, D., Hernández Felipe, D., Neuhaus, J.M. & Paris, N. (2001) Demonstration in yeast of the function of BP-80, a putative plant vacuolar sorting receptor. *Plant Cell*, **13**, 781–792.

Huttner, W.B. & Zimmerberg, J. (2001) Implication of lipid microdomains for membrane curvature, budding and fission. *Curr. Opin. Cell Biol.*, **13**, 478–484.

Jackson, C.L. & Casanova, J.E. (2000) Turning on ARF: the Sec 7 family of guanine nucleotide-exchange factors. *Trend Cell Biol.*, **10**, 60–67.

Jahn, R. & Südof, T.C. (1999) Membrane fusion and exocytosis. *Annu. Rev. Biochem.*, **68**, 863–911.

Johnson, K.F. & Kornfeld, S. (1992) The cytoplasmic tail of the mannose 6-phosphate/insulin-like growth factor-II receptor has two signals for lysosomal enzyme sorting in the Golgi. *J. Cell Biol.*, **119**, 249–257.

Kahn, R.A. & Gilman, A.G. (1986) The protein cofactor necessary for ADP-ribosylation of Gs by cholera toxin is itself a GTP binding protein. *J Biol. Chem.*, **261**, 7906–7911.

Kaiser, C. (2000) Thinking about p24 proteins and how transport vesicles select their cargo. *Proc. Natl. Acad. Sci. USA*, **97**, 3783–3785.

Kantheti, P., Qiao, X.X., Diaz, M.E., Peden, A.A., Meyer, G.E., Carskadon, S.L., Kapfhamer, D., Sufalko, D., Robinson, M.S., Noebels, J.L. & Burmeister, M. (1998) Mutation in the AP-3 δ in the mocha mouse links endosomal transport to storage deficiency in platelets, melanosomes, and synaptic vesicles. *Neuron*, **21**, 111–122.

Kappeler, F., Klopfenstein, D.R., Foguet, M., Paccaud, J.P. & Hauri, H.P. (1997) The recycling of ERGIC-53 in the early secretory pathway. *J. Biol. Chem.*, **272**, 31801–31808.

Keon, J.P.R., Jewit, S. & Hargreaves, J.A. (1995) A gene encoding γ-adaptin is required for apical extension growth in *Ustillago maydis*. *Gene*, **162**, 141–145.

Kirchhausen, T., Bonifacino, J.S. & Riezman, H. (1997) Linking cargo to vesicle formation – receptor tail interaction with coat proteins. *Curr. Opin. Cell Biol.*, **9**, 488–495.

Kirchhausen, T. (1999) Adaptors for clathrin-mediated traffic. *Annu. Rev. Cell Biol.*, **15**, 705–732.

Kirchhausen, T. (2000a) Three ways to make a vesicle. *Nature Rev.*, **1**, 187–198.

Kirchhausen, T. (2000b) Clathrin. *Annu. Rev. Biochem.*, **69**, 699–727.

Kirsch, T., Paris, N., Butler, J.M., Beevers, L. & Rogers, J.C. (1994) Purification and initial characterization of a potential plant vacuolar targeting receptor. *Proc. Natl. Acad. Sci. USA*, **91**, 3403–3407.

Kiyosue, T. & Shinozaki, K. (1995) Cloning of a carrot cDNA for a member of the family of ADP-ribosylation factors (ARFs) and characterization of the binding of nucleotides by its product after expression in *E. coli*. *Plant Cell Physiol.*, **36**, 849–856.

Kobayashi-Uehara, A., Shimosaka, E. & Handa, H. (2001) Cloning and expression analyses of cDNA encoding an ADP-ribosylation factor from wheat: tissue-specific expression of wheat Arf. *Plant Sci.*, **160**, 535–542.

Kuge, O., Hara Kuge, S., Orci, L, Ravazzola, M., Amherdt, M., Tanigawa, G., Wieland, F.T. & Rothman, J.E. (1993) Zeta-COP, a subunit of coatomer, is required for COP-coated vesicle assembly. *J. Cell Biol.*, **123**, 1727–1734.

Lanoix, J., Ouwendijk, J., Lin, C.C., Stark, A., Love, H.D., Osterman, J. & Nilson, T. (1999) GTP hydrolysis by arf-1 mediates sorting and concentration of Golgi resident enzymes into functional COPI vesicles. *EMBO J.*, **18**, 4935–4948.

LeBorgne, R., Griffiths, G. & Hoflack, B. (1996) Mannose 6-phosphate receptors and ADP-ribosylation factors cooperate for high affinity interaction of the AP-1 Golgi assembly proteins with membranes. *J. Biol. Chem.*, **271**, 2162–2170.

LeBorgne, R. & Hoflack, B. (1997) Mannose 6-phosphate receptors regulate the formation of clathrin-coated vesicles in the TGN. *J. Cell Biol.*, **137**, 335–345.

Lederkremer, G.Z., Cheng, Y., Petre, B.M., Vogan, E., Springer, S., Schekman, R., Walz, T. & Kirchausen, T. (2001) Structure of the Sec23p/24p and Sec13p/31p complexes of COPII. *Proc. Natl. Acad. Sci. USA*, **98**, 10704–10709.

Lippincott-Schwartz, J., Cole, N.B. & Donaldson, J.G. (1998) Building a secretory apparatus: role of ARF/COP in Golgi biogenesis and maintenance. *Histochem. Cell Biol.*, **109**, 449–462.

Liu, S.H., Towler, M.C., Chen, E., Chen, C.Y., Song, W., Apodaca, G. & Brodsky, F.M. (2001) A novel clathrin homolog that co-distributes with cytoskeletal components functions in the trans-Golgi network. *EMBO J.*, **20**, 272–284.

Lowe, M. & Kreis, T.E. (1995) *In vitro* assembly and disassembly of coatomer. *J. Biol. Chem.*, **270**, 31364–31371.

Lowe, M. & Kreis, T.E. (1996) *In vitro* assembly of coatomer, the COP-I coat precursor. *J. Biol. Chem.*, **271**, 30725–30730.

Ma, D., Zerangue, N., Lin, Y.F., Collins, A., Yu, M., Jan, Y.N. & Jan, L.Y. (2001) Role of ER export signals in controlling surface potassium channel numbers. *Science*, **291**, 316–319.

Maldonado-Mendoza, I.E. & Nessler, C.L. (1997) Molecular characterization of the AP19 gene family in *Arabidopsis thaliana*: components of the Golgi AP-1 clathrin assembly protein complex. *Plant Mol. Biol.*, **35**, 865–872.

Malhotra, V., Serafini, T., Orci, L., Shepherd, J.C. & Rothman, J.E. (1989) Purification of a novel class of coated vesicles mediating biosynthetic protein transport through the Golgi stack. *Cell*, **58**, 239–336.

Mallet, W.G. & Brodsky, F.M. (1996) A membrane-associated protein complex with selective binding to the clathrin coat adaptor AP1. *J. Cell Sci.*, **109**, 3059–3068.

Malsam, J., Gommel, D., Wieland, F.T. & Nickel, W. (1999) A role of for ADP-ribosylation factor in the control of cargo uptake during COPI-coated vesicle biogenesis. *FEBS Lett.*, **462**, 267–272.

Marks, M.S., Ohno, H., Kirchhausen, T. & Bonifacino, J.S. (1997) Protein sorting by tyrosine-based signals: adapting to the Ys and wherefores. *Trends Cell Biol.*, **7**, 124–128.

Marty, F. (1999) Plant vacuoles. *Plant Cell*, **11**, 587–599.

Marzioch, M., Henthorn, D.C., Herrmann, J.M., Wilson, R., Thomas, D.Y., Bergeron, J.J., Solari, R.C. & Rowley, A. (1999) Erp1p and Erp2p, partners for Emp24p and Erv25p in a yeast p24 complex. *Mol. Biol. Cell*, **10**, 1923–1938.

Matsuoka, K., Morimitsu, Y., Uchida, K. & Schekman, R. (1998a) Coat assembly and v-SNAREs concentration into synthetic COPII vesicles. *Mol. Cell*, **2**, 703–708.

Matsuoka, K., Orci, L., Amherdt, M., Bednarek, S.Y., Hamamoto, S., Schekman, R. & Yeung, T. (1998b) COPII-coated vesicle formation reconstituted with purified coat proteins and chemically defined liposomes. *Cell*, **93**, 263–275.

Matsuoka, K., Schekman, R., Orci, L. & Heuser, J.E. (2001) Surface structure of the COPII-coated vesicle. *Proc. Natl. Acad. Sci. USA*, **98**, 13705–13709.

Mellman, I. & Warren, G. (2000) The road taken: past and future foundations of membrane traffic. *Cell*, **100**, 99–112.

Memon, A.R., Clark, G.B. & Thompson, G.A. Jr. (1993) Identification of an ARF type low molecular mass GTP-binding protein in pea *Pisum sativum*. *Biochem. Biophys. Res. Commun.*, **193**, 809–813.

Memon, A.R., Hwang, S., Deshpande, N., Thompson, G.A. Jr. & Herrin, D.L. (1995) Novel aspects of regulation of a cDNA (ARF1) from *Chlamydomonas* with high sequence identity to animal ADP ribosylation factor 1. *Plant Mol. Biol.*, **29**, 567–577.

Meyer, C., Zizioli, D., Lausmann, S., Eskelinen, E.L., Haman, J., Saftig, P., von Figura, K. & Schu, P. (2000) μ1A-adaptin-deficient mice: lethality, loss of AP-1 binding and rerouting of mannose 6-phosphate receptors. *EMBO J.*, **19**, 2193–2203.

Movafeghi, A., Happel, N., Pimpl, P., Tai, G.-H. & Robinson, D.G. (1999) *Arabidopsis* Sec21p and Sec23p homologs. Probable coat proteins of plant COP-coated vesicles. *Plant Physiol.*, **119**, 1437–1445.

Muñiz, N., Nuoffer, C., Hauri, H.P. & Riezman, H. (2000) The Emp24 complex recruits a specific cargo molecule into endoplasmic reticulum-derived vesicles. *J. Cell Biol.*, **148**, 925–930.

Nebenführ, A. & Staehelin, L.A. (2001) Mobile factories: Golgi dynamics in plant cells. *Trends Plant Sci.*, **6**, 160–167.

Nickel, W. & Wieland, F. (1997) Biogenesis of COPI-coated transport vesicles. *FEBS Lett.*, **413**, 395–400.

Nickel, W., Malsam, J., Gorgas, K., Ravazzola, M., Jenne, N., Helms, J.B. & Wieland, F.T. (1998) Uptake by COPI-coated vesicles of both anterograde and retrograde cargo is inhibited by GTPγS in vitro. *J. Cell Sci.*, **111**, 3081–3090.

Nishimura, N. & Balch, W.E. (1997) A di-acidic signal required for selective export from the endoplasmic reticulum. *Science*, **277**, 556–558.

Orci, L., Glick, B.S. & Rothman, J.E. (1986) A new type of coated vesicular carrier that appears not to contain clathrin: its possible role in protein transport within the Golgi stack. *Cell*, **46**, 171–184.

Orci, L., Malhotra, V., Amherdt, M., Serafini, T. & Rothman, J.E. (1989) Dissection of a single round of vesicular transport: sequential intermediates for intercisternal movement in the Golgi stack. *Cell*, **56**, 357–368.

Orci, L., Stammes, M., Ravazzola, M., Amhert, M., Perrelet, A., Sollner, T.H. & Rothman, J.E. (1997) Bidirectional transport by distinct populations of COPI-coated vesicles. *Cell*, **90**, 335–349.

Orsel, J.G., Sincock, P.M., Krise, J.P. & Pfeffer, S.R. (2000) Recognition of the 300-kDa mannose 6-phosphate receptor cytoplasmic domain by 47-kDa tail-interacting protein. *Proc. Natl. Acad. Sci. USA*, **97**, 9047–9051.

Palmer, D.J., Helms, J.B., Beckers, C.J., Orci, L. & Rothman, J.E. (1993) Binding of coatomer to Golgi membranes requires ADP-ribosylation factor. *J. Biol. Chem.*, **168**, 12083–12089.

Paris, N., Stanley, C.M., Jones, R.L. & Rogers, J.C. (1996) Plant cells contain two functionally distinct vacuolar compartments. *Cell*, **85**, 563–572.

Paris, N., Rogers, S.W., Jiang, L., Kirsh, T., Beevers, L., Philips, T.E. & Rogers, J.C. (1997) Molecular cloning and further characterization of a probable plant vacuolar sorting receptor. *Plant Physiol.*, **115**, 29–39.

Pavel, J., Harter, C. & Wieland, F.T. (1998) Reversible dissociation of coatomer: functional characterization of a β/δ-coat protein subcomplex. *Proc. Natl. Acad. Sci. USA*, **95**, 2140–2145.

Pearse, B.M.F. (1975) Coated vesicles from pig brain purification and biochemical characterization. *J. Mol. Biol.*, **97**, 93–98.

Pelham, H.R.B. & Rothman, J.E. (2000) The debate about transport in the Golgi – two sides of the same coin? *Cell*, **102**, 713–719.

Peng, R., Grabowski, R., De Antoni, A. & Gallwitz, D. (1999) Specific interaction of the yeast cis-Golgi syntaxin Sed5p and the coat protein complex II component Sec24p of endoplasmic reticulum-derived transport vesicles. *Proc. Natl. Acad. Sci. USA*, **96**, 3751–3761.

Peng, R.W., De Antoni, A. & Gallwitz, D. (2000) Evidence for overlapping and distinct functions in protein transport of coat protein Sec24p family members. *J. Biol. Chem.*, **275**, 11521–11528.

Pepperkok, R., Whitney, J.A., Gomez, M. & Kreis, T.E. (2000) COPI vesicles accumulating in the presence of a GTP restricted ARF1 mutants are depleted of anterograde and retrograde cargo. *J. Cell Sci.*, **113**, 135–144.

Peyroche, A., Paris, S. & Jackson, C.L. (1996) Nucleotide exchange on Arf mediated by yeast Geal protein. *Nature*, **384**, 479–481.

Peyroche, A., Antonny, B., Robineau, S., Acker, J., Cherfils, J. & Jackson, C.L. (1999) Brefeldin A acts to stabilize an abortive ARF-GDP-Sec7 domain protein complex: involvement of specific residues of the Sec7 domain. *Mol. Cell*, **3**, 275–285.

Phillipson, B.A., Pimpl, P., Pinto daSilva, L.L., Crofts, A.J., Taylor, J.P., Movafeghi, A., Robinson, D.G. & Denecke, J. (2001) Secretory bulk flow of soluble proteins is efficient and COPII dependent. *Plant Cell*, **13**, 2005–2020.

Pimpl, P., Movafeghi, A., Coughlan, S., Denecke, J., Hillmer, S. & Robinson, D.G. (2000) *In situ* localization and *in vitro* induction of plant COPI-coated vesicles. *Plant Cell*, **12**, 2219–2235.

Poussu, A., Lohi, O. & Lehto, V.P. (2000) Vear, a novel Golgi-associated protein with VHS and gamma-adaptin "ear" domains. *J. Biol. Chem.*, **275**, 7176–7183.

Presley, J.F., Ward, T.H., Pfeifer, A.C., Siggia, E.D., Phair, R.D. & Lippincott-Schwartz, J. (2002) Dissection of COPI and Arf1 dynamics *in vivo* and role in Golgi membrane transport. *Nature*, **417**, 187–193.

Puertollano, R., Randazzo, P.A., Presley, J.F., Hartnell, L.M. & Bonifacino, J.S. (2001a) The GGAs promote ARF-dependent recruitment of clathrin to the TGN. *Cell*, **105**, 93–102.

Puertollano, R., Aguilar, R.C., Gorshkova, I., Crouch, R.J. & Bonifacino, J.S. (2001b) Sorting of mannose 6-phosphate receptors mediated by the GGAs. *Science*, **292**, 1712–1716.

Randazzo, P.A., Terui, T., Sturch, S., Fales, H.M., Ferrige, A.G. & Kahn, R.A. (1995) The myristoylated amino terminus of ADP-ribosylation factor 1 is a phospholipid- and GTP-sensitive switch. *J. Biol. Chem.*, **270**, 14809–14815.

Regad, F., Bardet, C., Tremousaygue, D., Moisan, A., Lescure, B. & Axelos, M. (1993) cDNA cloning and expression of *Arabidopsis* GTP-binding protein of the ARF family. *FEBS Lett.*, **316**, 133–136.

Regazzi, R., Ulrich, S., Kahn, R.A. & Wolheim, C.B. (1991) Redistrubition of ADP-ribosylation factor during stimulation of permeabilized cells with GTP analogues. *Biochem. J.*, **275**, 639–644.

Reinhard, C., Harter, C., Bremser, M., Brugger, B., Sohn, K., Helms, J.B. & Wieland, F. (1999) Receptor-induced polymerization of coatomer. *Proc. Natl. Acad. Sci. USA*, **96**, 1224–1228.

Ritzenthaler, C., Nebenführ, A., Movafeghi, A., Stussi-Garaud, C., Behnia, L., Pimpl, P., Staehelin, L.A. & Robinson, D.G. (2002) Revaluation of the effects of Brefeldin A on plant cells using tobacco BY-2 cells expressing Golgi-targeted GFP and COPI-antisera. *The Plant Cell*, **14**, 237–261.

Robinson, D.G., Hinz, G. & Holstein, S.E.H. (1998a) The molecular characterization of transport vesicles. *Plant Mol. Biol.*, **38**, 47–76.

Robinson, D.G., Bäumer, M., Hinz, G. & Hohl, I. (1998b) Vesicle transfer of storage proteins to the vacuole: the role of the Golgi apparatus and multivesicular bodies. *J. Plant Physiol.*, **152**, 659–667.

Robinson, M.S. & Bonifacino, J.S. (2001) Adaptor-related proteins. *Curr. Opin. Cell Biol.*, **13**, 444–453.

Roca, R., Stiefel, V. & Puigdomenech, P. (1998) Characterization of the sequence coding for the clathrin coat assembly protein AP17 (sigma 2) associated with the plasma membrane from *Zea mays* and constitutive expression of its gene. *Gene*, **208**, 67–72.

Rohn, W.M., Rouille, Y., Waguri, S. & Hoflack, B. (2000) Bi-directional trafficking between the trans-Golgi network and the endosomal/lysosomal system. *J. Cell Sci.*, **113**, 2093–2101.

Rothman, J.E. & Wieland, F.T. (1996) Protein sorting by transport vesicles. *Science*, **172**, 227–234.

Salama, N.R., Yeung, T. & Schekman, R. (1993) The Sec13p complex and reconstitution of vesicle budding from the ER with purified cytosolic proteins. *EMBO J.*, **12**, 4073–4082.

Sanderfoot, A.A., Ahmed, S.U., Marty-Mazars, D., Papoport, I., Kirchhausen, T., Marty, F. & Kaikhel, N.V. (1998) A putative vacuolar cargo receptor partially colocalizes with AtPEP12p on a pre-vacuolar compartment b in *Arabidopsis* roots. *Proc. Natl. Sci. USA*, **95**, 9920–9925.

Sanderfoot, A.A., Kovaleva, V., Zheng, H. & Raikhel, N.V. (1999) The t-SNARE AtVAMP3p resides on the prevacular compartment in *Arabidopsis* root cells. *Plant Physiol.*, **121**, 929–938.

Sanderfoot, A.A., Assad, F.F. & Raikhel, N. (2000) The *Arabidopsis* genome. An abundance of soluble N-ethylmaleimide-sensitive factor adaptors. *Plant Physiol.*, **124**, 1558–1569.

Scheel, J., Pepperkok, R., Lowe, M., Griffths, G. & Kries, T.E. (1997) Dissociation of coatner from membranes is required for brefeldin A induced transfer of Golgi enzymes of the endoplasmic reticulum. *J. Cell Biol.*, **137**, 319–333.

Scheele, U. & Holstein, S.E.H. (2002) Functional evidence for the identification of an *Arabidopsis* clathrin light chain polypeptide. *FEBS Lett.*, **514**, 355–360.

Schekman, R. & Orci, L. (1996) Coat proteins and vesicle budding. *Science*, **271**, 1526–1533.

Schimmöller, F., Singer-Krüger, B., Schröder, S., Krüger, U., Barlowe, C. & Riezman, H. (1995) The absence of Em24p, a component of ER-derived COPII-coated vesicles, causes a defect in transport of selected proteins to the Golgi. *EMBO J.*, **14**, 1329–1339.

Schmidt, S.L. (1997) Clathrin-coated vesicle formation and protein sorting an integrated process. *Annu. Rev. Biochem.*, **66**, 511–548.

Seaman, M.N., Sowerby, P.J. & Robinson, M.S. (1996) Cytosolic and membrane associated proteins involved in the recruitment of AP-1 adaptors onto the trans-Golgi network. *J. Biol. Chem.*, **271**, 25446–25451.

Serafini, T., Orci, L., Amherdt, M., Brunner, M., Kahn, R.A. & Rothman, J.E. (1991a) ADP-ribosylation factors a subunit of the coat of Golgi-derived COP-coated vesicle: a novel role for a GTP-binding protein. *Cell*, **67**, 239–253.

Serafini, T., Stenbeck, G., Brecht, A., Lottspeich, F., Orci, L., Rothman, J.E. & Wieland, F.T. (1991b) A coat subunit of Golgi-derived non-clathrin-coated vesicles with homology to the clathrin-coated vesicle coat protein beta-adaptin. *Nature*, **349**, 215–220.

Sevier, C.S., Weisz, O.A., Davis, M. & Machamer, C.E. (2000) Efficient export of the vesicular stomatitis virus G protein from the endoplasmic reticulum requires a signal in the cytoplasmic tail that includes both tyrosine-based and di-acidic motifs. *Mol. Biol. Cell*, **11**, 13–22.

Shinotsuka, C., Yoshida, Y., Kawamoto, K., Takatsu, H. & Nakayama, K. (2002) Overexpression of an ADP-ribosylation factor – guanine nucleotide exchange factor, BIG2, uncouples brefeldin-A induced adaptor protein-1 coat dissociation and membrane tubulation. *J. Biol. Chemistry*, **277**, 9468–9473.

Simmen, T., Schmidt, A., Hunziker, W. & Beermann, F. (1999) The tyrosinase tail mediates sorting to the lysosomal compartment in MDCK cells via a di-leucine and a tyrosine-based signal. *J. Cell Sci.*, **112**, 45–53.

Spang, A., Matsuoka, K., Hamamoto, S., Schekman, R. & Orci, L. (1998) Coatomer, Arf1p, and nucleotide are required to bud coat protein complex I-coated vesicles from large synthetic liposomes. *Proc. Natl. Acad. Sci. USA*, **95**, 11199–11204.

Springer, S. & Schekman, R. (1998) Nucleation of COPII vesicular coat complex by endoplasmic reticulum to Golgi vesicle SNAREs. *Science*, **281**, 698–700.

Springer, S., Spang, A. & Schekman, R. (1999) A primer on vesicle budding. *Cell*, **97**, 145–148.

Springer, S., Chen, E., Duden, R., Marzioch, M., Rowley, A., Hamamoto, S., Merchant, S. & Schekman, R. (2000) The p24 proteins are not essential for vesicular transport in *Saccharomyces cerevisiae*. *Proc. Natl. Acad. Sci. USA*, **97**, 4034–4039.

Stamnes, M.A. & Rothman, J.E. (1993) The binding of AP-1 clathrin adaptor particles to Golgi membranes requires ADP-ribosylation factor, a small GTP-binding protein. *Cell*, **73**, 999–1005.

Stamnes, M.A., Craighead, M.W., Hoe, M.H., Lampen, N., Geromanos, S., Tempst, P. & Rothman, J.E. (1995) An integral membrane component of coatomer-coated transport vesicles defines a family of proteins involved in budding. *Proc. Natl. Acad. Sci. USA*, **92**, 8011–8015.

Steinmann, T., Geldner, N., Grebe, M., Mangold, S., Jackson, C.L., Paris, S., Galweiler, L., Palme, K. & Jurgens, G. (1999) Coordinated polar localization of auxin efflux carrier PIN 1 by GNOM ARF GEF. *Science*, **286**, 316–318.

Stenbeck, G., Schreiner, R., Hermann, D., Auerbach, S., Lottspeich, F., Rothman, J.E. & Wieland, F.T. (1992) Gamma-COP, a coat subunit of non-clathrin-coated vesicles with homology to Sec21p. *FEBS Lett.*, **314**, 195–198.

Stepp, J.D., Huang, K. & Lemmon, S.K. (1997) The yeast adaptor protein complex, AP-3, is essential for the efficient delivery of alkaline phosphatase by the alternate pathway to the vacuole. *J. Cell Biol.*, **139**, 1761–1774.

Szopa, J. & Mueller-Roeber, B. (1994) Cloning and expression analysis of an ADP-ribosylation factor from *Solanum tuberosum* L. *Plant Cell Rep.*, **14**, 180–183.

Takatsu, H., Katoh, Y., Shiba, Y. & Nakayama, K. (2001) Golgi-localizing, gamma-adaptin ear homology domain, ADP-ribosylation factor-binding (GGA) proteins interact with acidic dileucine sequences within the cytoplasmic domains of sorting receptors through their Vps27p/Hrs/STAM (VHS) domains. *J. Biol. Chem.*, **276**, 28541–28545.

Teal, S.B., Hsu, V.W., Peters, P.J., Klausner, R.D. & Donaldson, J.G. (1994) An activating mutation in ARF1 stabilizes coatomer binding to Golgi membranes. *J. Biol. Chem.*, **269**, 3135–3138.

Teuchert, M., Schafer, W., Berghofer, S., Hoflack, B., Klenk, H.D. & Garten, W. (1999) Sorting of furin at the trans-Golgi network. Interaction of the cytoplasmic tail sorting signals with AP-1 Golgi-specific assembly proteins. *J. Biol. Chem.*, **274**, 8199–8207.

Togawa, A., Morinaga, N., Ogasawara, M., Moss, J. & Vaughan, M. (1999) Purification and cloning of a brefeldin A-inhibited guanine nucleotide-exchange protein for ADP-ribosylation guanine nucleotide-exchange protein for ADP-ribosylation factors. *J. Biol. Chem.*, **274**, 12308–12315.

Verma, D.P. (2001) Cytokinesis and building of the cell plate in plants. *Annu. Rev. Plant Physiol. Plant Mol. Biol.*, **52**, 751–784.

Verwoert, I.I.G.S., Brown, A., Slabas, A.R. & Stuije, A.R. (1995) A *Zea mays* GTP-binding protein of the ARF family complements an *Esherichia coli* mutant with a temperature-sensitive malonyl-coenzyme A: acyl carrier protein transacylase. *Plant Mol. Biol.*, **27**, 629–633.

Vostmeier, C. & Gallwitz, D. (2001) An acidic sequence of a putative yeast Golgi membrane protein binds COPII and facilitates ER export. *EMBO J.*, **20**, 6742–6750.

Vowels, J.J. & Payne, G.S. (1998) A dileucine-like sorting signal directs transport into an AP-3-dependent, clathrin-independent pathway to the yeast vacuole. *EMBO J.*, **17**, 4211.

Waters, M.G., Serafini, T. & Rothman, J.E. (1991) A cytosolic protein complex containing subunits of non-clathrin coated Golgi transport vesicles. *Nature*, **349**, 248–251.

Weissman, J.T., Plutner, H. & Balch, W.E. (2001) The mammalian guanine nucleotide exchange factor mSec12 is essential for activation of the Sar1 GTPase directing endoplasmic reticulum export. *Traffic*, **2**, 465–475.

Wieland, F. & Harter, C. (1999) Mechanisms of vesicle formation: insights from the COP system. *Curr. Opin. Cell Biol.*, **11**, 440–446.

Wen, C. & Greenwald, I. (1999) p24 proteins and quality control of LIN-12 and GLP-1 trafficking in *Caenorhabditis elegans*. *J. Cell Biol.*, **145**, 1165–1175.

Yamaji, R., Adamik, R., Takeda, K., Togawa, A., Pacheco-Rodriguez, G., Ferrans, V.J., Moss, J. & Vaughan, M. (2000) Identification and localization of two brefeldin A-inhibited guanine nucleotide-exchange proteins for ADP-ribosylation factors in a macromolecular complex. *Proc. Natl. Acad. Sci. USA*, **97**, 2567–2572.

Yoshihisa, T., Barlowe, C. & Schekman, R. (1993) Requirement for a GTPase-activating protein in vesicle budding from the endoplasmic reticulum. *Science*, **259**, 1466–1468.

Zhao, L., Helms, J.B., Brügger, B., Harter, C., Martoglio, B., Graf, R., Brunner, J. & Wieland, F.T. (1997) Direct and GTP-dependent interaction of ADP-ribosylation factor 1 with coater subunit β. *Proc. Natl. Acad. Sci. USA*, **94**, 4418–4423.

Zhao, L., Helms, J.B., Brunner, J. & Wieland, F.T. (1999) GTP-dependent binding of ADP-ribosylation factor to coatner in close proximity to the binding site fordilysine retrieval motifs and p23. *J. Biol. Chem.*, **274**, 14198–14203.

Zhu, Y., Doray, B., Poussu, A., Lehto, V.-P. & Kornfeld, S. (2001) Binding of GGA2 to the lysosomal enzyme sorting motif of the mannose 6-phosphate receptor. *Science*, **292**, 1716–1718.

4 Endomembrane and cytoskeleton interrelationships in higher plants

Chris Hawes, Claude Saint-Jore and Federica Brandizzi

4.1 Introduction

It is well accepted that compartmentalisation of the cytoplasm was a key event in the evolution of eukaryotic cells. This characteristic feature is exemplified by the subdivision of the cytosol into two phases, bounded by the membranes of the various organelles comprising the secretory system. The spatial organisation of the secretory organelles is in turn regulated by the various filaments and proteins of the cytoskeleton. The secretory system is responsible for the transport of macromolecules from sites of synthesis such as the endoplasmic reticulum (ER) and Golgi, to sites of action or deposition, for example the vacuolar system or cell surface. Likewise, molecules may enter the system via the endocytic pathway, although even today, we have little information on the physiological importance of this pathway in plants (Hawes *et al.*, 1995; Geldner *et al.*, 2001). It has been generally assumed that the transport of macromolecular cargo between the various endomembrane compartments is mediated by vesicle vectors which bud from donor compartments and fuse after targeting at specific receptor compartments (Sanderfoot & Raikhel, 1999). However, this view is not unanimously held and non-vesicular transport pathways have also been proposed (Brandizzi *et al.*, 2002; see also Chapter 2).

One consequence of the vesicle transport hypothesis is, concomitant with the transport of cargo molecules in the lumen of vesicles, is an inevitable flow of membrane in the cytoplasm and to and from the cell surface. In non-growing cells, it has to be assumed that there is a recycling of membrane at the cell surface via an endocytic mechanism that balances the inevitable loss of endomembrane to the plasma membrane (Steer & O'Driscoll, 1991; Geldner *et al.*, 2001).

The myriad of dynamic events undertaken by a fully functioning secretory pathway is in no way random but comprises what appears to be an extremely sophisticated and co-ordinated set of transport and targeting events. Besides regulating the biosynthesis and modification of cargo molecules, modulating their package into transport vectors, targeting and accepting them at the correct destination, the cell has also derived mechanisms to facilitate the transport steps, whilst maintaining the integrity of the intracellular environment. In this chapter, we consider the role of the cytoskeleton, both in transport of cargo through the endomembrane system and in maintaining the three-dimensional architecture of the constituent organelles, concentrating on the relationship between the Golgi apparatus and the ER.

4.2   Pharmacological experiments have indicated the importance of the actin cytoskeleton

In mammalian cells, it is generally accepted that microtubules play a major role in maintaining the three-dimensional organisation of the endomembrane system including the ER and the Golgi apparatus (Terasaki & Reese, 1994; Allan & Schroer, 1999; Thyberg & Moskalewski, 1999). Indeed, in the presence of the anti-microtubule agent nocodazole, the perinuclear organisation of the Golgi is lost and the organelle becomes distributed around the cytoplasm as a series of mini-stacks, reminiscent of the organisation of the plant Golgi system (Cole *et al.*, 1996). These mini-stacks are closely associated with transitional ER exit sites and arise because disruption of the microtubule cytoskeleton inhibits movement of pre-Golgi structures to the peri-nuclear Golgi region (Presley *et al.*, 1997). A redistribution of the existing Golgi membranes then occurs as Golgi components undergo a continuous cycling back to the ER (Miles *et al.*, 2001; Ward *et al.*, 2001), but forward membrane transport is inhibited which results in the formation of the mini-stacks. However, actin and associated proteins have also been shown to be involved in the organisation of the mammalian Golgi complex, to be associated with COPI vesicles and may be involved in the retro-grade transport of some proteins (Valderrama *et al.*, 2001; Fucini *et al.*, 2002; Luna *et al.*, 2002).

Historically, in plants, there is little evidence for a role of the microtubule cytoskeleton in mediating any aspect of the secretory pathway other than that of transport of vesicles to the developing cell plate. Therefore, it is the actin cytoskeleton that has been proposed to regulate the spatial organisation of the endomembranes, although prior to the advent of fluorescent protein technology the evidence for the function of an actin/myosin motor system has been scant. It has been shown, using the fluorochrome 3,3'-dihexyloxacarbocyanine iodide (DiOC$_6$) to label the ER in living cells, that actin may be involved in the re-modelling of the ER after disruption by low temperature or centrifugation (Quader *et al.*, 1987, 1989). Also, many ultrastructural reports, especially those employing freeze-substitution tech-nology, have described association of cortical ER with microfilament bundles (see Lichtscheidl & Hepler, 1996 for review).

A role of the actin cytoskeleton in the functioning of the Golgi apparatus has been difficult to establish. Actin disrupting agents such as the cytochalasins have been used to induce accumulations of secretory vesicles around Golgi stacks (Mollenhauer & Morré, 1976; Picton & Steer, 1981, 1983; Steer & O'Driscoll, 1991; see also Chapter 12), suggesting a function of actin filaments in secretory vesicle trafficking. Clustering of Golgi stacks after cytochalasin treatment has also been observed in maize root meristem cells using electron microscopy (Mollenhauer & Morré, 1976; Satiat-Jeunemaitre *et al.*, 1996) and after immunofluorescence labelling with the Golgi monoclonal antibody JIM 84 (Satiat-Jeunemaitre *et al.*, 1996; see Fig. 4.1). This data suggested that the Golgi apparatus is dependent on the integrity of the actin filament network for its spatial organisation.

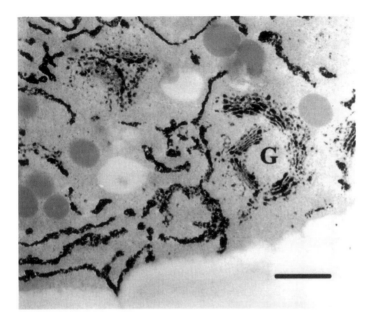

Figure 4.1 Maize root cap meristem cell showing clumping of Golgi stacks (G) after actin depolymer-isation with 20 μg/ml cytochalasin D for 1 h. Material was prepared using zinc iodide osmium impregnation technique to selectively highlight the endomembrane system. Bar = 500 nm.

It also seems likely that actin mediates transport in the endocytic pathway, as has been shown from experiments on the auxin efflux carrier PIN1 (Geldner *et al.*, 2001). In *Arabidopsis* roots, the PIN1 protein was shown to accumulate in vesicular compartments on treatment of the roots with BFA. These compartments were formed by vesiculation of the Golgi apparatus on drug treatment and it was suggested that they also accumulate PIN1 via the endocytic pathway. When roots were treated with cytochalasin D to disrupt the actin cytoskeleton, not only the internalisation of PIN into BFA compartments, but also the polar relocalisation of PIN1 to the plasma membrane on wash-out of BFA was inhibited. Thus, the cycling of the protein to and from the plasma membrane appears to be actin dependent, although some evidence was also presented for microtubule dependent delivery of the protein to the developing cell plate.

4.3 *In vivo* imaging of endomembrane organelles reveals dependence on the cytoskeleton

Although immunocytochemistry has revealed much about the organisation of the cytoskeleton of plant cells and of the various components of the endomembrane system, it was the advent of *in vivo* imaging that revolutionised the study of the dynamics of cellular components. Initially, endomembranes were imaged by the use of lipophilic dyes such as the carbocyanine $DiOC_6$ which stains the ER (Quader *et al.*, 1989), and has also been used in an attempt to identify the Golgi apparatus (Kawazu *et al.*, 1995). More recently the amphiphilic styryl dyes such as FM1-43 and FM4-64 which insert into the outer leaflet of the plasma membrane have been reported to highlight components of the putative endocytic pathway and the tonoplast (Carroll *et al.*, 1998; Emans *et al.*, 2002). Unfortunately, the ceramide dyes widely used to locate the Golgi apparatus in mammalian cells do not locate this organelle in plants (Fricker *et al.*, 2001). The cytoskeleton has also been investigated *in vivo* by the use of fluorescence analogue cytochemistry, whereby fluorochrome conjugated proteins such as fluoresceinated brain tubulin are microinjected into cells and incorporation into cytoskeletal arrays monitored (Hush *et al.*, 1994). A similar approach has been employed with the microinjection of rhodamine phalloidin into dividing plant cells to study the role of actin in phragmoplast formation (Molchan *et al.*, 2002).

The development of coelenterate fluorescent proteins as *in vivo* reporters has, however, revolutionised cell biology (Sullivan & Kay, 1999). Some of the earliest works utilising GFP expression in plants were based around targeting to the endoplasmic reticulum (Boevink *et al.*, 1996; Haseloff *et al.*, 1997). However, the importance of the cytoskeleton in controlling movement of the elements of the endomembrane system became clear when GFP was first targeted to the Golgi apparatus and to both the ER and Golgi apparatus in leaves of a tobacco species, *Nicotiana clevelandii* (Boevink *et al.*, 1998). The Golgi stacks in leaf epidermal cells were targeted by expressing the signal anchor sequence of a rat sialyl transferase (the transmembrane domain plus flanking amino acids) spliced to GFP (ST-GFP), and were shown to be highly motile exhibiting a variety of different movements. This phenomenon was subsequently confirmed in BY2 cells expressing a plant mannosidase GFP construct (Nebenführ *et al.*, 1999). When a GFP construct comprising the *Arabidopsis* homologue of the yeast K/HDEL receptor protein (*At*ERD2-GFP), which is responsible for recycling escaped ER proteins back from the Golgi, was expressed in tobacco leaves, both the ER and the Golgi fluoresced (Boevink *et al.*, 1998). Remarkably, the Golgi bodies showed two different sets of movements. Firstly, they were observed rapidly streaming over the ER in *trans*-vacuolar strands, and, secondly they could be seen moving in a more controlled fashion over the polygonal network of cortical ER tubules.

These studies also revealed that branching tubules from the ER network could exhibit directional growth and homotypically fuse with other pre-existing tubules. When such GFP-expressing leaves were fixed and the epidermal cells permeabilised, they could be stained with rhodamine phalloidin to reveal the actin cytoskeleton (Boevink *et al.*, 1998). The actin cytoskeleton essentially formed what appeared to be a random cortical network of actin cables, but when superimposed onto the polygonal ER network of the same cells, revealed by the expression of *At*ERD2-GFP, it was apparent that the ER tubules lay exactly over the actin cytoskeleton. Moreover, by reducing the digital signal in the confocal channel detecting ER/Golgi fluorescence, and leaving only Golgi fluorescence, it was obvious that the Golgi stacks also co-aligned with the actin. This data, taken along with the fact that actin depolymerising drugs, such as cytochalasin and latrunculin, inhibited Golgi and ER movement, has resulted in the conclusion that both ER tubule extension and Golgi movement are reliant on an active acto-myosin system (Boevink *et al.*, 1998; Brandizzi *et al.*, 2002).

A similar relationship between the actin cytoskeleton and the Golgi apparatus has also been reported in tobacco BY2 suspension culture cells (Nebenführ *et al.*, 1999; Saint-Jore *et al.*, 2002; see Fig. 4.2). In contrast, depolymerisation of micro-tubules had no appreciable effect on Golgi movements and immunostaining showed little relationship between the cortical microtubules and GFP-tagged Golgi (Nebenführ *et al.*, 1999; Saint-Jore *et al.*, 2002). Also, use of the myosin ATPase inhibitor 2,3-butanedione monoxime (BDM) also reversibly inhibited Golgi movement in BY2 cells (Nebenführ *et al.*, 1999). A conventional myosin II has been reported to be associated with the mammalian Golgi apparatus, although it is thought that this motor is not involved in Golgi movement *per se*, but in an actin-myosin mediate budding of vesicles from the *trans*-Golgi network (Stow *et al.*, 1998). Although we have no further information on the myosin(s) that may be driving Golgi movement in plant systems, analysis of the *Arabidopsis* genome has revealed 17 myosin sequences which all fall into two groups, class VIII and class XI, which exclusively contain plant or algal myosins (Reddy & Day, 2001). There-fore, it is not unreasonable to presume that one of these homologues drives Golgi

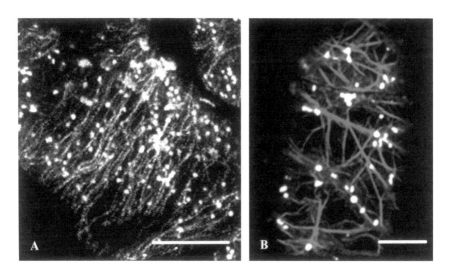

Figure 4.2 Tobacco BY2 cells expressing ST-GFP in the Golgi and immuno-stained for tubulin (A) or stained with rhodamine phalloidin for the actin cytoskeleton (B). Golgi do not appear to co-align with microtubules but do show a close association with the actin filament network. Bars = 10 and 5 μm. (Reproduced with permission from Saint-Jore *et al.*, 2002.)

movement in tobacco cells. Interestingly, in a range of cell types, peroxisomes have also been shown to perform similar movements over the actin cytoskeleton (Mathur *et al.*, 2002). However, based on pharmacological experiments, the authors suggested that the movement may not be myosin based, but dependent on an actin polymerisation-based process.

With the use of spectral variants of GFP it is possible to visualise concurrently two or more fluorescent organelles at the same time. Through the expression of an actin-targeted construct comprising the actin-binding domain of a mouse talin spliced to YFP concurrently with ST-GFP, it has been possible to confirm the association of actin with the Golgi in living leaf epidermal cells (Brandizzi *et al.*, 2002). Again, there was no obvious association between the cortical microtubular cytoskeleton and the mobile Golgi stacks in cells co-expressing GFP tubulin and ST-GFP. As expected, depolymerisation of the microtubule cytoskeleton with colchicine did not inhibit Golgi movement (Brandizzi *et al.*, 2002).

Another consequence of the cell having a motile Golgi apparatus concerns the integrity of the Golgi stack itself. Consisting of a stack of cisternal membranes which appear to be continually budding transport vesicles, the Golgi body appears to stay intact whilst moving. Thus, one can predict that there must be some physical elements responsible for maintaining its structure. Staehelin and Moore (1995) have proposed that individual stacks are embedded in or a surrounded by a structural Golgi matrix which may be responsible for preventing the stacks falling apart and for retaining intercisternal vesicles. Also, the presence of intercisternal elements which appear as filaments lying parallel to the cisternal membranes or as cross bridges between cisternae, has been well documented (Hawes *et al.*, 1996). It has been recently suggested that such elements may play a major role in establishing intercisternal bonds, especially between medial and *trans*-cisternae (Ritzenthaler *et al.*, 2002). Here it was reported that intercisternal filaments are present in all cisternae of BY2 Golgi stacks remaining after ten or more minutes of treatment with the secretory inhibitor brefeldin A (BFA, see also Chapter 12), suggesting that they may help confer some resistance to the retrograde redistribution of Golgi membranes into the ER on treatment with the drug.

The dearth of information on plant Golgi matrix proteins is however, in contrast to the situation in mammalian cells where a number of matrix proteins have been characterised (Nakamura *et al.*, 1997; Shorter *et al.*, 1999; Barr, 2002). This structure is now thought to be composed of dynamic arrays of protein complexes specific for the different Golgi cisternae which confer compartmental identity as well as being responsible for binding cisternae, tethering vesicles and interacting with Rab GTPases (Barr, 2002). The fact that the mammalian matrix is also thought to cycle in and out of the Golgi complex (Ward *et al.*, 2001) may reflect a key difference between the role of matrix proteins in conferring stability to a moving plant Golgi stack and maintaining the organisation of the more static mammalian Golgi complex.

The observation of Golgi movement in both leaf and suspension culture cells has resulted in considerable speculation as to their ability to receive secretory product and membrane from the ER whilst moving (Boevink *et al.*, 1998; Nebenführ *et al.*, 1999; Nebenführ & Staehelin, 2001). An extremely apt term *mobile factories* has been coined to describe the function of the plant Golgi in the production and distribution of secretory products (Nebenführ & Staehelin, 2001). Although it is not yet known whether the Golgi in all plant tissues exhibit such mobility, some extremely pertinent questions can now be posed regarding Golgi functioning. For instance, if Golgi stacks are moving how do they recruit membrane and secretory proteins from the ER and how do they manage to secrete or even receive membrane from the endocytic pathway when they are motile? Answers to such questions are not easily forthcoming, although various models have been proposed, especially with regards to ER to Golgi transport.

Boevink *et al.* (1998) on observing moving Golgi in leaf epidermal cells made the suggestion that perhaps the Golgi are travelling over the ER collecting secretory material from the surface, much like a vacuum cleaner or *Hoover* collects dirt from a carpet. It was suggested that the vectors for such an exchange could be vesicles travelling over a short range or even tubules emanating from the ER surface. One feature of Golgi movement within leaf cells and also observed in GFP labelled Golgi in BY2 cells is that the movement is not the same for all Golgi but extremely variable, exhibiting a range of speeds (from stationary to greater than 2 µm/sec) and even showing stop–start movements. Thus, it is unlikely that the force for moving Golgi stacks is simply generated by the shear forces created by cytoplasmic streaming. The fact that Golgi stacks exhibit stop-go movements in BY2 cells resulted in a hypothesis which stated that in order to receive material from the ER, Golgi stacks would have to come to a halt in close proximity to exit sites on the ER (Nebenführ *et al.*, 1999). This would require the generation of some form of extremely localised *stop signal* and presumably a similar *start* signal. Such a model is partially supported by the discovery that membrane protein transport from ER to Golgi can take place when Golgi stacks are static, and is discussed later (Brandizzi *et al.*, 2002). Interestingly, when Golgi movement is inhibited by the action of actin depolymerising drugs, Golgi sometimes clump together on small islands of cisternal ER which are formed at the three way junctions between tubules of the polygonal cortical network (Boevink *et al.*, 1998). It is therefore tempting to speculate that such junctions have a physiological significance and may even represent the elusive exit sites on the ER. Further discussion on the molecular basis of exit of material from the ER is given in Chapter 6.

To date, data on Golgi movement have come from a very limited number of cell types although our observations of Golgi in stable transformants of *Arabidopsis* indicate that in the majority of tissues Golgi show some movement. Likewise, we have little data on the quantity of membrane that must be transferred between the ER and Golgi in a cell that is not heavily secreting. In growing tissues, certainly the major activity of the Golgi is in the production of cell wall polysaccharides. The traffic from the ER may then simply be regulated by the quantity of membrane required for constructing new plasma membrane. In non-growing polysaccharide secreting cells, such as those of the root cap one could then envisage a situation whereby the balance in the membrane economy of the cell is regulated by endocytosis and very little membrane may be transported from the ER. To date, we have no *in vivo* data on ER–Golgi relationships in such cells.

4.4   Transport between ER and Golgi may be independent of the cytoskeleton: evidence from pharmacological and photobleaching studies

Whilst data from GFP-expressing cells have indicated that the actin cytoskeleton is responsible for supporting movement of individual Golgi stacks over the ER and around the cell, it appears that the cytoskeleton is not necessary for the transfer of material between the two organelles. In mammalian cells, microtubules and associated motor proteins are not only involved in the maintenance of Golgi structure but also in anterograde and retrograde protein and membrane transport between the ER and Golgi (Cole *et al.*, 1996; Thyberg & Moskalewski, 1999; Robertson & Allan, 2000). Considering the close association between the ER and Golgi in many plant cells, an analogous cytoskeletal involvement in transport between the two organelles could be predicted.

Saint-Jore *et al.* (2002) tested this hypothesis pharmacologically by investigating the effect of the secretory inhibitor Brefeldin A (BFA) on the Golgi apparatus of BY2 and tobacco leaf epidermal cells in the presence of various cytoskeletal inhibitors. BFA has been shown to induce the redistribution of Golgi-targeted GFP constructs into the ER in a reversible manner (Fig. 4.3A,B; Boevink *et al.*, 1998; Ritzenthaler *et al.*, 2002; Saint-Jore *et al.*, 2002; and further discussion on the molecular effects of Golgi-disrupting drugs such as BFA is given in Chapter 12). As would be expected, when the microtubule cytoskeleton was depolymerised by colchicine or oryzalin there was no effect on the BFA-induced redistribution of Golgi targeted ST-GFP into the ER. However, the same result was obtained when the actin cytoskeleton was disrupted by cytochalasin D or latrunculin B (Fig. 4.3C). Such experiments were also carried out in the presence of the protein synthesis inhibitor cycloheximide, indicating that fluorescence in the ER was not due to the synthesis of new GFP during the experimental period. These experiments indicated that there is no role for the cytoskeleton in retrograde transport, as induced by BFA, between the Golgi and ER.

Remarkably, Golgi bodies appear to reform from the ER, in BFA treated cells, on removal of the drug (Satiat-Jeunemaitre & Hawes, 1992; Boevink *et al.*, 1998; Saint-Jore *et al.*, 2002), a phenomenon that can also occur in cycloheximide treated cells (Saint-Jore *et al.*, 2002). Again this effect can be observed in the total absence of the cytoskeleton (Fig. 4.3D), indicating that forward transport from the ER may not require microtubules or actin (Saint-Jore *et al.*, 2002).

The conclusion from the experiments described above has been corroborated from a different approach to the study of transport in living, fluorescent protein-expressing cells, namely the use of a photobleaching strategy (Brandizzi *et al.*, 2002). Selective photobleaching followed by recovery measurements has proved to be an immensely powerful tool in the study of protein transport in and between the membranes of the secretory pathway (Lippincott-Schwartz *et al.*, 1998, 2001). The technique known as fluorescence recovery after photobleaching (FRAP) is based on the principle of photobleaching a predefined area of fluorescence with an

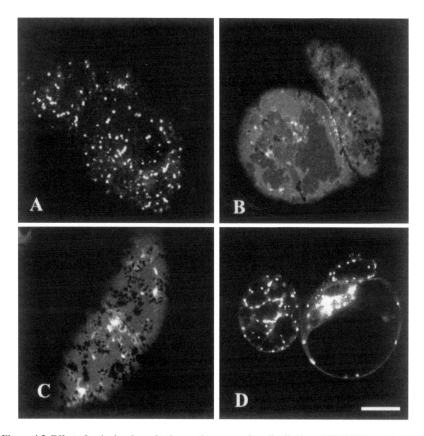

Figure 4.3 Effect of actin depolymerisation on the retrograde redistribution of ST-GFP from the Golgi into the ER on brefeldin A (BFA) treatment and recovery of Golgi after removal of BFA in BY2 cells. (A) Control cell expressing ST-GFP in the Golgi. (B) Cells treated with 180 μM BFA for 3 h. The majority of GFP fluorescence has redistributed into the ER which has been converted from a tubular network to large cisternae. (C) Cell treated as in B but with prior treatment for 1.5 h with 39.4 μM cytochalasin D to depolymerise the actin cytoskeleton. (D) Reformation of Golgi in BY2 cells after removal of BFA and incubation in fresh culture medium plus 39.4 μM cytochalasin D (reproduced with permission from Saint-Jore *et al.*, 2002). Bar = 10 μm. For details on the short term effects of BFA on cells see Chapter 12.

intense (but not damaging) pulse of laser light and observing and quantifying the rate of fluorescence recovery due to the movement of fluorescent molecules into the bleached area.

In leaf cells expressing various Golgi targeted fluorescent protein constructs, FRAP has been used to assess the flow of membrane protein between the ER and Golgi (Brandizzi *et al.*, 2002). Such experiments have yet to be carried out on moving Golgi, as inhibition of movement was necessary prior to photobleaching in order to quantify recovery rates. Single Golgi bodies expressing either *At*ERD2-GFP or ST-GFP were photobleached after inhibition of movement with

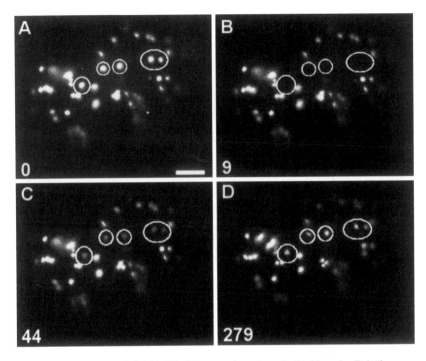

Figure 4.4 Photobleaching of Golgi in ST-GFP expressing tobacco leaf epidermal cells in the presence of cytochalasin D to inhibit Golgi movement. Five Golgi stacks (ringed in A) were selectively photobleached (B) using the Zeiss LSM 510 photobleaching software and recovery monitored over 279 seconds (C, D). Recovery of fluorescence indicated movement of fluorescent protein into the Golgi stacks (Brandizzi *et al.*, 2002). Bar = 5 µm.

latrunculin B. With both constructs, fluorescence in the bleached Golgi recovered to 80–90% of the level prior to bleaching within 5 minutes (Fig. 4.4). However, under conditions of energy depletion or treatment with *N*-ethyl-maleimide, which would be predicted not only to stop myosin-based movement but also fusion of membrane between ER and Golgi (Orci *et al.*, 1991), there was no recovery of fluorescence. These data indicated that upon depletion of fluorescence in the Golgi, the bleached molecules must have exchanged with fluorescent pools of the protein located elsewhere in the cell, possibly the ER. Thus, corroborating the suggestion of Saint-Jore *et al.* (2002) that actin is not required for anterograde ER to Golgi transport. Comparable experiments were carried out in the absence of both actin and microtubules and a similar recovery of fluorescence in the Golgi was recorded. The implications of these experiments are that the fluorescent chimeras located in the Golgi are not static but in a state of continual turnover, cycling from the ER to the Golgi and out of the Golgi to another compartment. Whether this is recycling to the ER, as has been suggested for mammalian Golgi membrane proteins (Miles *et al.*, 2001; Ward *et al.*, 2001), is yet to be established.

4.5 Summary

It is certainly the case that there are major differences in the organisation of the plant endomembrane system as compared to that in mammalian cells, with actin playing a far more prominent role than microtubules and their associated motors. In leaf cells, the cortical network of ER certainly appears to be modelled by extending tubules along a cortical actin framework, although the network does not break down on removal of the actin. Golgi bodies appear to track over the ER driven, presumably by a myosin motor interacting with the actin cytoskeleton. However, it appears that actin and Golgi movements are not needed for close range transport between ER and Golgi. At this moment, we do not know whether the transfer between ER and Golgi is faster when Golgi stacks are moving over the ER, or what the cytoskeletal requirements are in cell types where ER and Golgi are spatially separate. Likewise, the transport of secretory vesicles from the Golgi is most likely actin-driven, but we have no data on transport to the Golgi (one terminus of the endocytic pathway) or on the driving forces behind any inter-cisternal vesicle shuttle.

Acknowledgements

We acknowledge the Biotechnology and Biological Sciences Research Council for supporting much of the work described here, and Oxford Brookes University for a research studentship to CS-J. Figure 4.1 was kindly supplied by Béatrice Satiat-Jeunemaitre.

References

Allan, V.J. & Schroer, T.A. (1999) Membrane motors. *Curr. Opin. Cell Biol.*, **11**, 476–482.

Barr, F.A. (2002) The Golgi apparatus: going round in circles. *Trends in Cell Biol.*, **12**, 101–104.

Boevink, P., Santa Cruz, S.S., Hawes, C., Harris, N. & Oparka, K. (1996) Virus-mediated delivery of the green fluorescent protein to the endoplasmic reticulum of plant cells. *Plant J.*, **10**, 935–941.

Boevink, P., Oparka, K., Santa Cruz, S., Martin, B., Betteridge, A. & Hawes, C. (1998) Stacks on tracks: the plant Golgi apparatus traffics on an actin/ER network. *Plant J.*, **15**, 441–447.

Brandizzi, F., Snapp, E., Roberts, A., Lippincott-Schwartz, J. & Hawes, C. (2002) Membrane protein transport between the ER and Golgi in tobacco leaves is energy dependent but cytoskeleton independent: evidence from selective photobleaching. *The Plant Cell*, **14**, 1293–1309.

Carroll, A., Moyen, C., Van Kesteren, P., Tooke, F., Battey, N. & Brownlee, C. (1998) Ca^{2+}, annexins, and GTP modulate exocytosis from maize root cap protoplasts. *The Plant Cell*, **10**, 1267–1276.

Cole, N.B., Sciaky, N., Marotta, A., Song, J. & Lippincott-Schwartz, J. (1996) Golgi dispersal during microtubule disruption: regeneration of Golgi stacks at peripheral endoplasmic reticulum exit sites. *Mol. Biol. Cell*, **7**, 631–650.

Emans, N., Zimmermann, S. & Fischer, R. (2002) Uptake of a fluorescent marker in plant cells is sensitive to brefeldin A and wortmannin. *The Plant Cell*, **14**, 71–86.

Fricker, M., Parsons, A., Tlalka, M., Blancaflor, E., Gilroy, S., Meyer, A. & Plieth, C. (2001) Fluorescent probes for living plant cells, in *Plant Cell Biology* (eds. C. Hawes & B. Satiat-Jeunemaitre), Oxford University Press, Oxford, pp. 35–84.

Fucini, R.V., Chen, J.L., Sharma, C., Kessels, M.M. & Stamnes, M. (2002) Golgi vesicle proteins are linked to the assembly of an actin complex defined by mAbp1. *Mol. Biol. Cell*, **13**, 621–631.

Geldner, N., Frimi, J., Stierhof, Y.-D., Jürgens, G. & Palme, K. (2001) Auxin transport inhibitors block PIN1 cyclin and vesicle trafficking. *Nature*, **413**, 425–428.

Haseloff, J., Siemering, K.R., Prosher, D.C. & Hodge, S. (1997) Removal of a cryptic intron and sub-cellular localization of green fluorescent protein are required to mark transgenic *Arabidopsis* plants brightly. *Proc. Natl. Acad. Sci. USA*, **94**, 2122–2127.

Hawes, C., Crooks, K., Coleman, J. & Satiat-Jeunemaitre, B. (1995) Endocytosis in plants: fact or arte-fact? *Plant Cell Environ.*, **18**, 1245–1252.

Hawes, C.R., Faye, L. & Satiat-Jeunemaitre, B. (1996) The Golgi apparatus and pathways of vesicle trafficking, in *Membranes: Specialised Functions in Plant Cells* (eds. M. Smallwood, P.J. Knox & D.J. Bowles), BIOS, Oxford, pp. 337–365.

Hush, J.M., Wadsorth, P., Callaham, D.A. & Hepler, P.K. (1994) Quantification of microtubule dynam-ics in living plant cells using fluorescence redistribution after photobleaching. *J. Cell Sci.*, **107**, 775–784.

Kawazu, Y., Kawano, S. & Kuroiwa, T. (1995) Distribution of the Golgi apparatus in the mitosis of cultured tobacco cells as revealed by DiOC$_6$ fluorescence microscopy. *Protoplasma*, **186**, 183–192.

Lichtscheidl, I. & Hepler, P. (1996) Endoplasmic reticulum in the cortex of plant cells, in *Membranes: Specialised Functions in Plant Cells* (eds. M. Smallwood, P.J. Knox & D.J. Bowles), BIOS, Oxford, pp. 383–402.

Lippincott-Schwartz, J., Cole, N. & Presley, J. (1998) Unravelling Golgi membrane traffic with green fluorescent protein chimeras. *Trends Cell Biol.*, **8**, 16–47.

Lippincott-Schwartz, J., Snapp, E. & Kenworthy, A. (2001) Studying protein dynamics in living cells. *Nat. Rev. Mol. Cell Biol.*, **2**, 444–456.

Luna, A., Matas, O.B., Martínez-Menárguez, J.A., Mato, E., Duran, J.M., Ballesta, J., Way, M. & Egea, G. (2002) Regulation of protein transport from the Golgi complex to the endoplasmic reticulum by CDC42 and N-WASP. *Mol. Biol. Cell*, **13**, 866–879.

Mathur, J., Mathur, N. & Hülskamp, M. (2002) Simultaneous visualisation of peroxisomes and cytoskeletal elements reveals actin and not microtubule-based peroxisome motility in plants. *Plant Physiol.*, **128**, 1031–1045.

Miles, S., McManus, H., Forsten, K.E. & Storrie, B. (2001) Evidence that the entire Golgi apparatus cycles in interphase HeLa cells: sensitivity of Golgi matrix proteins to an ER exit block. *J. Cell Biol.*, **155**, 543–555.

Molchan, T.M., Valster, A. & Hepler, P.K. (2002) Actomyosin promotes cell plate alignment and late lateral expansion in *Tradescantia* stamen hair cells. *Planta*, **214**, 683–693.

Mollenhauer, H.H. & Morré, D.J. (1976) Cytochalasin B, but not colchicine inhibits migration of secre-tory vesicles in root tips of maize. *Protoplasma*, **87**, 39–48.

Nebenführ, A., Gallagher, L.A., Dunahay, T.G., Frohlick, J.A., Mazurkiewicz, A.M., Meehl, J.B. & Staehelin, L.A. (1999) Stop-and-go movements of plant Golgi stacks are mediated by the acto-myosin system. *Plant Physiol.*, **121**, 1127–1142.

Nakamura, N., Low, M., Levine, T.P., Rabouille, C. & Warren, G. (1997) The vesicle docking protein p115 binds GM130, a *cis*-Golgi matrix protein, in a mitotically regulated manner. *Cell*, **89**, 445–455.

Nebenführ, A. & Staehelin, L.A. (2001) Mobile factories: Golgi dynamics in plant cells. *Trends Plant Sci.*, **6**, 160–167.

Orci, L., Tagaya, M., Amherdt, M., Perrelet, A., Donaldson, J.G., Lippincott-Schwartz, J., Klausner, R.D. & Rothman, J.E. (1991) Brefeldin-A, a drug that blocks secretion, prevents the assembly of non-clathrin-coated buds on Golgi cisternae. *Cell*, **64**, 1183–1195.

Picton, J.M. & Steer, M.W. (1981) Determination of secretory vesicle production rates by dictyosomes in pollen tubes of *Tradescantia* using cytochalasin D. *J. Cell Sci.*, **49**, 261–272.

Picton, J.M. & Steer, M.W. (1983) The effect of cycloheximide on dictyosome activity in *Tradescantia* pollen tubes determined using cytochalasin D. *Eur. J. Cell Biol.*, **29**, 133–138.

Presley, J.F., Cole, N.B., Schroer, T.A., Hirschberg, K., Zaal, K.M.J. & Lippincott-Schwartz, J. (1997) ER-Golgi transport visualized in living cells. *Nature*, **389**, 81–85.

Quader, H., Hofmann, A. & Schnepf, E. (1987) Shape and movement of the endoplasmic reticulum in onion bulb epidermis cells: possible involvement of actin. *Euro. J. Cell Biol.*, **44**, 17–26.

Quader, H., Hofmann, A. & Schnepf, E. (1989) Reorganization of the endoplasmic reticulum in epidermal cells of onion bulb scales after cold stress: involvement of cytoskeletal elements. *Planta*, **177**, 273–280.

Reddy, A.S.N. & Day, I.S. (2001) Analysis of the myosins encoded in the recently completed *Arabidopsis thaliana* genome sequence. *Genome Biol.*, **2**, 0024.1–0024.17.

Ritzenthaler, C., Nebenführ, A., Movafeghi, A., Stussi-Garaud, C., Behnia, L., Pimple, P., Staehelin, L.A. & Robinson, D.G. (2002) Reevaluation of the effects of brefeldin A on plant cells using tobacco bright yellow 2 cells expressing Golgi-targeted green fluorescent protein and COP1 antisera. *Plant Cell*, **14**, 237–261.

Robertson, A. & Allan, V.J. (2000) Brefeldin A-dependent membrane tubule formation reconstituted *in vitro* is driven by a cell cycle-regulated microtubule motor. *Mol. Biol. Cell*, **11**, 941–955.

Saint-Jore, C.M., Evins, J., Batoko, H., Brandizzi, F., Moore, I. & Hawes, C. (2002) Redistribution of membrane proteins between the Golgi apparatus and endoplasmic reticulum in plants is reversible and not dependent on cytoskeletal networks. *The Plant J.*, **29**, 661–678.

Sanderfoot, A.A. & Raikhel, N.V. (1999) The specificity of vesicle trafficking: coat proteins and SNARES. *Plant Cell*, **11**, 629–641.

Satiat-Jeunemaitre, B. & Hawes, C. (1992) Redistribution of a Golgi glycoprotein in plant cells treated with Brefeldin A. *J. Cell Sci.*, **103**, 1153–1166.

Satiat-Jeunemaitre, B., Steele, C. & Hawes, C. (1996) Golgi membrane dynamics are cytoskeleton dependent: a study on Golgi stack movement induced by brefeldin A. *Protoplasma*, **191**, 21–33.

Shorter, J., Watson, R., Giannakou, M.-E., Clarke, M., Warren, G. & Barr, F. (1999) GRASP55, a second mammalian GRASP protein involved in the stacking of Golgi cisternae in a cell-free system. *The EMBO J.*, **18**, 4949–4960.

Staehelin, L.A. & Moore, I. (1995) The plant Golgi apparatus: structure, functional organisation and trafficking mechanisms. *Annu. Rev. Plant Physiol. Plant Mol. Biol.*, **46**, 261–288.

Steer, M.W. & O'Driscoll, D. (1991) Vesicle dynamics and membrane turnover in plant cells, in *Endocytosis, Exocytosis and Membrane Traffic in Plants* (eds C. Hawes, J.O.D. Coleman & D.E. Evans), CUP, Cambridge UK, pp. 128–142.

Stow, J.L., Fath, K.R. & Burgess, D.R. (1998) Budding roles for myosin II on the Golgi. *Trends Cell Biol.*, **8**, 138–141.

Sullivan, F.F. & Kay, S.A. (1999) *Green Fluorescent Proteins*, Academic Press, San Diego.

Terasaki, M. & Reese, T.S. (1994) Interactions among endoplasmic reticulum, microtubules, and retrograde movements of the cell surface. *Cell Motil. Cytoskeleton*, **29**, 291–300.

Thyberg, J. & Moskalewski, S. (1999) Role of microtubules in the organization of the Golgi complex. *Exp. Cell Res.*, **246**, 263–279.

Valderrama, F., Duran, J.M., Babia, T., Barth, H., Renau-Piqueras, J. & Egea, G. (2001) Actin microfilaments facilitate the retrograde transport from the Golgi complex to the endoplasmic reticulum in mammalian cells. *Traffic*, **2**, 717–726.

Ward, T.H., Polishchuck, R.S., Caplan, S., Hirschberg, K. & Lippincott-Schwartz, J.L. (2001) Maintenance of Golgi structure and function depends on the integrity of ER export. *J. Cell Biol.*, **155**, 557–570.

5 Intra-Golgi transport: escalator or bucket brigade?

Andreas Nebenführ

5.1 The problem

One of the great debates of current cell biology centres on a puzzling dilemma that lies at the heart of the Golgi apparatus: how can the structural and functional integrity of this organelle be maintained in the face of a constant flux of secretory products through its midst? This question relates to the different roles played by the Golgi in the secretory pathway. It is not only a station *en route* to the cell surface and vacuoles/lysosomes; it also serves as an important biosynthetic organelle that harbors a large number of enzymatic activities. These two functions impose two conflicting requirements upon the Golgi apparatus. On the one hand, the flow of secreted proteins, lipids, and carbohydrates has to proceed as smoothly as possible. On the other hand, the enzymes that operate on these molecules have to be retained within their cisternae. It is clear that these two processes can occur simultaneously only if there is a continuous and efficient interplay between mixing and sorting of substrates and enzymes. Mixing is necessary to ensure contact between enzymes and substrates, while sorting allows the substrates to move on and the enzymes to stay behind.

Cells are apparently able to solve this problem: several stable biochemical gradients across Golgi stacks have been described. A large number of enzymes have been shown to localize preferentially to subgroups of cisternae. Most of these studies have been carried out in animals (e.g. Rabouille *et al.*, 1995), but a few published reports suggest that this is also true for plants. For example, mammalian sialyltransferase is targeted to the same plant Golgi sub-compartment (*trans*) as in animal cells (Wee *et al.*, 1998). Similarly, the only plant Golgi enzyme that has been localized at the EM level, soybean α-1,2 mannosidase I, also targets to the same sub-compartment (*cis*) as its mammalian homologs (Nebenführ *et al.*, 1999). Indirect evidence for specific intra-Golgi localization of Golgi enzymes also comes from immunodetection of specific epitopes created in the Golgi stack (e.g. Zhang & Staehelin, 1992). In addition, the membrane lipid composition changes from *cis* to *trans*, with the *cis*-most cisternae being more like the ER and the *trans*-most cisternae resembling the composition of the plasma membrane (van Meer, 1998). How cells achieve this remarkable feat has been a matter of contentious debate for some time. In principle, several conceptual models can be proposed to solve this problem.

5.2 The models

In order to facilitate the discussion of the different models, it is helpful to define some of the terms used. The proteins that are carried in vesicles are referred to as *cargo* since they are thought to perform no specific function during transport. The two classes of macromolecules that are discussed as potential cargo molecules are the Golgi enzymes and the secretory proteins. In contrast, the proteins involved in mediating the shuttling of vesicles between cisternae are often called the *trafficking machinery*. They include the coat proteins (COPI, see Chapter 3) and targeting receptors (SNAREs, see Chapter 10). In recent years, an additional group of proteins that associate with the cytoplasmic side of Golgi membranes has been characterized, and these are usually referred to as *Golgi matrix*. *Anterograde transport* signifies the typical forward movement of secretory cargo from the ER via *cis*, medial, and *trans* Golgi to their final destinations. *Retrograde transport*, likewise, is transport in the opposite direction.

5.2.1 The cisternal progression/maturation model

The cisternal progression model is historically the oldest model of intra-Golgi transport (Grassé, 1957). It is based on the premise that secretory cargo never leaves individual cisternae. According to this model, new cisternae continuously form from ER-to-Golgi transport vesicles at the *cis* side of the Golgi. This addition of new cisternae at the forming face gradually displaces the older cisternae through the stack until they reach the *trans* side where they are consumed by packaging of secretory cargo into post-Golgi transport vesicles. Over time, it has become obvious that this simple model cannot explain the apparently stable distribution of Golgi enzymes in the stack. Therefore a countercurrent of recycling vesicles, that would move resident Golgi enzymes from the older to the younger cisternae, has been postulated (Fig. 5.1A; Schnepf, 1993; Bannykh & Balch, 1997; Glick *et al.*, 1997; Mironov *et al.*, 1997). At steady state, the rate of anterograde cisternal progression is predicted to match the rate of retrograde vesicle transport, and the distribution of Golgi enzymes would consequently be stable. This modified model has been termed *cisternal progression/maturation* so as to indicate the evolving composition of the cisternae as they progress through the Golgi stacks. An interesting aspect of this updated form of the model is that *cis* cisternae form by the fusion of anterograde ER-Golgi vesicles with retrograde intra-Golgi vesicles.

5.2.2 The vesicular shuttle model

The simplest way to explain the stable distribution of Golgi enzymes within a stack is to assume that the cisternae are stable, long-lived structures that house the various enzymatic activities. This implies that secretory products have to leave a cisterna in order to get to the next. In the vesicular shuttle model, this is accomplished by transport vesicles that move cargo in the anterograde direction (Farquhar &

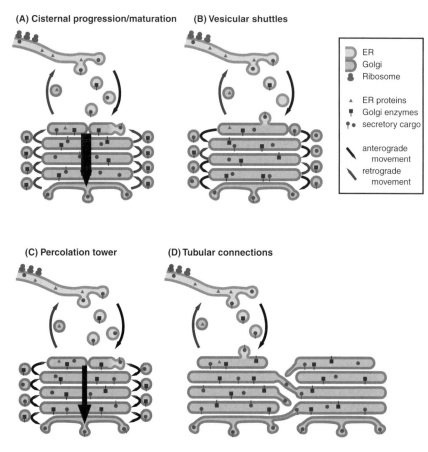

Figure 5.1 Schematic representation of the different models of intra-Golgi transport discussed in the text.

Palade, 1981). In order to accommodate emerging data that was not compatible with this strict form of the model, a second class of vesicles has been postulated that would recycle the escaped ER residents or Golgi enzymes back to their proper location in an earlier compartment of the secretory pathway (Farquhar & Palade, 1981). However, this retrograde transport would be expected to be a minor component of the total vesicular transport, since it is very likely that the majority of vesicles carries cargo in the anterograde direction (Fig. 5.1B).

5.2.3   The percolation tower model

Interestingly, the cisternal progression/maturation model and the vesicular shuttle model are not mutually exclusive. It is conceivable that cisternae move slowly through the stack while vesicles rapidly shuttle cargo *and* enzymes in both anterograde and retrograde directions (Fig. 5.1C; Pelham & Rothman, 2000). In this scenario, the distribution of Golgi enzymes is ensured by their differential affinity for the different *age classes* of cisternae and their resulting differential propensity to enter transport vesicles. Similarly, the anterograde transport of secretory products follows simply from the continual addition of new cargo at the *cis* side and simultaneous removal from the *trans* side. This model offers two distinct advantages: first, it does not require the sorting step necessary for the formation of separate anterograde and retrograde vesicles; second, it does not impose any directionality on the intra-Golgi vesicles, thus requiring fewer specific recognition steps, which is favored by the lack of Golgi-specific SNAREs (Pelham & Rothman, 2000).

5.2.4 The tubular connections model

An alternative model for anterograde cargo transport through stable cisternae assumes that Golgi cisternae can, on occasion, form direct tubular connections with other cisternae at different levels in the stack (Fig. 5.1D; Ayala, 1994; Weidman, 1995). These direct connections would eliminate the need for transport vesicles since cargo can move directly from one cisterna to another. This scenario obviously requires additional assumptions that would ensure the directionality of cargo transport, at the same time maintaining the biochemical identities of the connected cisternae (Mironov *et al.*, 1997).

5.3 The data

5.3.1 Rise and fall of cisternal progression

The oldest speculations on intra-Golgi transport date back almost half a century (Grassé, 1957) to a time when the involvement of this organelle in the secretory pathway was still an unproven hypothesis. The model was based solely on EM observations and represents the original cisternal progression model. Subsequent experiments with radioactive tracers confirmed the direction of cargo transport within the Golgi from *cis* to *trans* and were generally viewed as supporting the model (reviewed in Farquhar & Palade, 1981), although alternative interpretations were usually not pursued.

The strongest evidence in favor of the cisternal progression model came from the works on scale-producing algae (reviewed in Melkonian *et al.*, 1991). These organisms synthesize large scales in the lumen of their Golgi cisternae from where they are delivered to the cell surface in membrane-bound transport carriers. The scales typically are too large to fit into the small intra-Golgi vesicle shuttles and consequently are found only within the cisternae. An extreme example of this is provided by the unicellular alga *Pleurochrysis scherfellii*, where a single scale of up to 2 μm diameter is assembled per cisterna (Brown Jr. and Romanivicz, 1976).

The dramatic size of these scales has even allowed the direct observation of their transport to the cell surface in living cells (Brown Jr., 1971).

During the 1970s, more sophisticated experimental protocols began to reveal shortcomings of this simplistic cisternal progression model. In particular, the apparently stable differential composition of the cisternae within a stack seemed incompatible with the maturation of cisternae as they changed from *cis* to medial to *trans*. These concerns led to the formulation of an alternative model of intra-Golgi transport that assumed stable cisternae and shuttling transport vesicles (Farquhar & Palade, 1981). This model received strong support when it became possible to reconstitute elements of intra-Golgi transport *in vitro* (Balch *et al.*, 1984). These experiments, took advantage of a series of new developments that allowed researchers to follow the progress through the Golgi by biochemical means. In particular, incomplete glycosylation of a foreign protein (the G-protein of vesicular stomatitis virus, or VSV-G) in a cell line that lacked *N*-acetyl glumosaminyl transferase I could rapidly be restored by fusion with a wild-type cell line, suggesting transfer of the VSV-G protein from one Golgi system to the other (Rothman *et al.*, 1984b). Importantly, this effect could be reconstituted with isolated membranes from the two cell lines, which was interpreted as dissociative (i.e. vesicle-mediated) transport between independent Golgi stacks (Rothman *et al.*, 1984a).

This biochemical approach resulted in the identification of a number of proteins necessary for vesicle formation and fusion, such as COPI coats, Arf1 GTPase, and NSF (*N*-ethyl maleimide-sensitive factor), thereby greatly enhancing our understanding of the mechanisms of vesicular transport (Rothman & Wieland, 1996). At the same time, however, it has to be cautioned that the cell-free assays could not formally distinguish between anterograde movement of secretory cargo and retrograde movement of modifying enzymes. While the data were clearly suggestive of anterograde vesicular shuttles, they could in principle also be interpreted in terms of recycling Golgi enzymes. A first indication that COPI vesicles indeed may play a role in retrograde transport came from work in yeast that demonstrated that COPI proteins interacted with the recycling signal on ER membrane proteins and were necessary for their return to the ER (Letourneur *et al.*, 1994).

5.3.2 Renaissance of progression

A renewed interest in cisternal progression was sparked by new data that demonstrated (again) that bulky secretory products can traverse the Golgi stack without leaving the cisterna they reside in. The first such report was in a detailed analysis of the unicellular green alga *Scherffelia dubia* (Becker *et al.*, 1995). This organism can produce large number of scales in a short time in response to experimental deflagellation (McFadden & Melkonian, 1986). Of the more than 10000 scales observed in 90 Golgi stacks of *Scherffelia*, not a single one was found in the abundant peri-Golgi vesicles (Becker *et al.*, 1995). Similar observations were later made in mammalian fibroblasts that produce large collagen complexes. In this case, the large electron-dense aggregates of procollagen also were seen only within the confines of cisternae and never in detached vesicles or tubules (Bonfanti *et al.*, 1998). These results argued strongly in favor of the cisternal progression model, and the data from mammalian cells helped to convince animal cell biologists of this. At this time, it also became apparent that the vesicles that surround Golgi cisternae could recycle resident Golgi enzymes back to their proper location on the conveyor belt of cisternae (reviewed in Glick & Malhorta, 1998). This concept can be simulated mathematically by assuming differential affinity of Golgi residents for a single type of recycling vesicles. Using an iterative computational approach, it is thus possible to generate a stable gradient of enzyme distribution (Glick *et al.*, 1997). This simple model has been modified recently to include a variable affinity of Golgi enzymes for recycling vesicles. In particular, it has been proposed that an enzyme's propensity to enter recycling vesicles increases dramatically when certain parameters within the cisterna reach a critical value during maturation (e.g. pH, or membrane composition). This *triggered recycling* makes the model much more robust and able to tolerate (simulated) the overexpression of individual Golgi enzymes (Weiss & Nilsson, 2000). While this theoretical model better matches the situation found *in vivo*, it remains to be seen whether its basic assumptions can be confirmed in living cells. Taken together, these theoretical papers have provided a hypothetical mechanism that allows the dynamic maintenance of biochemical gradients in the Golgi and has thereby removed one of the major objections to the original progression model.

5.3.3 What's in a vesicle?

A central difference between the cisternal progression/maturation and vesicular shuttle models is the composition of the intra-Golgi vesicles (compare Figs. 5.1A and B). The former model assumes that these vesicles predominantly contain Golgi enzymes that are recycled to younger cisternae. In contrast, the latter model predicts that most vesicles would contain cargo travelling in the anterograde direction. The percolation tower model, on the other hand, implies similar numbers of molecules for anterograde and retrograde cargo. While the directionality of vesicle movement is very difficult to determine experimentally, it is possible to examine the contents of intra-Golgi vesicles using several techniques. Over the last few years, a number of publications have addressed this issue and provided a wealth of data that allows us to draw tentative conclusions.

It is generally assumed that vesicle formation at the Golgi is driven by recruitment of the COPI coat (see Chapter 3). Consequently, one group of experiments has used a biochemical approach to determine the composition of purified COPI vesicles. In order to facilitate isolation of these vesicles the coat was stabilized by the addition of GTPγS which prevents uncoating. These GTPγS-stabilized vesicles contained few Golgi enzymes, but also little anterograde cargo (Sönnichsen *et al.*, 1996). This dilemma was solved when it was discovered that GTP hydrolysis is required for proper selection of cargo by the coat proteins (Pepperkok *et al.*, 2000). In fact, COPI vesicles generated under these conditions did contain significant amounts of retrograde cargo, but still very little anterograde cargo (Lanoix *et al.*, 2001).

Interestingly, using this experimental approach it is possible to distinguish two different classes of COPI vesicles that are enriched in either *cis*-Golgi proteins or medial-Golgi proteins, suggesting that they formed at different levels within the stacks (Lanoix *et al.*, 2001). These results corroborate previous data from yeast that show that COPI is necessary for retrograde transport to the ER (Letourneur *et al.*, 1994). Thus, biochemical as well as genetic data favors the idea that COPI vesicles are involved in retrograde but not anterograde transport.

Another approach to assessing the composition of intra-Golgi vesicles is by immunoelectron microscopy. In this case, the presence of specific proteins in peri-Golgi vesicles is tested with cross-reacting antibodies in thin sections prepared for electron microscopy. Initial studies using this approach seemed to confirm the vesicular shuttle model. Two populations of COPI-positive vesicles were described that contained either anterograde (VSV-G and proinsulin) or retrograde cargo (KDEL receptor) (Orci *et al.*, 1997) while Golgi enzymes were found in lower concentrations (Orci *et al.*, 2000a). It has been suggested that these two types of vesicles may perform distinct functions in intra-Golgi and Golgi-ER retrograde transports (Orci *et al.*, 2000b). However, a more recent detailed analysis of Golgi stacks and their surrounding vesicles suggests that this initial conclusion may be wrong. A careful double-labeling study revealed that Golgi enzymes (mannosidase II) are more likely than anterograde cargo (VSV-G) to be present in peri-Golgi vesicles, although the latter were not completely excluded (Martínez-Menárguez *et al.*, 2001). Similarly, COPI vesicles that form at the ER-Golgi intermediate compartment found in mammalian cells exclude anterograde cargo and instead preferentially harbor proteins that are recycled to the ER (Martínez-Menárguez *et al.*, 1999).

The apparent discrepancy found in these morphological studies is difficult to resolve and may be a result of the enormous complexity of the Golgi apparatus. Complete three-dimensional (3D) tomographic reconstruction of large parts of Golgi complexes in mammalian cells reveals a bewildering multitude of membrane compartments of complex morphology and diverse, often unpredicted, interactions (Ladinsky *et al.*, 1999; Marsh *et al.*, 2001). Given this complicated topology it is clearly difficult to unequivocally identify the compartment that is labeled (or not labeled) by the antibody. Double labeling for separate markers as well as careful interpretation of unlabeled membranes appears to be essential. Although one of the studies described (Martínez-Menárguez *et al.*, 2001) follows this recipe, more experiments will be necessary to resolve this dispute. For example, to address the question of cargo enrichment in COPI vesicles, it will be necessary to identify not only the vesicles but also their donor compartments since the composition of COPI vesicles can be expected to vary depending on the compartment on which they form (compare Lanoix *et al.*, 2001).

5.3.4   Where's the anterograde cargo?

Further strong evidence in favor of the cisternal progression model comes from a comparison of large and small cargo molecules that are expressed in the same cell (Mironov *et al.*, 2001). By combining temperature-sensitive mutants of VSV-G (tagged with green fluorescent protein, GFP) and an elaborate scheme of temperature shifts in a cell type that can be stimulated to produce procollagen, it is possible to create synchronized waves and pulses of both types of secretory cargo that enter the Golgi stack in a finely controllable way. The central findings of this study are that both large and small cargo molecules traverse the stack at the same rapid rate and that neither of them enters peri-Golgi vesicles (Mironov *et al.*, 2001). These results clearly establish that cisternal progression is fast enough to account for the secretory transport rates reported in the past. They also discount peri-Golgi vesicles as playing a significant role in anterograde transport of secretory products (at least for some types of molecules), thereby keeping in doubt the percolation model of intra-Golgi transport. Intriguingly, one set of experiments employing very small pulses of VSV-G-GFP protein revealed no lateral mobility of secretory cargo within the Golgi complex (Mironov *et al.*, 2001). This is in striking contrast to the unrestricted mobility of Golgi enzymes within the interconnected Golgi apparatus of these cells (Cole *et al.*, 1996). Thus, it appears that secretory products are confined to their individual stack within the mammalian Golgi ribbon and are excluded not only from the peri-Golgi vesicles but also from the tubular connections between the corresponding cisternae of adjacent stacks of the Golgi ribbon.

5.3.5   *Plants do it too*

While these recent studies highlight the high degree of sophistication in experimental approaches possible today in mammalian cell systems, it should not be overlooked that plant researchers have continued to provide indirect evidence in favor of the cisternal progression/maturation model. For example, in studying the biogenesis of dense vesicles that mediate transport to the protein storage vacuole in pea cotyledons, it was found that protein aggregates form already at the *cis* side of the Golgi (Hillmer *et al.*, 2001). As these aggregates move through the Golgi stack, they are always contained within membranes that are continuous with the Golgi cisternae, suggesting that these *dense buds* never dissociate and reach the *trans* Golgi by cisternal progression (Hillmer *et al.*, 2001). These data are in apparent disagreement with similar observations of artificial protein aggregates in animal cells. In this case, no continuities between the membranes of the Golgi and those surrounding the aggregates, the so-called megavesicles, were found (Volchuk *et al.*, 2000). A possible solution to this dilemma may lie in a 3D tomographic reconstruction of the megavesicles since it is possible that narrow connections to adjacent cisternae may not be visible in the serial thin sections.

Another line of evidence comes from studies of the effects of the drug brefeldin A (BFA) on tobacco BY-2 suspension culture cells (Ritzenthaler *et al.*, 2002). Treatment of BY-2 cells with BFA leads to a rapid loss of *cis* cisternae from Golgi stacks. Interestingly, these medial-*trans* Golgi remnants are still labeled by the *cis*-Golgi marker GmMan1-GFP (soybean mannosidase I-green fluorescent protein, Nebenführ *et al.*, 1999). The disappearance of fluorescently labeled Golgi stacks instead coincides with the loss of Golgi cisternae of *trans*-like appearance. The simplest interpretation of this phenomenon is that the *cis*-Golgi marker protein cannot recycle back to younger cisternae due to the BFA-induced block in COPI formation (see Chapter 12) and therefore progresses within the (former) *cis* cisternae as they mature into *trans*-like appearance (Ritzenthaler *et al.*, 2002).

5.3.6   Tubule or not tubule – that is the question

A fairly new concept in intra-Golgi transport is the idea of having direct membrane continuities between different levels of adjacent Golgi stacks (Fig. 5.1D; Weidman, 1995). These tubular connections could, in principle, allow (small) anterograde cargo to move through the Golgi without entering transport vesicles. So far, evidence for these connections has been sketchy, and recent detailed 3D tomographic reconstructions of Golgi apparatus in high-pressure frozen/freeze-substituted mammalian cells revealed no evidence of them (Ladinsky *et al.*, 1999; Marsh *et al.*, 2001). However, a most recent 3D analysis of pancreatic beta cells stimulated for maximal insulin production suggests that such inter-cisternal connections can exist at least under certain conditions (B. Marsh, personal communication). The proposed tubules between neighboring stacks probably cannot exist in plant cells where Golgi stacks travel as individual units and do not form the ribbon found in mammalian cells (Boevink *et al.*, 1998; Nebenführ *et al.*, 1999). In this case, the connections would have to form within a single stack (Weidman, 1995), but there is no evidence for such tubules in plants to date.

One prediction of the tubular connections model is that secretory cargo uses these membrane continuities to travel from one cisterna to another (Weidman, 1995). However, given the high mobility of the Golgi enzymes within the membrane (Cole *et al.*, 1996), it is difficult to envision how the distinct biochemical composition of adjacent cisternae could be maintained if these direct membrane continuities exist. This problem is exacerbated by the recent finding that anterograde cargo exhibits very little lateral mobility within the Golgi complex (Mironov *et al.*, 2001). Larger secretory products are also most likely excluded from narrow tubules. Thus, given the dearth of direct evidence in favor of frequent inter-cisternal connections and the overwhelming amount of data that is difficult to reconcile with this model, it seems doubtful whether tubular connections can play a significant role in intra-Golgi transport.

5.4   The solution?

Given the large number of publications which have accumulated in recent years that provide experimental evidence for various aspects of the cisternal progression/ maturation model, it is now an inescapable conclusion that cisternal progression does occur (Pelham & Rothman, 2000; Pelham, 2001; Barr, 2002). A direct consequence of this conclusion is that Golgi enzymes have to be recycled back to younger cisternae, a feat most likely accomplished by COPI vesicles. While this appears to be a full vindication of the earliest ideas of intra-Golgi traffic (Grassé, 1957), it has to be cautioned that the alternative modes of transport cannot be conclusively ruled out. Indeed, it is possible that COPI vesicles can also move in the anterograde direction to deliver certain kinds of secretory products to the next cisterna. Only a better understanding of the molecular mechanisms involved in vesicle targeting within the Golgi can be expected to provide a definitive answer. The new-found consensus that Golgi cisternae are not stable entities, however, provides a unifying conceptual framework with which we can interpret new results.

5.5 The implications

One line of research that has been pursued for a number of years is aimed at identi-fying the targeting signals and mechanisms that allow modifying enzymes to main-tain their specific positions within the Golgi (Opat *et al.*, 2001). An interesting aspect of Golgi enzyme recycling is that it does not appear to rely on any sorting receptors. In other words, Golgi enzymes seem to be able to interact directly with the machinery necessary for vesicle formation to ensure efficient recycling. The two major hypotheses that have influenced our thinking in this matter are the kin-recognition model and the membrane thickness model. The kin-recognition model postulates that enzymes that reside in the same cisterna can interact with each other and therefore are sorted together (Nilsson *et al.*, 1993). The membrane thickness model assumes that the thickness of the lipid bilayer, which increases in the Golgi from *cis* to *trans*, determines the positioning of the proteins since *cis*-Golgi enzymes tend to have shorter transmembrane domains than *trans*-Golgi enzymes (Bretscher & Munro, 1993). Both of these models were formulated within the context of the vesicular shuttle model, i.e. assuming stable cisternae which retain their enzymes, but they can be modified to fit the recycling requirement of the cisternal progression/maturation model (Füllekrug & Nilsson, 1998).

However, recent publications suggest that proteins of the Golgi matrix may also play a role in targeting Golgi enzymes. In particular, members of the p24 family of membrane proteins that are implicated in ER-Golgi traffic were found to form complexes with the Golgi matrix proteins GRASP55 and GRASP65 (Barr *et al.*, 2001). GRASP55 can interact with the small GTPase Rab2 and appears to be necessary for the maintenance of both Golgi structure and membrane traffic (Short *et al.*, 2001). A similar dual role has been proposed for GRASP65 (Barr *et al.*, 1997; Nakamura *et al.*, 1997). These data are suggestive of an intimate link between stacking of Golgi cisternae and membrane traffic through the stack. It will be interesting to see whether this connection can be confirmed and what its role in the targeting of Golgi enzymes is.

One consequence of the cisternal progression/maturation model is that the *cis*-most cisterna of a Golgi stack is formed from both anterograde ER-to-Golgi transport vesicles and retrograde intra-Golgi recycling vesicles. This implies that new cisternae are 'growing' until they reach a certain size and the next cisterna becomes initiated. It is not known what determines when a cisterna is 'large enough', but it seems reasonable to speculate that the next older cisterna can act as a template. Another problem arises in plant cells where Golgi stacks can travel through the cytoplasm (Boevink *et al.*, 1998; Nebenführ *et al.*, 1999), and the arrival of ER-to-Golgi vesicles may not represent a uniform, continuous stream. Do stacks under these conditions sometimes form smaller cisternae, or are there feedback mechanisms to ensure that always a complete cisterna is formed (Nebenführ & Staehelin, 2001)? This question relates to the way the membrane exchange between ER and Golgi is organized in cells with a high level of intracellular motility (see Chapter 4). Nothing is currently known about the spatio–temporal relationship of ER export sites and Golgi stacks in plants, but data from yeast suggests that there may be a functional interaction between the two organelles. The Golgi apparatus in the budding yeast *Saccharomyces cerevisiae* is dispersed into individual cisternae that do not form stacks. In contrast, the closely related species *Pichia pastoris* forms Golgi stacks like most other eukaryotes. Interestingly, *P. pastoris* contains discrete ER export sites adjacent to Golgi stacks whereas transport vesicle formation occurs all over the ER in *S. cerevisiae* (Rossanese *et al.*, 1999). This correlation has been interpreted as representing a functional constraint on the places of Golgi formation at least in yeast cells (Rossanese *et al.*, 1999), but possibly also in mammalian cells (Hammond & Glick, 2000). It remains to be seen whether this also holds true for plant cells with their mobile Golgi stacks.

An even more fundamental question raised by the cisternal progression/maturation model is whether the Golgi can be considered a stable organelle at all. Given the continuous flux of its major distinguishing feature, the stacked cisternae, the question arises of what makes the Golgi the Golgi? How can the structural and functional integrity of this organelle be maintained when all its components appear to be just transport intermediates? Should the Golgi be considered a dynamic outgrowth of the ER, or does it contain structural elements that define its identity and that never become associated with another organelle? This also relates to questions on how the number of Golgi stacks in a cell can increase (only by fission, or also by *de novo* formation?), and to the fate of the Golgi during mitosis and cytokinesis (where it essentially disappears from mammalian, but not plant cells) or during experiments that disrupt normal membrane flow (such as BFA, where Golgi membranes fuse with the ER) (Warren & Wickner, 1996; Lippincott-Schwartz *et al.*, 2000; Rossanese & Glick, 2001). The new data supporting the cisternal progression/maturation model seem to favor a dynamic definition of the Golgi that essentially postulates a self-organizing principle of balanced anterograde and retrograde transport, thus leading to a stable steady state (Mistelli, 2001). This interpretation appears to be supported by experiments that demonstrate cycling of Golgi enzymes through the ER (e.g. Zaal *et al.*, 1999). However, data on Golgi matrix proteins seem to contradict this view and instead argue for a constant scaffold that acts as an anchor for Golgi membranes (e.g. Seemann *et al.*, 2000). This issue clearly is the next great debate that surrounds the Golgi apparatus and, given the history of Golgi research, we can expect many unexpected findings.

References

Ayala, J. (1994) Transport and internal organization of membranes: vesicles, membrane networks and GTP-binding proteins. *J. Cell Sci.*, **107**, 753–763.

Balch, W.E., Dunphy, W.G., Braell, W.A. & Rothman, J.E. (1984) Reconstitution of the transport of protein between successive compartments of the Golgi measured by the coupled incorporation of N-acetylglucosamine. *Cell*, **39**, 405–416.

Bannykh, S.I. & Balch, W.E. (1997) Membrane dynamics at the endoplasmic reticulum-Golgi interface. *J. Cell Biol.*, **138**, 1–4.

Barr, F.A. (2002) The Golgi apparatus: going round in circles? *Trends Cell Biol.*, **12**, 101–104.

Barr, F.A., Preisinger, C., Kopajtich, R. & Körner, R. (2001) Golgi matrix proteins interact with p24 cargo receptors and aid in their efficient retention in the Golgi apparatus. *J. Cell Biol.*, **155**, 885–891.

Barr, F.A., Puype, M., Vandekerckhove, J. & Warren, G. (1997) GRASP65, a protein involved in the stacking of Golgi cisternae. *Cell*, **91**, 253–262.

Becker, B., Bölinger, B. & Melkonian, M. (1995) Anterograde transport of algal scales through the Golgi complex is not mediated by vesicles. *Trends Cell Biol.*, **5**, 305–307.

Boevink, P., Oparka, K., Sant Cruz, S., Martin, B., Betteridge, A. & Hawes, C. (1998) Stacks on tracks: the plant Golgi apparatus traffics on an actin/ER network. *Plant J.*, **15**, 441–447.

Bonfanti, L., Mironov, A.A. Jr., Martínez-Menárguez, J.A., Martella, O., Fusella, A., Baldassarre, M., Buccione, R., Geuze, H.J., Mironov, A.A. & Luini, A. (1998) Procollagen traverses the Golgi stack without leaving the lumen of cisternae: evidence for cisternal maturation. *Cell*, **95**, 993–1003.

Bretscher, M.S. & Munro, S. (1993) Cholesterol and the Golgi apparatus. *Science*, **261**, 1280–1281.

Brown, R.M. Jr. (1971) Movement of the protoplast and function of the Golgi-apparatus in the algae. *Publ. Wiss. Film Sekt. Biol.*, **7**, 361.

Brown, R.M. Jr. & Romanivicz, D.K. (1976) Biogenesis and structure of Golgi-derived cellulosic scales in *Pleurochrysis*. I. Role of the endomembrane system in scale assembly and exocytosis. *Appl. Polymer Symp.*, **28**, 537–585.

Cole, N.B., Smith, C.L., Sciaky, N., Terasaki, M., Edidin, M. & Lippincott-Schwartz, J. (1996) Diffusional mobility of Golgi proteins in membranes of living cells. *Science*, **273**, 797–801.

Farquhar, M.G. & Palade, G.E. (1981) The Golgi apparatus (complex) – (1954–1981) – from artifact to center stage. *J. Cell Biol.*, **91**, 775–1035.

Füllekrug, J. & Nilsson, T. (1998) Protein sorting in the Golgi complex. *Biochim. Biophys. Acta*, **1404**, 77–84.

Glick, B.S., Elston, T. & Oster, G. (1997) A cisternal maturation mechanism can explain Golgi asymmetry. *FEBS Lett.*, **414**, 177–181.

Glick, B.S. & Malhorta, V. (1998) The curious state of the Golgi apparatus. *Cell*, **95**, 883.

Grassé, P.P. (1957) Ultrastructure, polarité et reproduction de l'appareil de Golgi. *Comptes Rendus de l'Académie des Sci.*, **245**, 1278–1281.

Hammond, A.T. & Glick, B.S. (2000) Dynamics of transitional endoplasmic reticulum sites in vertebrate cells. *Mol. Biol. Cell*, **11**, 3013–3030.

Hillmer, S., Movafeghi, A., Robinson, D.G. & Hinz, G. (2001) Vacuolar storage proteins are sorted in the *cis*-cisternae of the pea cotyledon Golgi apparatus. *J. Cell Biol.*, **152**, 41–50.

Ladinsky, M.S., Mastronarde, D.N., McIntosh, J.R. & Howell, K.E. (1999) Golgi stucture in three dimensions: functional insights from the normal rat kidney cell. *J. Cell Biol.*, **144**, 1–16.

Lanoix, J., Ouwendijk, J., Stark, A., Szafer, E., Cassel, D., Dejgaard, K., Weiss, M. & Nilsson, T. (2001) Sorting of Golgi resident proteins into different subpopulations of COPI vesicles: a role for ArfGAP. *J. Cell Biol.*, **155**, 1199–1212.

Letourneur, F., Gaynor, E.C., Hennecke, S., Demolliere, C., Duden, R., Emr, S.D., Riezman, H. & Cosson, P. (1994) Coatomer is essential for retrieval of dilysine-tagged proteins to the endoplasmic reticulum. *Cell*, **79**, 1199–1207.

Lippincott-Schwartz, J., Roberts, T.H. & Hirschberg, K. (2000) Secretory protein trafficking and organelle dynamics in living cells. *Annu. Rev. Cell Dev. Biol.*, **16**, 557–289.

Marsh, B.J., Mastronarde, D.N., Buttle, K.F., Howell, K.E. & McIntosh, J.R. (2001) Organellar relationships in the Golgi region of the pancreatic beta cell line, HIT-T15, visualized by high resolution electron tomography. *PNAS*, **98**, 2399–2406.

Martínez-Menárguez, J.A., Geuze, H.J., Slot, J.W. & Klumperman, J. (1999) Vesicular tubular clusters between the ER and Golgi mediate concentration of soluble secretory proteins by exclusion from COPI-coated vesicles. *Cell*, **98**, 81–90.

Martínez-Menárguez, J.A., Prekeris, R., Oorschot, V.M.J., Scheller, R., Slot, J.W., Geuze, H.J. & Klumperman, J. (2001) Peri-Golgi vesicles contain retrograde but not anterograde proteins consistent with the cisternal progression model of intra-Golgi transport. *J. Cell Biol.*, **155**, 1213–1224.

McFadden, G.I. & Melkonian, M. (1986) Golgi apparatus activity and membrane flow during scale biogenesis in the green flagellate *Scherffelia dubia* (Prasinophyceae). I. Flagellar regeneration. *Protoplasma*, **130**, 186–198.

Melkonian, M., Becker, B. & Becker, D. (1991) Scale formation in algae. *J. Electron Microsc. Tech.*, **17**, 165–178.

Mironov, A.A., Beznoussenko, G.V., Nicoziani, P., Martella, O., Trucco, A., Kweon, H.-S., Di Giandomenico, D., Polishchuk, R.S., Fusella, A., Lupetti, P., Berger, E.G., Geerts, W.J.C., Koster, A.J., Burger, K.N.J. & Luini, A. (2001) Small cargo proteins and large aggregates can traverse the Golgi by a common mechanism without leaving the lumen of cisternae. *J. Cell Biol.*, **155**, 1225–1238.

Mironov, A.A., Weidman, P. & Luini, A. (1997) Variations on the intracellular transport theme: maturing cisternae and trafficking tubules. *J. Cell Biol.*, **138**, 481–484.

Mistelli, T. (2001) The concept of self-organization in cellular architecture. *J. Cell Biol.*, **155**, 181–185.

Nakamura, N., Lowe, M., Levine, T.P., Rabouille, C. & Warren, G. (1997) The vesicle docking protein p115 binds GM130, a *cis*-Golgi matrix protein, in a mitotically regulated manner. *Cell*, **89**, 445–455.

Nebenführ, A., Gallagher, L., Dunahay, T.G., Frohlick, J.A., Masurkiewicz, A.M., Meehl, J.B. & Staehelin, L.A. (1999) Stop-and-go movements of plant Golgi stacks are mediated by the acto-myosin system. *Plant Physiol.*, **121**, 1127–1141.

Nebenführ, A. & Staehelin, L.A. (2001) Mobile factories: Golgi dynamics in plant cells. *Trends Plant Sci.*, **6**, 160–167.

Nilsson, T., Slusarewicz, P., Hoe, M.H. & Warren, G. (1993) Kin recognition: a model for the retention of Golgi enzymes. *FEBS Lett.*, **330**, 1–4.

Opat, A.S., van Vliet, C. & Gleeson, P.A. (2001) Trafficking and localization of resident Golgi glycosylation enzymes. *Biochimie*, **83**, 763–773.

Orci, L., Amherdt, M., Ravazzola, M., Perrelet, A. & Rothman, J.E. (2000a) Exclusion of Golgi residents from transport vesicles budding from Golgi cisternae in intact cells. *J. Cell Biol.*, **150**, 1263–1270.

Orci, L., Ravazzola, M., Volchuk, A., Engel, T., Gmachl, M., Amherdt, M., Perrelet, A., Söllner, T.H. & Rothman, J.E. (2000b) Anterograde flow of cargo across the Golgi stack potentially mediated via bidirectional "percolating" COPI vesicles. *Proc. Natl. Acad Sci. USA*, **97**, 10400–10405.

Orci, L., Stamnes, M., Ravazzola, M., Anherdt, M., Perrelet, Alain, Söllner, T.H. & Rothman, J.E. (1997) Bidirectional transport by distinct populations of COPI-coated vesicles. *Cell*, **90**, 335–349.

Pelham, H.R.B. (2001) Traffic through the Golgi apparatus. *J. Cell Biol.*, **155**, 1099–1101.

Pelham, H.R.B. & Rothman, J.E. (2000) The debate about transport in the Golgi – two sides of the same coin? *Cell*, **102**, 713–719.

Pepperkok, R., Whitney, J.A., Gomez, M. & Kreis, T.E. (2000) COPI vesicles accumulating in the presence of a GTP restricted Arf1 mutant are depleted of anterograde and retrograde cargo. *J. Cell Sci.*, **113**, 135–144.

Rabouille, C., Hui, N., Hunte, F., Kieckbusch, R., Berger, E.G., Warren, G. & Nilsson, T. (1995) Mapping the distribution of Golgi enzymes involved in the construction of complex oligosaccharides. *J. Cell Sci.*, **108**, 1617–1627.

Ritzenthaler, C., Nebenführ, A., Movafeghi, A., Stussi-Garaud, C., Behnia, L., Pimpl, P., Staehelin, L.A. & Robinson, D.G. (2002) Reevaluation of the effects of Brefeldin A on plant cells using tobacco bright yellow 2 cells expressing Golgi-targeted green fluorescent protein and COPI antisera. *Plant Cell*, **14**, 237–261.

Rossanese, O.W. & Glick, B.S. (2001) Deconstructing Golgi inheritance. *Traffic*, **2**, 589–596.

Rossanese, O.W., Soderholm, J., Bevis, B.J., Sears, I.B., O'Connor, J., Williamson, E.K. & Glick, B.S. (1999) Golgi structure correlates with transitional endoplasmic reticulum organization in *Pichia pastoris* and *Saccharomyces cerevisiae*. *J. Cell Biol.*, **145**, 69–81.

Rothman, J., Miller, R. & Urbani, L. (1984a) Intercompartmental transport in the Golgi complex is a dissociative process: facile transfer of membrane protein between two Golgi populations. *J. Cell Biol.*, **99**, 260–271.

Rothman, J., Urbani, L. & Brands, R. (1984b) Transport of protein between cytoplasmic membranes of fused cells: correspondence to processes reconstituted in a cell-free system. *J. Cell Biol.*, **99**, 248–259.

Rothman, J.E. & Wieland, F.T. (1996) Protein sorting by transport vesicles. *Science*, **272**, 227–234.

Schnepf, E. (1993) Golgi apparatus and slime secretion in plants: the early implications and recent models of membrane traffic. *Protoplasma*, **172**, 3–11.

Seemann, J., Jokitalo, E.J. & Warren, G. (2000) The role of the tethering proteins p115 and GM130 in transport through the Golgi apparatus *in vivo*. *Mol. Biol. Cell*, **11**, 635–645.

Short, B., Preisinger, C., Körner, R., Kopajtich, R., Byron, O. & Barr, F.A. (2001) A GRASP55-rab2 effector complex linking Golgi structure to membrane traffic. *J. Cell Biol.*, **155**, 877–883.

Sönnichsen, B., Watson, R., Clausen, H., Misteli, T. & Warren, G. (1996) Sorting by COPI-coated vesicles under interphase and mitotic conditions. *J. Cell Biol.*, **134**, 1411–1425.

van Meer, G. (1998) Lipids of the Golgi membrane. *Trends Cell Biol.*, **8**, 29–33.

Volchuk, A., Amherdt, M., Ravazzola, M., Brügger, B., Rivera, V.M., Clackson, T., Perrelet, A., Söllner, T.H., Rothman, J.E. & Orci, L. (2000) Megavesicles implicated in the rapid transport of intra-cisternal aggregates across the Golgi stack. *Cell*, **102**, 335–348.

Warren, G. & Wickner, W. (1996) Organelle inheritance. *Cell*, **84**, 395–400.

Wee, E.G.-T., Sherrier, D.J., Prime, T.A. & Dupree, P. (1998) Targeting of active sialyltransferase to the plant Golgi apparatus. *Plant Cell*, **10**, 1759–1768.

Weidman, P.J. (1995) Anterograde transport through the Golgi complex: do Golgi tubules hold the key? *Trends Cell Biol.*, **5**, 302–305.

Weiss, M. & Nilsson, T. (2000) Protein sorting in the Golgi apparatus: a consequence of maturation and triggered sorting. *FEBS Lett.*, **486**, 2–9.

Zaal, K.J.M., Smith, C.L., Polishchuck, R.S., Altan, N., Cole, N.B., Ellenberg, J., Hirschberg, K., Presley, J.F., Roberts, T.H., Siggia, E., Phair, R.D. & Lippincott-Schwartz, J. (1999) Golgi membranes are absorbed into and reemerge from the ER during mitosis. *Cell*, **99**, 589–601.

Zhang, G.F. & Staehelin, L.A. (1992) Functional compartmentation of the Golgi apparatus of plant cells. Immunocytochemical analysis of high-pressure frozen- and freeze-substituted sycamore maple suspension culture cells. *Plant Physiol.*, **99**, 1070–1083.

6 Retrograde transport from the Golgi

Jürgen Denecke

6.1 Introduction

Proteins leave the endoplasmic reticulum (ER) via a process generally described as *vesicle budding*, in which specific adaptor molecules mediate cross talk between the cargo molecules that are to be transported and a group of coat proteins in the cytosol. The latter are recruited to the membranes, assemble into macromolecular networks or lattices and impose a force on the membrane, governed essentially by the protein–protein interactions of coat components and adaptors, that eventually lead to the formation of tubules or vesicles. These will then shed the coats and fuse with each other or with the *cis*-Golgi cisternae, thus providing a constant flux of membranes and cargo from the ER to the Golgi. Golgi-derived membrane budding occurs in a similar way, except that the coat components are different and more diverse, mainly because Golgi-derived transport carriers can have at least three different destinations, one of which is back to the ER. The importance of this latter route in preventing the escape or possibly detrimental deposition of certain molecules to distal locations is the subject of this chapter, which aims at illustrating the major findings that led to our current understanding of the *cis*-Golgi compartment in plants. The open questions remaining, as well as the contradictory findings or incomplete models will be critically analysed and discussed in the light of recent technical advances that should permit the frontiers to be moved further towards a better understanding of this very exciting centrepiece of the plant secretory pathway.

6.2 Why should proteins and membranes recycle back to the ER?

Retrograde transport from the Golgi apparatus cannot be discussed without first introducing anterograde transport towards the Golgi and the cargo molecules following this route. Although actual ER-derived COPII vesicles have not been visualised in plants, indirect evidence for COPII-dependent transport to the Golgi in plants was obtained using a variety of *in vivo* approaches in which manipulation of key components of the COPII vesicle budding machinery led to a different behaviour of transport markers. These include the co-expression of dominant-negative mutants of the GTPases Rab1 and Sar1 (Batoko *et al.*, 2000; Takeuchi *et al.*, 2000; Phillipson *et al.*, 2001) and overexpression of the guanosine nucleotide exchange factor Sec12 (Phillipson *et al.*, 2001), which led to the inhibition of anterograde transport of cargo molecules. In addition, components of the COPII coat with significant homology to yeast and mammalian isoforms are encoded by sequences found in the *Arabidopsis* genome- and expressed sequence tags (EST)-databases and were detected immunologically in plant tissues (Movafeghi *et al.*, 1999). It has not been established in plants whether COPII-coated structures result in vesicle formation or tubular ER extensions. It also remains to be analysed whether they actually fuse with the *cis*-Golgi, or whether they become the *cis*-Golgi after homotypic fusion and subsequent maturation through selective retrograde transport. Regardless of the model, ER export involves a constant influx of membranes and protein on the *cis*-side of the Golgi.

Recent evidence suggests that the COPII-dependent route does support a significant amount of lumenal bulk flow (Phillipson *et al.*, 2001). The fact that the ER chaperone calreticulin leaves the ER frequently and returns via Golgi-derived COPI vesicles (Pimpl *et al.*, 2000; Phillipson *et al.*, 2001) illustrates the dynamic nature of the early secretory pathway and highlights the crucial role of retrograde transport to rescue valuable components lost from the ER. Clearly, retrograde transport of calreticulin cannot be marginal because the presence of the HDEL motif gives rise to up to 100-fold higher protein levels compared to truncated calreticulin (Crofts *et al.*, 1999). Degradation of the truncated calreticulin lacking its HDEL motif is thought to occur in a post-ER compartment, because inhibition of ER export via Sec12 over-expression increases the stability, but not the synthesis rate of the protein (Phillipson *et al.*, 2001). Wild type calreticulin is largely dependent on its HDEL motif to escape degradation in post-ER compartments and it is likely that the same is true for the ER chaperone BiP and other abundant reticuloplasmins.

The relevance of retrograde transport should not only be discussed with respect to the recycling of ER residents that have leaked out with the bulk flow, but also in regard to the recycling of the transport machinery, such as the protein sorting receptors (Klumperman, 2000). The earliest documentation of this principle arose from the mannose-6-phosphate receptor which binds to its ligands in the Golgi, releases its ligands at lower pH in the prelysosome (Hoflack *et al.*, 1987; Griffiths *et al.*, 1988), and is then recycled to the *trans*-Golgi where it can bind to new ligands and start a transport cycle again (Kornfeld & Mellman, 1989; Kornfeld, 1992). A similar principle has been established for the HDEL receptor that was shown to recycle between the *cis*-Golgi and the ER to retrieve ER residents that escape with the bulk flow (Pelham, 1988; Townsley *et al.*, 1993).

The recycling of the membranes themselves is at least as important as that of the transport machinery to explain homeostasis and a constant membrane surface of the end-locations in the secretory pathway, the plasma membranes and the vacuolar membranes. It is possible that homeostasis is not maintained during some developmental processes or responses to stress which involve an upregulation of secretory activity, a general increase in the cell surface, cell division, tip-growth or the deposition of storage proteins during seed development, but in general, membranes need to be constantly recycled, and one major route leads from the Golgi back to the ER.

Using fluorescent ER and Golgi markers, it was elegantly established that there is close contact between ER and Golgi (Boevink *et al.*, 1998). This discovery suggests that ER-derived vesicles, if they exist, will have a very short period of existence *in vivo* and reach (or give rise to) the Golgi cisternae quickly. Alternatively, COPII-shaped membrane tubules that make contact with the Golgi would not be ruled out by these observations. From a protein transport perspective, transient tubular connections between ER and Golgi would elegantly explain the results supporting the bulk flow principle (Phillipson *et al.*, 2001). Bulk flow of soluble cargo would not be predicted to be very efficient if restricted to the central lumen of typical COPII vesicles isolated *in vitro* (Barlowe *et al.*, 1994). This is because of their small size of 50 nm, which leaves only little volume in the centre if membrane spanning cargo occupies a 10–15 nm zone on the inner membrane surface together with their sorting receptors. Instead, tubular connections could support far more bulk flow than their volume alone, simply because solutes would be expected to have a higher mobility than the membrane carrier and would simply flow through the tube. It should also be noted that in contrast to COPII vesicles, it has been relatively easy to visualise Golgi-derived COPI vesicles in the electron microscope (Pimpl *et al.*, 2000), and this discrepancy may very well be due to actual differences between COPII- and COPI-dependent transport. Vesiculation *in vitro* is by no means contradictory to tubular transport *in vivo*, because it is well established in the mechanics of fluids that tubular fluid streams vesiculate automatically. This instability is entirely dependent on the velocity and viscosity of the fluid medium, a property redily utilised in cell sorters or ink-jet printers. *In vitro* generation of COPII-coated structures may thus form vesicles in the absence of a target membrane. This may not necessarily occur *in vivo*, and if tubular connections between ER and Golgi do exist, retrieval of ER residents will have to be very efficient and thus restrictive, excluding bulk flow in the retrograde direction, and nothing would better suit than a COPI vesicle which can be visualised *in vivo*, generated *in vitro*, and shown to contain large quantities of calreticulin (Pimpl *et al.*, 2000). To demonstrate the true nature of COPII-dependent ER-export structures, serial sections and systematic EM analysis followed by 3D reconstitution will probably be the only way forward. Although technically difficult and time-consuming, it will actually show us what we must otherwise deduce indirectly from biochemical experiments.

A more philosophical point is the issue of the need for a Golgi apparatus. Does the Golgi apparatus exist as a consequence of ER export and the need for sorting to different compartments? Or is it an entity that exists as such and receives ER material to process it further? This question may seem irrelevant, but it is the foundation to the question regarding the two major models describing the Golgi apparatus, the cisternal progression model and the vesicular transport model (Pelham & Rothman, 2000). Indeed, as clathrin-coated vesicles are exported from the *trans*-Golgi apparatus for transport in the direction of the lytic vacuole, one could ask why this would not be possible directly from the ER? It seems very unlikely that the subtle pH difference between the ER and the Golgi, for which unambiguous data are elusive in plants, would restrict vacuolar cargo-binding to the Golgi apparatus. Is Golgi maturation necessary for anterograde Golgi export and could the underlying mechanism be part of a quality control mechanism to ensure that ER residents and ER-export machinery have been properly removed prior to further transport events? This could very well be the case, and provide an interesting challenge for future research.

6.3 Retrograde transport of soluble proteins

In plants, it has been shown that a variety of C-terminal peptides can act as vacuolar sorting signals, all following the general consensus (K/H/R)DEL, and further variants, a topic which has been extensively reviewed over the last decade. It remains interesting that some ER residents consistently use the HDEL signal throughout the plant kingdom whereas others consistently use KDEL (Hadlington & Denecke, 2000). Following the discovery of the *Arabidopsis* homologue of the HDEL receptor in yeast (ERD2) (Lee *et al.*, 1993), the transport of this putative receptor was indirectly studied via fusions to the reporter green fluorescent protein (GFP). These studies have established that ERD2-GFP is predominantly co-localised with Golgi markers and it re-distributes to the ER upon treatment with the drug Brefeldin A (BFA) (Boevink *et al.*, 1998). It has not been shown whether ERD2 or ERD2-GFP binds to the expected ligands. However, it was possible to demonstrate that the receptor can be saturated *in vivo* using dosage experiments with the secretory marker α-amylase tagged with a C-terminal HDEL motif (Phillipson *et al.*, 2001). This protein is fully retained until a certain threshold level is reached after which secretion starts and increases with dosage. Secretion of α-amylase-HDEL can reach similar levels to those observed for un-tagged α-amylase when an equilibrium is reached between *de novo* synthesis of α-amylase-HDEL and escape from its receptor. This is due to the fact that α-amylase is equally stable in the ER and in the culture medium and can be measured with a quantitative assay. The newly established cargo system can now be utilised to test if ERD2 overexpression increases the ER retention capacity. If successful, it may establish an *in vivo* activity assay for this receptor molecule.

There is still a gap in the model of how ERD2 can return to the Golgi in a way that is sufficiently fast to overtake bulk flow of ER residents. This would certainly not be possible if the tubular transport model is adopted, unless a second fast-track ER-export route is employed by the receptor. Regardless of the type of eukaryotic cell under investigation, it is impossible to explain all the observations made so-far with merely one route from ER to Golgi and one back to ER. For instance, blocking ER export by overproduction of Sec12 inhibited α-amylase secretion but with the same or even higher efficiency the weak secretion of α-amylase-HDEL (Phillipson *et al.*, 2001). If HDEL-ligands and their receptors leave the ER in the same COPII-dependent transport route, then such results would not be obtained. If ERD2 is prevented from returning to the Golgi in the retrograde direction, retrograde transport would be inhibited as well, and α-amylase-HDEL would not be affected much. In fact, the opposite effect was observed: in other words blocking COPII transport with Sec12 overproduction did not compromise retrograde transport (Phillipson *et al.*, 2001). Unravelling the ER-export route followed by the HDEL-receptor will thus be an important issue to understand retrograde transport, due to the intricate way in which the two transport routes must be interconnected.

The possibility of a COPII-independent ER-export pathway has been raised recently based on the observation that inhibition of COPII-dependent transport affects not all soluble proteins (Törmäkangas *et al.*, 2001). In addition, the visual-isation of unusually large ER-export carriers in some plant tissues, clearly surpass-ing the size of a typical COPII vesicle in yeast (Hara-Nishimura *et al.*, 1998; Toyooka *et al.*, 2000), suggests that ER export in plants may be more complicated than previously assumed. Likewise, it has become clear from research in mamma-lian cells that COPI transport does not possess a monopoly on retrograde transport to the ER, as illustrated by the rab6-dependent route followed by some toxins (Girod *et al.*, 1999; Storrie *et al.*, 2000). Although evidence for such a pathway is currently restricted to the identification of rab6-related plant cDNAs and genomic sequences with predicted coding regions, it is only a matter of time until this path-way will be studied in plants. The conclusion from this is that ER-Golgi transport may be controlled by a binary transport system, in which proteins are transported with different efficiencies.

In order for transport receptors to work properly, the recycling of receptors back to the organelle of ligand binding is at least as important or even more than the transport of the ligands themselves. This has important consequences for experi-mental approaches to study protein transport. If recycling of components is essen-tial for homeostasis, it is likely that inhibition of one transport route may lead to the collapse of its matching retrograde route and vice versa. For instance, the fact that trans-dominant negative mutants of COPII components such as the GTPases Sar1p and Rab1p cause accumulation of cargo molecules in the ER lumen (Batoko *et al.*, 2000; Takeuchi *et al.*, 2000; Phillipson *et al.*, 2001) may be unexpected in the first instance. Either approach should prevent COPII vesicle fusion with the Golgi, but not prevent COPII vesicle budding itself. In other words, one would have expected the COPII vesicles to accumulate in the cytosol, by analogy to the accumulation of COPII vesicles *in vitro* using the non-hydrolysable GTP homologue GTPγS (Barlowe *et al.*, 1994), and since COPII vesicles are at least 6-fold smaller than the wavelength of visible light, fluorescence microscopy should have revealed a weak cytosolic stain. However, the long-term effect of interfering with COPII vesicle fusion with the Golgi may be that essential machinery for vesicle budding is lost from the ER and not recycled from the Golgi via COPI-mediated transport. Hence, if applied for long times, it is likely that both COPI and COPII vesicle budding are inhibited. This would lead to accumulation of cargo molecules in the ER lumen, as observed (Batoko *et al.*, 2000; Takeuchi *et al.*, 2000; Phillipson *et al.*, 2001). A recent addition to the available tools to manipulate ER-Golgi trafficking is the GTPase rab2 (Cheung *et al.*, 2002). A tobacco pollen-specific rab2 (Ntrab2) was shown to localise to the Golgi of elongating pollen tubes, which was prevented using dominant-negative mutations in Ntrab2 which also led to a block in the normal localisation of Golgi-resident as well as plasmalemma and secreted proteins. The mutant proteins also inhibited pollen tube growth and were shown to operate in a pollen tube specific fashion, suggesting that other rab2 isoforms must be present to regulate ER-Golgi traffic in vegetative tissues. This work beautifully illustrates the crucial role of ER-Golgi traffic and the pleiotropic effects that can be observed, considering the general notion that rab2 should really be involved in retrograde traffic from the Golgi.

A number of seemingly conflicting reports about the effects of BFA or other transport inhibitors could also be explained if the possible dependence of COPII transport on regular COPI transport is taken into consideration (Ritzenthaler *et al.*, 2002). In this study, the authors provide evidence for the cisternal maturation model for the Golgi apparatus. This model predicts that anterograde intra-Golgi transport occurs via simple progression of stacks driven by selective retrograde vesicle transport which progressively changes the composition of the cisternae (Vitale & Denecke, 1999). The molecular target of BFA is the guanosine nucleotide exchange factor for ARF1p, resulting in a stabilisation of an abortive complex with the GDP-bound form of ARF1p at the Golgi membrane (Chardin & McCormick, 1999; Peyroche *et al.*, 1999). This leads to the inhibition of COPI vesicle formation, which is also documented in plants and includes an inhibition of coatomer recruitment *in vitro* (Pimpl *et al.*, 2000). BFA also promotes a re-distribution of Golgi proteins into the ER and a progressive loss of Golgi cisternae from *cis-* to *trans-*direction *in vivo* (Boevink *et al.*, 1998; Ritzenthaler *et al.*, 2002; Saint-Jore *et al.*, 2002), suggesting that in the long term, ER export may also be inhibited.

The much disputed role of COPI vesicles in anterograde intra-Golgi transport (Pelham & Rothman, 2000) is also questioned by a number of recent findings that suggest a predominant role of COPI vesicles in retrograde transport only (Martínez-Menárguez *et al.*, 2001; Mironov *et al.*, 2001; Pelham, 2001). Moreover, the vicinity of a vesicle in the periphery of the Golgi does not prove a Golgi origin, nor does it confirm its destination. One can only speculate from their cargo composition, which is extremely difficult to quantify using immunocytochemistry as illustrated by contradictory findings (Balch *et al.*, 1994; Orci *et al.*, 1997; Martínez-Menárguez *et al.*, 1999, 2001). But of course, the presence of calreticulin and HDEL-tagged α-amylase in plant COPI vesicles (Pimpl *et al.*, 2000) does by no means exclude the possibility that a second COPI population exists with anterograde cargo. Further experiments with a greater variety of cargo molecules will be necessary to describe the system better in plants.

In conclusion, great care must be taken when interpreting the results of experiments, and a combination of approaches will always be better than the mere sum of the individual approaches together. A further complication arises from the definition of vesicular cargo molecules and its distinction from vesicular machinery (Klumperman, 2000). For instance, SNAREs or sorting receptors should not be regarded as cargo molecules, because they are clearly part of the vesicle transport machinery. Soluble proteins always constitute true cargo molecules, whereas membrane-spanning proteins can either be cargo or machinery, and it is important to distinguish between the two while interpreting published findings.

6.4   From where does retrieval of soluble proteins occur?

Several reports present evidence for COPI-dependent transport from the Golgi in plant cells (Movafeghi *et al.*, 1999; Contreras *et al.*, 2000; Pimpl *et al.*, 2000). From these findings it appears as if retrograde transport between the Golgi and the ER in plants is likely to occur in a very similar way as described for mammalian cells, except for the lack of the *intermediate compartment*. Despite hard efforts over the last 10 years, evidence for such a compartment has remained elusive in plant cells via any biochemical or microscopical approach, and it seems to be a feature shared with the yeast *Saccharomyces cerevisiae*. Several lines of evidence suggest that the *cis*-Golgi acts as the main compartment from which proteins carrying the typical ER retention motifs are retrieved. Using the straightforward assessment of the presence of endo-H sensitive high mannose *N*-linked glycans versus the complex endo-H sensitive forms, it could be shown that KDEL-tagged phaseolin does not penetrate far into the Golgi. The same has been shown for assembly defective phaseolin which also remains in the endo-H sensitive form (Pedrazzini *et al.*, 1997). This could be due to co-retrieval via association with BiP, which possesses the HDEL motif. In both cases, retrieval must occur from a Golgi compartment in which these processing enzymes are absent or inactive. Wild type assembly-competent phaseolin is ER-export competent and acquires endo-H resistance. The abundant ER protein calreticulin also exhibits exclusive endo-H sensitive glycans (Crofts *et al.*, 1999), although it does prove to be ER-export competent (Phillipson *et al.*, 2001). The difficulty with this approach is that it is not easy to pinpoint the location in the Golgi where glycan processing occurs, because immunocytochemistry does not report on enzyme activity, whereas biochemical assays with cell fractionation do not permit reproducible separation of *cis*, medial and *trans*-Golgi cisternae. More direct and thus much more meaningful evidence arose from direct electron microscopy, which revealed a clear statistically meaningful concentration of HDEL and KDEL proteins in the *cis*-Golgi (Bauly *et al.*, 2000; Phillipson *et al.*, 2001). This is perhaps the best evidence currently available to demonstrate the role of the *cis*-Golgi in the retrograde transport back to the ER. If retrieval of HDEL or KDEL proteins is possible from medial or *trans*-Golgi cisternae in plants it will depend on the use of proteins which have competence for these glycan modifications and which have directly been detected in the ER or the nuclear envelope. Such experiments are yet to be carried out.

6.5   Retrograde transport of membrane spanning proteins

Membrane proteins may not be transported via similar kinetics and cargo sorting mechanisms as soluble proteins because they are restricted to the limited surface of the membrane itself. When spanning the membranes, they will also have contact with both sides of the membrane of the secretory pathway and can potentially interact with cytosolic and lumenal components of the transport machinery. Although the ER retention of membrane-spanning proteins is much less under-stood in plants, the first results are emerging and provide interesting data. Experiments in which the transmembrane domain and cytosolic tail of the ER resident protein calnexin were fused to the C-terminus of invertase (a soluble secreted protein translocated to the ER via a typical N-terminal signal peptide) revealed that a lysine residue near the C-terminus of calnexin is important for ER-retention of the fusion protein (Barrieu & Chrispeels, 1999). The mutant molecule in which the lysine residue was substituted by glutamine progressed further through the secretory system, yielding a proteolytic fragment containing invertase in the vacuoles, whilst the microsomal fraction contained a membrane-anchored full length fusion protein.

Slightly different results were obtained with the tomato Cf-9 disease resistance gene. Unlike calnexin, the Cf-9 gene-product encodes a type I membrane protein carrying a typical cytosolic dilysine motif known as the KKXX signal, with XX referring to any amino-acids in the two most C-terminal positions of the coding region. A GFP fusion to the transmembrane domain and cytosolic tail of Cf-9 was targeted to the ER, whereas mutation of the KKXX motif to NNXX led to secretion of a proteolytic fragment devoid of the transmembrane domain (Benghezal *et al.*, 2000).

The difference in the observed end-location may be a property of the lumenal reporter after cleavage, or reflect a difference in the location that can be reached after escaping from the retention machinery. Recent evidence suggests that the length of the transmembrane domain can determine either ER-retention, Golgi retention or transport to the plasma membrane (Brandizzi *et al.*, 2002). While varying the length of the transmembrane domain from 23 to either 20 or 17 amino acids, the authors could demonstrate that the first was localised to the plasma membrane, the second was retained in the Golgi whilst the shortest domain conferred retention in the ER. Subtle differences between the transmembrane domains of calnexin and the Cf-9 gene product may not be apparent from studying the primary sequences, but could yet account for the differences observed in the two former studies. Apart from this, it is difficult to compare these results due to the different nature of the lumenal reporter protein. It was not clear where proteolysis occurs, and the proteins in the end-location no longer contained the transmembrane domains (Barrieu & Chrispeels, 1999; Benghezal *et al.*, 2000). But ignoring the results on the final destination, both reports imply positively charged residues near the C-terminus of the cytosolic tail as an important constituent of the ER retention signal. Whether such *di-lysine* motifs are retrieved from the Golgi and interact with the COPI transport machinery in plants remains to be established, and in addition to a signal-mediated retention in the ER, there could very well be an additional mechanism which relies on the exclusion from anterograde transport and relies on the length of the transmembrane domain (Brandizzi *et al.*, 2002).

Besides actual cargo molecules, research on the localisation of sorting machinery has also been carried out. As discussed before, a GFP-fusion protein with the *Arabidopsis thaliana* ERD2 gene product could be detected in the Golgi apparatus and was shown to be re-localised to the ER upon treatment with BFA (Boevink *et al.*, 1998). Similarly, it could be shown that ERD2-GFP remains trapped in the ER upon co-expression of the trans-dominant negative mutant of the GTPase Sar1 which is restricted to its GTP-bound form (Takeuchi *et al.*, 2000). This result would suggest that both ERD2-GFP and bulk flow occur via COPII-dependent transport. Also the *Arabidopsis thaliana* Rer1 homologue (Sato *et al.*, 1999) would appear to be transported in a COPII-dependent fashion, as deduced from GFP-fusion protein assays (Takeuchi *et al.*, 2000). The latter protein was shown to be essential for the retrieval from the Golgi of the Sec12, Sec71 and Sec63 gene products in yeasts (Sato *et al.*, 1995, 1997). Inhibition of Golgi-derived COPI transport via co-expression of GTP- or GDP-restricted forms of the GTPase ARF1 cause re-distribution of ERD2-GFP to the ER. The GDP-restricted form of ARF1 also caused ER retention of the Golgi marker sialyltransferase (ST)-GFP (Lee *et al.*, 2002). Since interference with ARF1 should target the Golgi, ER-retention must be indirect, possibly by inhibiting COPII transport. Most interestingly, GFP-Rer1 behaved completely different and remained associated with a Gogi-like structure upon treatment with the GTP- and GDP-restricted ARF1 or BFA (Sato *et al.*, 1999; Takeuchi *et al.*, 2000, 2002). The latter is a surprising result and may be another suggestion that Golgi-derived retrograde transport is perhaps not restricted to COPI-dependent transport. It was discussed before that the retrograde COPI route has recently been found to be complemented by a different, rab6-dependent route (Girod *et al.*, 1999; Storrie *et al.*, 2000). If two distinct pathways lead from the Golgi to the ER, the simple need to recycle the required machinery would almost predict that there should also be two routes leading from the ER to the Golgi. Evidence for COPII-independent transport has been obtained for soluble proteins in plants, and it seems to occur at a very fast rate (Törmäkangas *et al.*, 2001). This seems to correspond well with evidence for divergent ER-export pathways in yeast to sort GPI-anchored proteins from other protein cargo (Muniz *et al.*, 2001).

6.6   Conclusions

It appears that research on the plant secretory pathway has opened up an enormous wealth of opportunities, now that biochemical, genetic and microscopical approaches can be combined to answer more difficult questions. But whilst this is an advantage, it also becomes clear that care has to be taken to distinguish primary targets of drugs and trans-dominant negative mutants from secondary, long term effects. The simple fact that ARF1 controls COPI transport, but will also influence COPII transport indirectly due to the dependence of these transport steps on each other, provides an example out of many to illustrate how many other observations such as those obtained with the drug BFA may have to be constantly re-interpreted, while further data emerge.

Acknowledgements

This work was supported by grants from the Biotechnology and Biological Sciences Research Council (BBSRC) and the European Union (Grant nrs. CHRX-CT94-0590).

References

Balch, W.E., McCaffery, J.M., Plutner, H. & Farquhar, M.G. (1994) Vesicular stomatitis virus glycoprotein is sorted and concentrated during export from the endoplasmic reticulum. *Cell*, **76**, 841–852.

Barlowe, C., Orci, L., Yeung, T., Hosobuchi, M., Hamamoto, S., Salama, N., Rexach, M.F., Ravazzola, M., Amherdt, M. & Schekman, R. (1994) COPII: a membrane coat formed by Sec proteins that drive vesicle budding from the endoplasmic reticulum. *Cell*, **77**, 895–907.

Barrieu, F. & Chrispeels, M.J. (1999) Delivery of a secreted soluble protein to the vacuole via a membrane anchor. *Plant Phys.*, **120**, 961–968.

Batoko, H., Zheng, H.Q., Hawes, C. & Moore, I. (2000) A rab1 GTPase is required for transport between the endoplasmic reticulum and Golgi apparatus and for normal Golgi movement in plants. *Plant Cell*, **12**, 2201–2218.

Bauly, J.M., Sealy, I.M., Macdonald, H., Brearley, J., Droge, S., Hillmer, S., Robinson, D.G., Venis, M.A., Blatt, M.R., Lazarus, C.M. & Napier, R.M. (2000) Overexpression of auxin-binding protein enhances the sensitivity of guard cells to auxin. *Plant Phys.*, **124**, 1229–1238.

Benghezal, M., Wasteneys, G.O. & Jones, D.A. (2000) The C-terminal dilysine motif confers endoplasmic reticulum localization to type I membrane proteins in plants. *Plant Cell*, **12**, 1179–1201.

Boevink, P., Oparka, K., Santa Cruz, S., Martin, B., Betteridge, A. & Hawes, C. (1998) Stacks on tracks: the plant Golgi apparatus traffics on an actin/ER network. *Plant J.*, **15**, 441–447.

Brandizzi, F., Frangne, N., Marc-Martin, S., Hawes, C., Neuhaus, J.M. & Paris, N. (2002) The destination for single-pass membrane proteins is influenced markedly by the length of the hydrophobic domain. *Plant Cell*, **14**, 1077–1092.

Chardin, P. & McCormick, F. (1999) Brefeldin A: the advantage of being uncompetitive. *Cell*, **97**, 153–155.

Cheung, A.Y., Chen, C.Y., Glaven, R.H., de Graaf, B.H., Vidali, L., Hepler, P.K. & Wu, H.M. (2002) Rab2 GTPase regulates vesicle trafficking between the endoplasmic reticulum and the Golgi bodies and is important to pollen tube growth. *Plant Cell*, **14**, 945–962.

Contreras, I., Ortiz-Zapater, E., Castilho, L.M. & Aniento, F. (2000) Characterization of COP I coat proteins in plant cells. *Biochem. Biophys. Res. Commun.*, **273**, 176–182.

Crofts, A.J., Leborgne-Castel, N., Hillmer, S., Robinson, D.G., Phillipson, B., Carlsson, L.E., Ashford, D.A. & Denecke, J. (1999) Saturation of the endoplasmic reticulum retention machinery reveals anterograde bulk flow. *Plant Cell*, **11**, 2233–2248.

Girod, A., Storrie, B., Simpson, J.C., Johannes, L., Goud, B., Roberts, L.M., Lord, J.M., Nilsson, T. & Pepperkok, R. (1999) Evidence for a COP-I-independent transport route from the Golgi complex to the endoplasmic reticulum. *Nat. Cell Biol.*, **1**, 423–430.

Griffiths, G., Hoflack, B., Simons, K., Mellman, I. & Kornfeld, S. (1988) The mannose 6-phosphate receptor and the biogenesis of lysosomes. *Cell*, **52**, 329–341.

Hadlington, J.L. & Denecke, J. (2000) Sorting of soluble proteins in the secretory pathway of plants. *Curr. Opin. Plant. Biol.*, **3**, 461–468.

Hara-Nishimura, I., Shimada, T., Hatano, K., Takeuchi, Y. & Nishimura, M. (1998) Transport of storage proteins to protein storage vacuoles is mediated by large precursor-accumulating vesicles. *Plant Cell*, **10**, 825–836.

Hoflack, B., Fujimoto, K. & Kornfeld, S. (1987) The interaction of phosphorylated oligosaccharides and lysosomal enzymes with bovine liver cation-dependent mannose 6-phosphate receptor. *J. Biol. Chem.*, **262**, 123–129.

Klumperman, J. (2000) Transport between ER and Golgi. *Curr. Opin. Cell Biol.*, **12**, 445–449.

Kornfeld, S. (1992) Structure and function of the mannose 6-phosphate/insulinlike growth factor II receptors. *Annu. Rev. Biochem.*, **61**, 307–330.

Kornfeld, S. & Mellman, I. (1989) The biogenesis of lysosomes. *Annu. Rev. Cell Biol.*, **5**, 483–525.

Lee, H.I., Gal, S., Newman, T.C. & Raikhel, N.V. (1993) The *Arabidopsis* endoplasmic reticulum retention receptor functions in yeast. *Proc. Natl. Acad. Sci. USA*, **90**, 11433–11437.

Lee, M.H., Min, M.K., Lee, Y.J., Jin, J.B., Shin, D.H., Kim, D.H., Lee, K.H. & Hwang, I. (2002) ADP-ribosylation factor 1 of *Arabidopsis* plays a critical role in intracellular trafficking and maintenance of endoplasmic reticulum morphology in *Arabidopsis*. *Plant Phys.*, **129**, 1507–1520.

Martínez-Menárguez, J.A., Geuze, H.J., Slot, J.W. & Klumperman, J. (1999) Vesicular tubular clusters between the ER and Golgi mediate concentration of soluble secretory proteins by exclusion from COPI-coated vesicles. *Cell*, **98**, 81–90.

Martínez-Menárguez, J.A., Prekeris, R., Oorschot, V.M., Scheller, R., Slot, J.W., Geuze, H.J. & Klumperman, J. (2001) Peri-Golgi vesicles contain retrograde but not anterograde proteins consistent with the cisternal progression model of intra-Golgi transport. *J. Cell Biol.*, **155**, 1213–1224.

Mironov, A.A., Beznoussenko, G.V., Nicoziani, P., Martella, O., Trucco, A., Kweon, H.S., Di Giandomenico, D., Polishchuk, R.S., Fusella, A., Lupetti, P., Berger, E.G., Geerts, W.J., Koster, A.J., Burger, K.N. & Luini, A. (2001) Small cargo proteins and large aggregates can traverse the Golgi by a common mechanism without leaving the lumen of cisternae. *J. Cell Biol.*, **155**, 1225–1238.

Movafeghi, A., Happel, N., Pimpl, P., Tai, G.H. & Robinson, D.G. (1999) *Arabidopsis* Sec21p and Sec23p homologs. Probable coat proteins of plant COP-coated vesicles. *Plant Phys.*, **119**, 1437–1446.

Muniz, M., Morsomme, P. & Riezman, H. (2001) Protein sorting upon exit from the endoplasmic reticulum. *Cell*, **104**, 313–320.

Orci, L., Stamnes, M., Ravazzola, M., Amherdt, M., Perrelet, A., Sollner, T.H. & Rothman, J.E. (1997) Bidirectional transport by distinct populations of COPI-coated vesicles. *Cell*, **90**, 335–349.

Pedrazzini, E., Giovinazzo, G., Bielli, A., de Virgilio, M., Frigerio, L., Pesca, M., Faoro, F., Bollini, R., Ceriotti, A. & Vitale, A. (1997) Protein quality control along the route to the plant vacuole. *Plant Cell*, **9**, 1869–1880.

Pelham, H.R. (1988) Evidence that luminal ER proteins are sorted from secreted proteins in a post-ER compartment. *EMBO J.*, **7**, 913–918.

Pelham, H.R. (2001) Traffic through the Golgi apparatus. *J. Cell Biol.*, **155**, 1099–1101.

Pelham, H.R. & Rothman, J.E. (2000) The debate about transport in the Golgi – two sides of the same coin? *Cell*, **102**, 713–719.

Peyroche, A., Antonny, B., Robineau, S., Acker, J., Cherfils, J. & Jackson, C.L. (1999) Brefeldin A acts to stabilize an abortive ARF-GDP-Sec7 domain protein complex: involvement of specific residues of the Sec7 domain. *Mol. Cell*, **3**, 275–285.

Phillipson, B.A., Pimpl, P., Crofts, A.J., Taylor, J.P., Movafeghi, A., Robinson, D.G. & Denecke, J. (2001) Secretory bulk flow of soluble proteins is COPII dependent. *Plant Cell*, **13**, 2005–2020.

Pimpl, P., Movafeghi, A., Coughlan, S., Denecke, J., Hillmer, S. & Robinson, D.G. (2000) *In situ* localization and *in vitro* induction of plant COPI-coated vesicles. *Plant Cell*, **12**, 2219–2236.

Ritzenthaler, C., Nebenfuhr, A., Movafeghi, A., Stussi-Garaud, C., Behnia, L., Pimpl, P., Staehelin, L.A. & Robinson, D.G. (2002) Reevaluation of the effects of brefeldin A on plant cells using tobacco bright yellow 2 cells expressing Golgi-targeted green fluorescent protein and COPI antisera. *Plant Cell*, **14**, 237–261.

Saint-Jore, C.M., Evins, J., Batoko, H., Brandizzi, F., Moore, I. & Hawes, C. (2002) Redistribution of membrane proteins between the Golgi apparatus and endoplasmic reticulum in plants is reversible and not dependent on cytoskeletal networks. *Plant J.*, **29**, 661–678.

Sato, K., Nishikawa, S. & Nakano, A. (1995) Membrane protein retrieval from the Golgi apparatus to the endoplasmic reticulum (ER): characterization of the RER1 gene product as a component involved in ER localization of Sec12p. *Mol. Biol. Cell*, **6**, 1459–1477.

Sato, K., Sato, M. & Nakano, A. (1997) Rer1p as common machinery for the endoplasmic reticulum localization of membrane proteins. *Proc. Natl. Acad. Sci. USA*, **94**, 9693–9698.

Sato, K., Ueda, T. & Nakano, A. (1999) The *Arabidopsis thaliana* RER1 gene family: its potential role in the endoplasmic reticulum localization of membrane proteins. *Plant Mol. Biol.*, **41**, 815–824.

Storrie, B., Pepperkok, R. & Nilsson, T. (2000) Breaking the COPI monopoly on Golgi recycling. *Trends Cell Biol.*, **10**, 385–391.

Takeuchi, M., Ueda, T., Yahara, N. & Nakano, A. (2002) Arf1 GTPase plays roles in the protein traffic between the endoplasmic reticulum and the Golgi apparatus in tobacco and *Arabidopsis* cultured cells. *Plant J.*, **31**, 499–515.

Takeuchi, M., Ueda, T., Sato, K., Abe, H., Nagata, T. & Nakano, A. (2000) A dominant negative mutant of Sar1 GTPase inhibits protein transport from the endoplasmic reticulum to the Golgi apparatus in tobacco and *Arabidopsis* cultured cells. *Plant J.*, **23**, 517–525.

Törmäkangas, K., Hadlington, J.L., Pimpl, P., Hillmer, S., Brandizzi, F., Teeri, T.H. & Denecke, J. (2001) A vacuolar sorting domain may also influence the way in which proteins leave the endoplasmic reticulum. *Plant Cell*, **13**, 2021–2032.

Townsley, F.M., Wilson, D.W. & Pelham, H.R. (1993) Mutational analysis of the human KDEL receptor: distinct structural requirements for Golgi retention, ligand binding and retrograde transport. *EMBO J.*, **12**, 2821–2829.

Toyooka, K., Okamoto, T. & Minamikawa, T. (2000) Mass transport of proforma of a KDEL-tailed cysteine proteinase (SH-EP) to protein storage vacuoles by endoplasmic reticulum-derived vesicle is involved in protein mobilization in germinating seeds. *J. Cell Biol.*, **148**, 453–463.

Vitale, A. & Denecke, J. (1999) The endoplasmic reticulum-gateway of the secretory pathway. *Plant Cell*, **11**, 615–628.

7 Protein modifications in the Golgi apparatus

Ken Matsuoka

7.1 Outline of protein modifications in the secretory pathway

Proteins synthesized by the membrane-bound polysome are translocated through the endoplasmic reticulum (ER) membrane or inserted into the ER membrane. Many of these proteins are transported through the secretory pathway, during which they are modified. These post-translational modifications include disulfide bridge formation, glycosylation, sulfation, lipid anchoring, covalent and non-covalent attachment of co-factors, and proteolytic processing. Some of these modifications, such as the glycosylation of asparagine-residues (N-glycosylation) and disulfide bridge formation, occur in the ER. Such modifications are conserved in higher eukaryotes, whereas the other modifications which occur in the ER and in the secretory organelles are not evolutionarily conserved in eukaryotes.

In this review, current knowledge of the post-translational protein modification in the plant Golgi apparatus is summarized in comparison with mammalian and fungal systems, and the possible modification mechanisms, which can be expected from a genomic analysis, are also discussed.

7.2 Protein glycosylation

7.2.1 Overview of the glycosylation

Attachment of a sugar moiety to proteins can occur by two distinct mechanisms. One is an enzymatic glycosylation, and the other is a glycation that is a result of a chemical reaction without an enzyme. Glycation of serum proteins in the blood of diabetes patient is common, due to the presence of a highly reactive aldehyde-type sugar, such as glucose (Uchida, 2000). Because glycation is an oxidative protein modification with aldose, this reaction may occur in the oxidative environment of the whole plant, including the cell wall. However, few reports on protein glycation in plants have been published, possibly due to a lack of interest in this reaction.

In contrast, there are many reports on protein glycosylation in plants. Until now, three types of glycosylation are known in eukaryotes: N-glycosylation, O-glycosylation, and the recently found C-mannosylation that is currently found in a couple of mammalian proteins (Van den Steen *et al.*, 1998). Among these glycosylations, the most common form is N-glycosylation, which starts in the ER by the co-translational transfer of a precursor oligosaccharide to specific asparagine residues of the nascent polypeptide chain. Processing of this oligosaccharide into hybrid- or complex-type N-glycans occurs in the ER and the Golgi apparatus, during the transport through the secretory pathway. At the end of the maturation, some plant N-glycans have structures that differ typically from those found in their mammalian and yeast counterparts. The unique feature of plant N-glycan is the presence of $\beta(1,2)$-xylose and $\alpha(1,3)$-fucose residues. Because this characteristic has been discussed in depth in a recent review (Lerouge *et al.*, 1998), the nature and role of such modifications are not discussed here.

The mechanism of O-glycosylation, i.e. the attachment of sugars to the hydroxyl residues in proteins, is not well understood, especially in plants. One of the most significant reasons for this is that this type of glycosylation is very divergent amongst the eukaryotes. In mammalian cells as well as in plants, O-glycoslyation occurs both in the secretory pathway and in cytosol. The most common O-glyco-sylation in the secretory pathway in animals is the glycosylation of serine and threonine residues with a mucin-type glycan (Carraway & Hull, 1989). The synthesis of mucin-type glycans is initiated by the peptidyl GalNAc-transferases localized in the Golgi apparatus (Carraway & Hull, 1989). In contrast, the most abundant O-glycan in yeast and fungi are homomannans (Ernst & Prill, 2001). This glycan is attached to the proteins in the ER by protein mannosyltransferases (Ernst & Prill, 2001). Recently there have been several reports that many archae and some Gram-positive bacteria have N- and O-glycosylated proteins (Schaffer *et al.*, 2001), although little is known about the mechanisms of their glycosylations.

O-glycosylation in the secretory system of higher plants is more complex than these organisms. Several different classes of O-glycan linkages are present in higher plant secretory proteins. These include arabinans and arabinogalactans that are attached to hydroxylated proline residues, and galactose and mucin-type gly-cans attached to serine (and possibly threonine) residue(s). These O-glycosylated proteins are predominantly found in the extracellular matrix, although some vacuolar proteins are also O-glycosylated (Matsuoka *et al.*, 1995; Kishimoto *et al.*, 1999).

7.2.2   O-glycosylation of Ser/Thr residues

Some of the serine residues in secretory and vacuolar proteins contain O-glycans. In the case of the cell wall matrix glycoprotein extensin, some of the serine residues are O-glycosylated with a monosaccharide galactose (Lamport *et al.*, 1973; Cho & Chrispeels, 1976) in addition to the extensive oligoarabinosylation of hydroxyproline residues (discussed below). Similar galactosylation seems to occur in some vacuolar proteins. For example, the isoforms of sporamin, which is a sweet potato storage protein, have a glycan with terminal galactose attached to the 39th serine residue of its precursor and is recognized only by galactose-specific lectins (Matsuoka *et al.*, 1995).

In addition to the monogalactosylation, mammalian mucin-type glycosylation can also be found in plants. A subset of Golgi membrane proteins from rice cells are recognized by peanut agglutinin and UEA-I, which recognize β-Gal(1-3)GalNAc and α-linked fucose residues, respectively (Mitsui *et al.*, 1990). This recognition is sensitive to alkaline hydrolysis and *O*-glycanase (Kimura *et al.*, 1993). Because brefeldin A, which blocks the transport of proteins from the endoplasmic reticulum to the Golgi apparatus (Stahelin & Driouich, 1997), inhibited the formation of the *O*-linked sugar chains of these Golgi-membrane associated proteins, it is likely that the mucin-type glycan is attached to proteins in the Golgi apparatus. A recent report on the analysis of sugar chain of glutelin, which is a storage protein transported to protein-storage vacuole through the Golgi apparatus in rice endosperm (Matsuoka & Bednarek, 1998), has revealed that this protein has an alkaline-sensitive *O*-glycan of Galβ-1,3GalNAc structure (Kishimoto *et al.*, 1999). This structure is known as the core 1 structure of the mammalian mucin.

In mammalian cells, control of mucin-type O-glycosylation involves an initiation step followed by a processing step. The initiation step is carried out by a large family of homologous UDP-GalNAc:polypeptide GalNAc-transferases (Van den Steen *et al.*, 1998). The processing step involves elongation, branching, and terminal modification of the *O*-glycans (Van den Steen *et al.*, 1998). Interestingly, however, no close homolog of mammalian UDP-GalNAc:polypeptide GalNAc-transferases can be found in the *Arabidopsis* genome database. It is therefore possible that the enzyme catalyzing the addition of initial GalNAc to serine (or threonine) residues in the plant Golgi belongs to a class distinct from the mammalian enzymes.

In contrast, *Arabidopsis* has at least 10 homologs of mammalian UDP-GlcNAc:Gal1-3GalNAc 6GlcNAc-transferase, which are involved in the synthesis of the core 2 branch of the mucin (Fig. 7.1). These *Arabidopsis* homologs have a primary structure typical of type II transmembrane proteins. The possible transmembrane region of these proteins consists of 19–23 AAs. In contrast, type II transmembrane proteins those might be localized in the ER, such as ER mannosidase, which is involved in the trimming of *N*-glycan (At1g51590), has a transmembrane region of 16 AAs. It is generally assumed that the thickness of the membrane bilayer increases in the later compartment of the secretory pathway. Furthermore, it has been proposed that the length of the transmembrane region that corresponds to the thickness of the membrane determines the localization of the protein in the secretory pathway when a protein does not have a specific signal for retention (Munro, 1998). A recent analysis of the localization of proteins in plant cells fits this hypothesis (Brandizzi, 2002). In this regard, it is interesting to note that most of the members in these *Arabidopsis* UDP-GlcNAc:Gal1-3GalNAc 6GlcNAc-transferase homologs with relatively short transmembrane region (At1g53100, At3g03690, At3g24040, At3g15350) have shorter cytosolic regions than the ones having longer transmembrane regions. In addition, some of these homologs with relatively long cytosolic region (At4g03340, AT1g03520, At5g39990, At5g15050) have a YXXΦ sequence that corresponds to the tyrosine-based internalization motif (Bonifacino & Dell'Angelica, 1999). Therefore, it is possible that these proteins are localized to the Golgi apparatus or other late secretory organelle where glycan

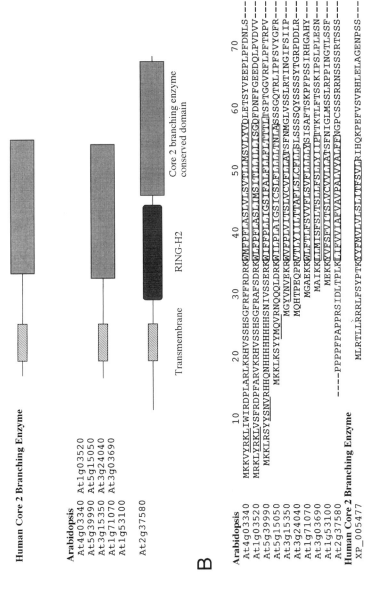

Figure 7.1 *Arabidopsis* homologs of mammalian Core 2 branching enzyme. (A) Schematic illustration of the human Core 2 branching enzyme and *Arabidopsis* homologs. (B) N-terminal transmembrane region and surrounding sequences of the human Core 2 branching enzyme and *Arabidopsis* homologs. Possible transmembrane regions are boxed.

trimming takes place. Future work will hopefully reveal the role that these gene products play in the modification of protein-attached glycans in plants.

7.2.3   Proline hydroxylation and glycosylation of hydroxylated proline

An O-glycosylation unique to plants is the glycosylation of hydroxyproline (Hyp) residues. These include the oligoarabinosylation found in extensins and Hyp-rich glycoproteins (HRGPs), and the addition of arabinogalactan to the arabinogalactan proteins. Although these proteins are rich in glycosylated Hyps, there are a number of cell wall proteins that contain non-glycosylated Hyp residues, e.g. extensins and HRGPs. In addition, some vacuolar proteins also contain Hyp residues. For example, class I chitinase of tobacco contains a clustered Hyp without glycosylation (Sticher *et al.*, 1992), whereas the sporamin precursor, when expressed in tobacco cell, becomes glycosylated at a particular proline residue after hydroxylation (Matsuoka *et al.*, 1995).

One approach to characterize the glycosylation mechanism of Hyp residues is to define the sequence motif that is required for this modification. Comparison of the glycosylated sequences implies the conserved sequence for glycosylation. Highly conserved peptide motifs containing contiguous Hyp residues are found in the glycosylated region of the extensins. In contrast, non-contiguous Hyp residues surrounded by Ala, Ser or Thr residues are the common features of the arabino-galactosylated proteins (Table 7.1). Thus, the Hyp contiguity hypothesis predicts the arabinosylation of contiguous Hyp residues (Kieliszewski & Lamport, 1994). In contrast, galactosylation can occur with non-contiguous Hyp residues surrounded by amino acids with small side chains.

However, further experimental evaluation is required to determine the sequence motif of O-glycosylation, because the glycosylation of Hyp is a complex mechanism with two separate steps, namely, proline hydroxylation and glycosylation. The enzyme, which catalyze the first step is idiomatically called *prolyl hydroxylase*, although it is actually an oxygenase that introduce oxygen between the hydrogen and carbon of peptidyl proline. Substrate specificity of the plant prolyl hydroxylases is not well defined, although it has been shown that poly(L-proline) is an excellent substrate (Tanaka *et al.*, 1981). In aqueous solution this peptide assumes a three-residue per turn extended helix, designated poly-proline II. Initially, this structure was designated as to define the substrate of this enzyme. However, recent

Table 7.1 Hyp O-glycosylation sites and surrounding sequences

Protein	Sequence	Glucan
Arabinogalactan protein	SPSOSOSOSO	Arabinogalactan
Gum arabic glycoprotein	DSOSOTOTAOO	Arabinogalactan
Sporamin	TTHEOASSET	Arabinogalactan
Extensin	OOOVYK	Oligoarabinose

Amino acids were written in single letter code. O, hydroxyproline. O, glycosylated hydroxyproline.

characterization of an *Arabidopsis* prolyl hydroxylase (AAC64297), which is a monomeric and possibly type II transmembrane protein of about 30 kDa, revealed that this *Arabidopsis* enzyme is capable of hydroxylating non-clustered proline residues as well (Hieta & Myllyharju, 2002). Peptide sequences that can be hydroxylated *in planta* somewhat differ from those hydroxylated by animal enzymes because plants do not hydroxylate transgenic collagen (Ruggiero *et al.*, 2000).

There are five genes that encode the close relatives of characterized prolyl hydroxylases in the *Arabidopsis* genome (Hieta & Myllyharju, 2002). Two of them (AAB80790 and AAF88161) also appear to be type II proteins with relatively long hydrophobic region of 23 and 24 amino acids long respectively. In these cases, the N-terminal most cytosolic region contains KXK or related sequences (KSK and KLR). The KXK sequence is known to interact with coatomer, which is a major coat protein of COPI vesicle trafficking through the Golgi apparatus (Jackson *et al.*, 1993). Although this motif was identified in type I transmembrane proteins, it may be possible that such motif may contribute to the Golgi localization of these possible prolyl hydroxylases.

The next step of the glycosylation is the transfer of glycan from the donor substrate to the acceptor Hyp residue. To date, two classes of sugar donors are known to be utilized for O-glycosylation. One is sugar nucleotides, which are used for mammalian O-glycosylation (Carraway & Hull, 1989). The other is lipid-linked sugars, which are used for O-glycosylation in yeast (Ernst & Prill, 2001). In plants, lipid-linked acidic arabinosyl oligosaccharides are transferred to HRGP acceptors (Bolwell, 1986). One possible intermediate for the attachment of arabinogalactan to protein is a lipid-linked glycan with about 15 sugar residues and a galactose reducing end (Hayashi & Maclachlan, 1984). Because polysaccharides in arabinogalactan protein contains repetitive glycan units (Bacic *et al.*, 1996), the block transfer of preassembled lipid-linked oligosaccharide to build polysaccharides in arabinogalactan proteins can explain the repetitive nature of the glycan side chain of this class. Although we cannot rule out the possibility that a sugar-nucleotide is used the formation of a peptide-sugar until the arabinogalactan transferase has been identified.

7.3   Protein tyrosine sulfation

Protein tyrosine O-sulfation is one of the most frequent post-translational modifications in mammalian secretory proteins (Kehoe & Bertozzi, 2000). In contrast, only one sulfated polypeptide from plant has been reported to date. This is the precursor to phytosulfokine (PSK), which is a peptide growth factor found in many higher plants (Matsubayashi *et al.*, 2001).

Phytosulfokine was first characterized as a growth factor found in a conditioned medium that stimulates the cell growth of low-density *Asparagus* cell cultures (Matsubayashi & Sakagami, 1996). This factor is a pentapeptide with two sulfated tyrosines. Sulfation is essential for the function of this hormone. Analysis of the

EST sequence has revealed that the PSK precursor is encoded as a preproprotein with a signal peptide at its N-terminus. The precursor contains an N-terminal propeptide between the signal peptide and mature hormone. A several amino acid extension after the hormone-coding region is found in all the precursors. Many possible PSK precursors are found in the EST/genome database and in all cases the precursor structures are similar although there are significant differences in the length of the N-terminal propeptide (Fig. 7.2). Because of the many different plant species which respond to this peptide and because many plant species have genes that can encode this hormone, it is likely that the protein sulfation activity is widely distributed in the higher plant kingdom (Matsubayashi *et al.*, 2001).

In mammalian cells, the sulfation reaction is catalyzed by tyrosylprotein sulfotransferase (TPST), a membrane-bound enzyme localized to the *trans*-Golgi network (Kehoe & Bertozzi, 2000). This protein uses 3'-phosphoadenosine 5'-phosphosulfate as a sulfate donor. The substrate polypeptide contains acidic residues at the N-terminal region relative to the sulfation site (Niehrs *et al.*, 1990). Human and mouse TPST cDNAs are type II transmembrane proteins of 370 amino acid residues with an apparent molecular mass of 54 kDa (Ouyang *et al.*, 1998). In spite of significant advances in the study of mammalian TPST, little is known about TPST in higher plants because the *Arabidopsis* genome does not encode a close homolog to mammalian TPST.

However, using synthetic peptides, which constitute parts of the PSK precursor, as substrates, Sakagami and co-workers have recently showed that TPST activity is present in microsomal membranes prepared from several monocot and dicot plant cells (Hanai *et al.*, 2000). They also showed that acidic amino acid residues adjacent and closely located to the N-terminus of the sulfation site are important for sulfation. This substrate specificity resembles that of the mammalian enzyme. An analysis of the localization of this enzyme by subcellular fractionation indicated that the activity was almost exclusively recovered in the Golgi-enriched fraction (Hanai *et al.*, 2000). Therefore, tyrosine sulfation in the secretory pathway in plants can occur in the Golgi apparatus as in the case in mammalian cells.

7.4   Proteolytic processing of proteins

7.4.1   Kexin/subtilase family proteases

It is well known that the Kex2p protease in yeast, mammal furin, and related proteases are the processing proteases that convert hormone precursors to mature peptide hormones. These *convertases* are serine proteases belonging to the subtilisin superfamily. These enzymes recognize a common sequence motif (K/R)-(X)n-(K/R), where n=0, 2, 4 or 6 and X is any amino acid but usually not a Cys (Seidah & Chretien, 1999). The cleavage site of these enzymes is the C-terminus of the second basic amino acid. Kex2p and furin are type I transmembrane proteins with a large catalytic domain at the lumenal side. These proteases are localized to the

```
Pinus taeda AW981986p              MFCGGSVRQPAKNMLSFIFAILLLTTVTSI----RPLDKGGPRNSRNSIV-DSELFVKEVLPIDDALNKVQRIDGEETCQKSEDEEECLNRRSL-AAHTDYITTQHHNSP
Zea mays AI712273p         MARRADCDGARGGARGPPAGVVTVMVLAAALAVLLLASSSS-----KTAPVASAARDDPSAAAAAVTSSRDLQNDGSAAAAEGKGKEKECGANDEDECMMRRTL-AAHTDYITTQQHHG
Oryza sativa    T02919     MRPTGRRSSPPVAAALALLLLVLPFFSHCASAA------ELVLQDGATGNGDEVSELMGAAEEEAAGLC--EEGNEECVERRMLRDAHLDYITTQKRNRP
Zea mays AI665040p                 MARRATVMVLAAALAVLLLASSSS-----KTAPVASAARDDPSAAAAAVTSSRDLQNDGSAAAAEGKGKEKECGANDEDECMMRRTL-AAHTDYITTQQHHG
Arabidopsis thaliana AAC32433      MANVSALLTIALLLCSTLMCTAR------PEPAISISTTAAD--------PCNMEKKIEGKLDDMHMVDENCGA--DDEDCLMRRTL-VAHTDYITTQKKKHP
Arabidopsis thaliana g3445203      MANVSALLTIALLLCSTLMCTA-----RPEPAISISITTAADP--------CNMEKKIEGKLDDMHMVDENCGA--DDEDCLMRRTL-VAHTDYITTQKKKHP
Gossypium arboreum BE052169p       MAKLASLFILTLLLVSTLSFSFAA---RSGPAFPNDSPAKTQSQGTTTD---------EIEQSEDRCEGV-GEDECLMRRTL-AAHLDYITTQKQKP
Oryza sativa    AU068854p          MAARTVAVAAALAVLLIFAASSATVAMAGRPTPTTSLDEEAAQAAAQ----------SEIGGGCKEGEGEEECLARRTL-TAHTDYITTQQHHN
Lycopersicon esculentum AW442998p  MSKANTSFFFIILLLCFALSYAS-----RPAPAFHEASLNI----------DHHQDHVRESKQVANEESCNGG-QDEECLERRNL-AAHLDYITTQNQNP
Arabidopsis thaliana g6723423      MGKFTTIFIMALL-LCSTLTVAA-----RLTPTTTTALSRE---------NSVKEIEGDKVEEESCNGI-GEEECLIRRSL-VLHTDYITTQNHKP
Glycin max AW423604p               MSKVVTLFTLALL-LSFNLIHAS-----RPNPSLNV----------VSSSHEDVAATKEIDEESCEB--GTEECLIRRTL-AAHVDYITTQKHKPKP
Mesembryanthemum.crystallinum BE131082p MSKLTTLLVIALLVCSITLINAG-----RPNPTSLI----------NEGKETEHAEMDENESCQG-LNDEECLMRRTL-VAHTDYITTQHHNP
Asparagus officinalis BAB20706     MSSKAITLLLIALLFSLSLAQAA-------RPLQPADSTK---------SVHVIPEKVHDEACEG-VGEEECLMRRTL-TAHVDYITTQDHNP
```

Figure 7.2 Example of the PSK precursors found in the genome database. Bold YITTQ sequence corresponding to the PSKα. Underlined acidic aminoacids are required for sulfation. Possible processing site for the kexin-like protease (RR) is written in italic.

trans-Golgi network (Van de Ven *et al.*, 1993). Some other proteases belong to this family, such as neuroendocrine convertases, do not contain membrane-spanning regions. These convertases are known to localize secretory glanules (Itoh *et al.*, 1996).

Rogers and co-workers recently tested the presence of such protease activity in plant cells by expressing an artificial substrate in the secretory pathway (Jiang & Rogers, 1999). They found that a processing protease activity, which recognizes a dibasic signal, is detectable in the *trans*-Golgi network where a class of vacuolar proteins is sorted to the vacuole. Interestingly, such dibasic amino acids (Arg-Arg) are found between the acidic motif and mature hormone part of the PSK precursor (Fig. 7.2). Therefore, it is possible that kexin/subtilase class of processing enzymes is present in the secretory pathway in plants. These putative processing proteases seems to be localized in an organelle that is a downstream from where the plant sulfotransferase localizes.

Although *Arabidopsis* does not appear to encode close relatives of Kex2p, furin, and neuroendocrine convertases, there are many possible genes for proteases that belong to subtilisin family. Recently, some of these subtilisin-type proteases have been found to be essential for plant development. The *SDD1* gene of *Arabidopsis*, which encodes a member of this family, is required for the control of cell lineage that leads to the formation of stomatal guard cells (Berger & Altmann, 2000). The *ALE1* gene that encodes another member of this family, is responsible for epidermal surface formation in *Arabidopsis* embryos and juvenile plants (Tanaka *et al.*, 2001). The N-terminus of the ALE1p precursor contains a long hydrophobic stretch that appears to be a type II transmembrane anchor. Many other subtiliase homologs appear to contain the usual signal peptide with 10–15 hydrophobic stretch. Because the length of the hydrophobic stretch of ALE1p is 19 amino acids, it is possible that this enzyme is localized to the Golgi apparatus or *trans*-Golgi network. Future immunocytochemical analysis along with the identification of substrates will reveal the cellular role and action of these proteases on the plant development.

7.4.2   *S1P/SKI-1 homologs in plants*

Recently, a new class of mammalian subtilisin-kexin-like convertases, called Site-1 protease/subtilisin kexin isozyme-1 (S1P/SKI-1), was identified (Sakai *et al.*, 1998; Seidah & Chretien, 1999). Their structure is closer to pyrolysin, a thermostable secretory protease in the archaea (Voorhorst *et al.*, 1996) than to mammalian pro-hormone convertases. It exhibits a specificity for cleavage at the C-terminus of the motif (R/K)-X-X-(L,T). S1P/SKI-1 is synthesized in the ER as an inactive precursor that is activated autocatalytically by the removal of an NH_2-terminal propeptide (Espenshade *et al.*, 1999). The activated S1P/SKI-1 is located in the Golgi apparatus. The substrate of this protease, a *sterol regulatory element binding protein* (SREBP) precursor, is an ER membrane protein, which has the ability to sense the level of sterol in the ER. Upon shortage of sterols in the ER, the SREBP precursor is transported to the Golgi, processed by S1P/SKI-1 and then by another protease

S2P. The resulting protein released into the cytosol by these processing reactions is the transcription factor SREBP, which is transported to the nucleus and which activates the sterol-regulated gene expression.

In *Arabidopsis* there is a gene encoding a close homolog to the S1P/SKI-1 (At5g19660). However, *Arabidopsis* does not appear to have a homolog to SREBP. Substrates of S1P/SKI-1 are not only membrane proteins e.g. the SREBP precursor, but also secretory protein precursors such as the brain-derived neurotrophic factor precursor. This neurotrophic factor precursor is processed to the mature form at the *trans*-Golgi network and/or immature secretory vesicles (Mowla *et al.*, 2001). It is therefore possible that the plant homolog of S1P/SKI-1 functions as a processing enzyme for soluble secretory or vacuolar proteins in the late secretory/vacuolar-targeting pathway. In this regard, it is interesting to note that the N-terminal sequence from the processing site of sporamin precursor when expressed in tobacco, Arg-Leu-Pro-Thr (Matsuoka *et al.*, 1990), fits the recognition motif of S1P/SKI-1. Because this processing occurs after the sorting to the vacuole (Matsuoka *et al.*, 1995), it will be interesting to analyze the intracellular localization and the substrate specificity of the tobacco homolog of S1P/SKI-1.

7.5   Concluding remarks

In this review, our knowledge on the post-translational processings of proteins in the plant Golgi apparatus has been summarized. However, unlike the enormous amounts of reports in mammalian systems, only little information is available on plants. Although plants secrete many enzymes containing several different co-factors, nothing is known about the mechanism of the introduction of such co-factors into proteins as yet. The production of useful mammalian proteins in plants is one of the important goals in plant biotechnology. Because proper modification and processing are essential for the expression of functional proteins, further understanding of protein processing in plants will be important not only to understand the plant itself but also to utilize plants as protein-producing factories.

References

Bacic, A., Du, H., Stone, B.A. & Clarke, A.E. (1996) Arabinogalactan proteins: a family of cell-surface and extracellular matrix plant proteoglycans. *Essays Biochem.*, **31**, 91–101.

Berger, D. & Altmann, T. (2000) A subtilisin-like serine protease involved in the regulation of stomatal density and distribution in *Arabidopsis thaliana*. *Genes Dev.*, **14**, 1119–1131.

Bolwell, G.P. (1986) Microsomal arabinosylation of polysaccharide and elicitor-induced carbohydrate-binding glycoprotein in French bean. *Phytochemistry*, **25**, 1807–1813.

Bonifacino, J.S. & Dell'Angelica, E.C. (1999) Molecular bases for the recognition of tyrosine-based sorting signals. *J. Cell Biol.*, **145**, 923–926.

Brandizzi, F., Frangne, N., Marc-Martin, S., Hawes, C., Neuhaus, J.-M. & Paris, N. (2002) The destination for single-pass membrane proteins is influenced markedly by the length of the hydrophobic domain. *Plant Cell*, **14**, 1077–1092.

Carraway, K.L. & Hull, S.R. (1989) O-glycosylation pathway for mucin-type glycoproteins. *Bioessays*, **10**, 117–121.

Cho, Y.-P. & Chrispeels, M.J. (1976) Serine-O-galactosyl linkages in glycopeptides from carrot cell walls. *Phytochemistry*, **15**, 165–169.

Ernst, J.F. & Prill, S.K. (2001) O-glycosylation. *Med. Mycol.*, **39**, 67–74.

Espenshade, P.J., Cheng, D., Goldstein, J.L. & Brown, M.S. (1999) Autocatalytic processing of site-1 protease removes propeptide and permits cleavage of sterol regulatory element-binding proteins. *J. Biol. Chem.*, **274**, 22795–22804.

Hanai, H., Nakayama, D., Yang, H., Matsubayashi, Y., Hirota, Y. & Sakagami, Y. (2000) Existence of a plant tyrosylprotein sulfotransferase: novel plant enzyme catalyzing tyrosine O-sulfation of preprophytosulfokine variants *in vitro*. *FEBS Lett.*, **470**, 97–101.

Hayashi, T. & Maclachlan, G. (1984) Biosynthesis of pentosyl lipids by pea membranes. *Biochem. J.*, **217**, 791–803.

Hieta, R. & Myllyharju, J. (2002) Cloning and characterization of a low-molecular weight proryl-4-hydroxylase from *Arabidopsis thaliana*. *J. Biol. Chem.*, **277**, 23965–23971.

Itoh, Y., Tanaka, S., Takekoshi, S., Itoh, J. & Osamura, R. (1996) Prohormone convertases (PC1/3 and PC2) in rat and human pancreas and islet cell tumors: subcellular immunohistochemical analysis. *Pathol. Int.*, **46**, 726–737.

Jackson, M., Nilsson, T. & Peterson, P.A. (1993) Retrieval of transmembrane proteins to the endoplasmic reticulum. *J. Cell Biol.*, **121**, 317–333.

Jiang, L. & Rogers, J.C. (1999) Functional analysis of a Golgi-localized Kex2p-like protease in tobacco suspension culture cells. *Plant J.*, **18**, 23–32.

Kehoe, J.W. & Bertozzi, C.R. (2000) Tyrosine sulfation: a modulator of extracellular protein–protein interactions. *Chem. Biol.*, **7**, R57–R61.

Kieliszewski, M.J. & Lamport, D.T.A. (1994) Extensin: repetitive motifs, functional sites, post-translational codes, and phylogeny. *Plant J.*, **5**, 157–172.

Kimura, S., Yamada, M., Igaue, I. & Mitusi, T. (1993) Structure and function of the Golgi-complex in rice cells – characterization of Golgi membrane-glycoproteins. *Plant Cell Physiol.*, **34**, 855–863.

Kishimoto, T., Watanabe, M., Mitsui, T. & Mori, H. (1999) Glutelin basic subunits have a mammalian mucin-type O-linked disaccharide side chain. *Arch. Biochem. Biophys.*, **370**, 271–277.

Lamport, D.T.A., Katona, L. & Roerig, S. (1973) Galactosylserine in Extensin. *Biochem. J.*, **133**, 125–131.

Lerouge, P., Cabanes-Macheteau, M., Rayon, C., Fischette-Laine, A.C., Gomord, V. & Faye, L. (1998) N-glycoprotein biosynthesis in plants: recent developments and future trends. *Plant Mol. Biol.*, **38**, 31–48.

Matsubayashi, Y. & Sakagami, Y. (1996) Phytosulfokine, sulfated peptides that induce the proliferation of single mesophyll cells of *Asparagus officinalis* L. *Proc. Natl. Acad. Sci. USA*, **93**, 7623–7627.

Matsubayashi, Y., Yang, H. & Sakagami, Y. (2001) Peptide signals and their receptors in higher plants. *Trends Plant Sci.*, **6**, 573–577.

Matsuoka, K. & Bednarek, S. (1998) Protein transport within the plant cell endomembrane system: an update. *Curr. Opin. Plant Biol.*, **1**, 463–469.

Matsuoka, K., Matsumoto, S., Hattori, T., Machida, Y. & Nakamura, K. (1990) Vacuolar targeting and post-translational processing of the precursor to the sweet potato tuberous root storage protein in heterologous plant cells. *J. Biol. Chem.*, **265**, 19750–19757.

Matsuoka, K., Watanabe, N. & Nakamura, K. (1995) O-glycosylation of a precursor to a sweet potato protein, sporamin, expressed in tobacco cells. *Plant J.*, **8**, 877–889.

Mitsui, T., Kimura, S. & Igaue, I. (1990) Isolation and characterization of Golgi membranes from suspension-cultured cells of rice (*Oryza sativa* L.). *Plant Cell Physiol.*, **31**, 15–25.

Mowla, S.J., Farhadi, H.F., Pareek, S., Atwal, J.K., Morris, S.J., Seidah, N.G. & Murphy, R.A. (2001) Biosynthesis and post-translational processing of the precursor to brain-derived neurotrophic factor. *J. Biol. Chem.*, **276**, 12660–12666.

Munro, S. (1998) Localization of proteins to the Golgi apparatus. *Trends Cell Biol.*, **8**, 11–15.

Niehrs, C., Kraft, M., Lee, R.W. & Huttner, W.B. (1990) Analysis of the substrate specificity of tyrosylprotein sulfotransferase using synthetic peptides. *J. Biol. Chem.*, **265**, 8525–8532.

Ouyang, Y., Lane, W.S. & Moore, K.L. (1998) Tyrosylprotein sulfotransferase: purification and molecular cloning of an enzyme that catalyzes tyrosine O-sulfation, a common posttranslational modification of eukaryotic proteins. *Proc. Natl. Acad. Sci. USA*, **95**, 2896–2901.

Ruggiero, F., Exposito, J.Y., Bournat, P., Gruber, V., Perret, S., Comte, J., Olagnier, B., Garrone, R. & Theisen, M. (2000) Triple helix assembly and processing of human collagen produced in transgenic tobacco plants. *FEBS Lett.*, **469**, 132–136.

Sakai, J., Rawson, R.B., Espenshade, P.J., Cheng, D., Seegmiller, A.C., Goldstein, J.L. & Brown, M.S. (1998) Molecular identification of the sterol-regulated luminal protease that cleaves SREBPs and controls lipid composition of animal cells. *Mol. Cell*, **2**, 505–514.

Schaffer, C., Graninger, M. & Messner, P. (2001) Prokaryotic glycosylation. *Proteomics*, **1**, 248–261.

Seidah, N.G. & Chretien, M. (1999) Proprotein and prohormone convertases: a family of subtilases generating diverse bioactive polypeptides. *Brain Res.*, **848**, 45–62.

Stahelin, L.A. & Driouich, A. (1997) Brefeldin A effects in plants: are different Golgi responses caused by different sites of action? *Plant Physiol.*, **114**, 401–403.

Sticher, L., Hofsteenge, J., Milani, A., Neuhaus, J.-M. & Meins, F. Jr., (1992) Vacuolar chitinases of tobacco: a new class of hydroxyproline-containing proteins. *Science*, **257**, 655–657.

Tanaka, H., Onouchi, H., Kondo, M., Hara-Nishimura, I., Nishimura, M., Machida, C. & Machida, Y. (2001) A subtilisin-like serine protease is required for epidermal surface formation in *Arabidopsis* embryos and juvenile plants. *Development*, **128**, 4681–4689.

Tanaka, M., Sato, K. & Uchida, T. (1981) Plant prolyl hydroxylase recognizes poly(L-proline) II helix. *J. Biol. Chem.*, **256**, 11397–11400.

Uchida, K. (2000) Role of reactive aldehyde in cardiovascular diseases. *Free Radic. Biol. Med.*, **28**, 1685–1696.

Van de Ven, W.J., Roebroek, A.J. & Van Duijnhoven, H.L. (1993) Structure and function of eukaryotic proprotein processing enzymes of the subtilisin family of serine proteases. *Crit. Rev. Oncog.*, **4**, 115–136.

Van den Steen, P., Rudd, P.M., Dwek, R.A. & Opdenakker, G. (1998) Concepts and principles of O-linked glycosylation. *Crit. Rev. Biochem. Mol. Biol.*, **33**, 151–208.

Voorhorst, W.G., Eggen, R.I., Geerling, A.C., Platteeuw, C., Siezen, R.J. & Vos, W.M. (1996) Isolation and characterization of the hyperthermostable serine protease, pyrolysin, and its gene from the hyperthermophilic archaeon *Pyrococcus furiosus*. *J. Biol. Chem.*, **271**, 20426–20431.

8 Sorting of lytic enzymes in the plant Golgi apparatus

Liwen Jiang and John C. Rogers

8.1 Introduction

All eukaryotic cells contain an endomembrane system for the secretory pathway that is comprised of functionally distinct, membrane-bounded organelles. The central components are endoplasmic reticulum (ER) and Golgi apparatus, from which membranes and proteins are directed to vacuoles or lysosomes via intermediate prevacuolar compartments/endosomes, or to the plasma membrane. Because each organelle is defined by its limiting membrane, transport of proteins between the organelles occurs in vesicles with a unique vesicle for each interval between one organelle and another. Secretory proteins enter the pathway in the ER. After folding into the proper three-dimensional conformation in the ER lumen, proteins destined for other organelles in the secretory pathway are exported to the Golgi apparatus (Schekman & Orci, 1996). The current model of protein sorting in the secretory pathway considers the Golgi apparatus, in particular the late Golgi or *trans*-Golgi network (TGN), to be the central organelle where the destination of soluble proteins is determined. Some proteins lacking specific targeting information follow the default pathway and enter vesicles that fuse with the plasma membrane to release their contents to the cell exterior. Relatively, little is known of the mechanisms by which proteins are directed into these secretory vesicles. In contrast, specific structural features on a protein, *targeting determinants*, may be recognized in the Golgi apparatus by mechanisms that direct the protein into another pathway.

In plant cells, the sorting process is complex because two separate pathways diverge from the Golgi apparatus to two different vacuole destinations (Jiang & Rogers, 1999c; Neuhaus & Rogers, 1998; Okita & Rogers, 1996). One pathway carries proteins to the protein storage vacuole (PSV) via dense vesicles or their equivalents (see Chapter 9). Targeting determinants that have been identified for proteins sorted into the PSV pathway are contained within C-terminal propeptides (Matsuoka & Neuhaus, 1999), and aggregation that begins in the *cis*-Golgi appears to be part of the sorting process (Hillmer *et al.*, 2001). Results from previous studies, where overexpression of proteins destined for the PSV pathway resulted in their secretion (Frigerio *et al.*, 1998; Neuhaus *et al.*, 1994), indicated the possibility that receptors might participate in the sorting process. Recent studies have demonstrated that RMR proteins, integral membrane proteins that traffic from Golgi to

PSVs (Jiang *et al.*, 2000), bind specifically to C-terminal propeptides in a manner that correlates well with the ability of the propeptides to function in sorting to the PSV (see Section 8.5, below). Thus, RMR proteins may be a type of sorting receptor specific for the PSV pathway. It was previously thought that mechanisms for sorting to PSVs were unique to plant cells, but recent results from studies of traffic of procathepsin L are consistent with a similar pathway in mammalian cells (Ahn *et al.*, 2002; Yeyeodu *et al.*, 2000).

The second pathway from Golgi to a vacuole in plant cells shares features in common with other eukaryotes. Animal, yeast, and plant cells have in common a lysosome/vacuole within the secretory pathway that maintains an acidic pH and functions as a degradative compartment; to distinguish this compartment from the PSV, it is identified as the lytic vacuole (Okita & Rogers, 1996). A common mechanism is responsible for delivering soluble proteins to the lysosome/lytic vacuole in cells of these three organisms. An integral membrane receptor protein binds cargo proteins at the relatively neutral pH in the TGN and causes them to be selected into clathrin-coated vesicles for delivery to an endosome/prevacuolar compartment; there the presence of an environment with acidic pH causes the release of cargo proteins from the receptor, which is then recycled back to the TGN. Subsequent fusion between the endosome/prevacuolar compartment and the lysosome/lytic vacuole will deliver the cargo proteins to the lysosome/lytic vacuole (Robinson & Hinz, 1997).

8.2 General features of the lytic vacuole pathway

8.2.1 Identification of organelles in the pathway

When sections of plant cells are viewed under transmission electron microscope, numerous organelles defined by a single membrane and an apparently empty lumen will be observed. The identity of these structures usually cannot be determined from their morphology. Thus, on a morphological basis, it is rather difficult to define a plant cell vacuole precisely, in a way that distinguishes it from other membrane-bounded structures such as prevacuolar compartments/endosomes. Size is helpful only in that, if it is large (i.e. >4 μm) it is likely to be a vacuole, although ring-like organelles that are endosomes in *Arabidopsis* protoplasts, as judged by the presence of the Rab protein Ara6 (Ueda *et al.*, 2001), may be several microns in diameter. For that reason we as well as others have worked to identify proteins that are present on vacuoles with specific functions (Jauh *et al.*, 1999; Jiang *et al.*, 2001; Jiang & Rogers, 2001; Neuhaus & Rogers, 1998; Paris *et al.*, 1996), reasoning that protein markers would provide biochemical definitions of function.

One relatively reliable marker for vacuoles is the presence of tonoplast intrinsic proteins (TIPs) (Hara-Nishimura & Maeshima, 2000; Maurel, 1997) in their tonoplasts. In general, different TIP isoforms are associated with specific vacuole functions (Jauh *et al.*, 1999; Neuhaus & Rogers, 1998; Paris *et al.*, 1996) (Fig. 8.1).

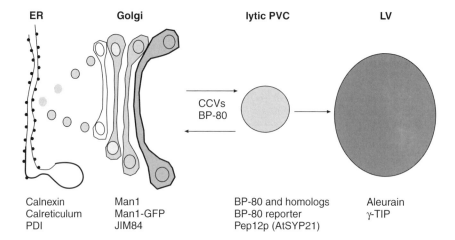

Figure 8.1 The lytic pathway and its components in plant cells. Shown is an oversimplified representation of protein trafficking along the lytic pathway from the endoplasmic reticulum (ER) to Golgi to lytic prevacuolar compartment (PVC) to lytic vacuole (LV). The established markers that have been used to identify these individual organelles within the lytic pathway are indicated below each organelle. Traffic between Golgi and PVC is mediated by clathrin-coated vesicles (CCVs) and involves a vacuolar sorting receptor BP-80; missorted proteins can be recruited back to Golgi apparatus from the PVC by a process that might require the cytoplasmic tail of the receptor (Jiang & Rogers, 1998). LV is the final destination of target proteins that contain a barley cysteine protease aleurain and a tonoplast intrinsic protein γ-TIP (Holwerda *et al.*, 1990; Jauh *et al.*, 1999; Paris *et al.*, 1996). Man1 (mannosidase I) and Man1-GFP fusion are *cis*-Golgi markers (Nebenführ *et al.*, 1999; Li *et al.*, 2002); JIM84 is a monoclonal antibody directed against Lewis α-containing *N*-glycans locating to *trans*-Golgi (Fitchette *et al.*, 1999; Satiat-Jeunemaitre & Hawes, 1992); BP-80 and homologs (AtELP and VSR) are markers for PVCs in confocal immunofluorescence (Li *et al.*, 2002) that are colocalized with AtPep12p.

A potential advantage of using TIP isoforms as markers for vacuoles is the fact that they are present in organelles that range from very large to very small; the latter were interpreted to be vacuole precursors (Jauh *et al.*, 1999). A second advantage is the fact that α-, γ-, and δ-TIPs were not detected in prevacuolar organelles for either lytic vacuoles (Jiang & Rogers, 1998; Paris *et al.*, 1997), or for the compound-type of PSV in certain root tip cells and in seed embryos (Jiang *et al.*, 2000, 2001). A potential disadvantage is the lack of information regarding their possible distribution in endosomes. In unpublished results from confocal immunofluorescence experiments, we have found that labeling for vacuolar pyrophosphatase (V-PPase) (Maeshima & Yoshida, 1989; Sarafian *et al.*, 1992) always colocalized with and was limited to membranes that also labeled for α-, γ-, or δ-TIP in barley and pea root tip cells (Jauh *et al.*, 1999). Thus, the presence of V-PPase and a TIP isoform may prove to be a reliable way to define a vacuole membrane (with the exception of the PSV membrane in seed embryos (Jiang *et al.*, 2001)), but the utility of this definition remains to be tested by other laboratories.

Figure 8.1 presents our current understanding of the organization of the pathway to the lytic vacuole, and identifies some of the established markers for individual organelles in this pathway. For our studies, a lytic vacuole is defined as an organelle whose membrane is marked by the presence of V-PPase and γ-TIP, and we have shown that γ-TIP is associated with vacuoles containing aleurain, a cysteine protease that also serves to mark lytic vacuoles (Jauh *et al.*, 1999; Paris *et al.*, 1996). Some lytic vacuoles can be induced to store vegetative storage proteins whose accumulation and mobilization are regulated by metabolic and environmental cues. Lytic vacuoles that can accumulate vegetative storage proteins are marked by the presence of δ- plus γ-TIPs, or by δ-TIP alone (Jauh *et al.*, 1998, 1999). To dissect the lytic vacuole pathway, it has been important to have available markers for the different organelles within the secretory endomembrane system. Antibodies to resident ER proteins, such as BiP, calnexin, calreticulin, and protein disulfide isomerase have served to identify ER (Staehelin, 1997). Antibody localization of the Golgi apparatus has been more problematic, but the monoclonal antibody JIM84, which recognizes a carbohydrate epitope that is predominantly localized to the Golgi (Fitchette *et al.*, 1999; Satiat-Jeunemaitre & Hawes, 1992), has been very useful, and recently an antibody to Golgi mannosidase I (Nebenführ *et al.*, 1999) has been developed (Li *et al.*, 2002). An alternative approach to visualizing ER and Golgi in studies of protein traffic has been to express fusions between green fluorescent protein or red fluorescent protein and resident proteins of those organelles (Boevink *et al.*, 1998; Kim *et al.*, 2001; Nebenführ *et al.*, 1999; Wee *et al.*, 1998). Between the Golgi apparatus and the lytic vacuole is a prevacuolar compartment; this organelle is discussed in detail below.

8.2.2   Soluble proteins sorted into the lytic pathway and their sorting determinants

Generally, three types of vacuolar targeting determinants have been described in plant proteins (Neuhaus & Rogers, 1998). C-terminal and internal sorting determinants in general function in the protein storage vacuole pathway and are considered in Chapter 9.

The sequence-specific vacuolar sorting determinant was defined from studies using two different proteins: sweet potato sporamin and barley aleurain. Sporamin is an abundant vegetative storage protein in sweet potato tubers (Maeshima *et al.*, 1985) that functions as a serine protease inhibitor (Yeh *et al.*, 1997). Prosporamin is transported through the Golgi apparatus where it acquires *O*-linked glycans (Matsuoka *et al.*, 1995), and then to a vacuole where the precursor is processed into mature sporamin by removal of its 16 amino acid N-terminal propeptide (Matsuoka *et al.*, 1990). The propeptide is essential and sufficient for vacuolar targeting (Matsuoka & Nakamura, 1991), and elegant analyses of the sequence specificity of targeting identified the Ile and Leu residues in the motif, NPIRL, as particularly important (Matsuoka & Nakamura, 1999; Nakamura *et al.*, 1993).

A second example of sequence-specific sorting came from studies of barley aleurain, a cysteine protease that is synthesized as a proenzyme (proaleurain) and transported to a post-Golgi compartment where it is processed to the mature form;

proteolytic processing requires an acidic pH (Holwerda *et al.*, 1990). Aleurain was localized by electron microscopic immunocytochemistry (immunoEM) to a vacuole that was morphologically and physically distinct from protein storage vacuoles in barley aleurone cells (Holwerda *et al.*, 1990). The N-terminal region of proaleurain is essential for vacuolar targeting in tobacco suspension culture cells, and the replacement of the motif NPIRP greatly reduced proper targeting (Holwerda *et al.*, 1992). Thus the four amino acids, NPIR, played a major role in vacuolar targeting of both proteins in tobacco suspension culture cells. The presence of an essential conserved motif from two different proteins indicated that the sequence might be recognized by a receptor that would direct the proteins into the lytic vacuole pathway. The vacuolar targeting functions of the prosporamin and proaleurain sequences have been subsequently used by attaching the sequences to green fluorescent protein, and then demonstrating that the chimeric proteins, when expressed transiently or in transgenic plants, localized to vacuoles (Di Sansebastiano *et al.*, 2001; Kim *et al.*, 2001). The studies of Di Sansebastiano *et al.* (1998, 2001) indicated that the proaleurain sequence directed the chimeric protein to an acidified vacuole that was separate from a vacuole with neutral pH having the characteristics of a protein storage vacuole. The reader should consult an excellent review for other sequences related to the prosporamin and proaleurain vacuolar sorting determinants (Matsuoka & Neuhaus, 1999).

The mechanisms by which the barley aspartic proteinase, phytepsin (Runeberg-Roos *et al.*, 1991), is targeted to vacuoles are much more complex. Results from immunofluorescence studies indicated that the protein was present in both lytic vacuoles and protein storage vacuoles (Paris *et al.*, 1996). The proenzyme undergoes two processing events to yield the mature enzyme, and the removal of an N-terminal propeptide and of a plant-specific, internal saposin domain (Guruprasad *et al.*, 1994). Interestingly, the N-terminal propeptide contains the sequence NPLR immediately preceding the site of cleavage from the mature enzyme (Runeberg-Roos *et al.*, 1991). Its similarity to the functional NPIR motif in proaleurain and prosporamin raised the possibility that it might participate in sorting the enzyme into the lytic vacuole pathway, while the saposin domain was suggested as a possible sorting determinant for the protein storage vacuole pathway (Paris *et al.*, 1996). Recently, however, results from studies where various deletion constructs were expressed in tobacco cells demonstrated that deletion of the saposin domain resulted in essentially complete secretion of the protein from cells (Törmäkangas *et al.*, 2001). This observation might suggest that the NPLR motif is not functionally significant. However, the complexities of the system limit our ability to interpret this result. The wild type enzyme is exported from the ER in a COPII-dependent manner and acquired complex modifications to *N*-glycans in the Golgi apparatus. However, surprisingly, the mutant form lacking the saposin domain was secreted by a pathway that appeared not to involve COPII (Törmäkangas *et al.*, 2001), and no evidence for Golgi transit was presented. Thus, it is possible that the saposin deletion mutant exited the ER by a pathway that bypassed the Golgi apparatus. If so, a sorting mechanism to recognize the NPLR motif in the Golgi would not have had access to the protein. We must conclude that the mechanism by which phytepsin is directed to lytic vacuoles has not yet been identified, and that vacuolar sorting mediated by the saposin domain probably occurs in the Golgi complex by mechanisms that are yet to be elucidated.

8.3 Evidence for receptor-mediated sorting: the roles of VSR proteins in plant cells

8.3.1 Identification of pea BP-80 and homologs from other plants

Since the NPIR motif was of major importance in the vacuolar sorting determinants of both proaleurain and prosporamin, it was hypothesized that the motif was recognized by a transmembrane sorting receptor within the Golgi apparatus; binding to the receptor would result in the soluble proteins being packaged into, presumably (based on parallels with mammalian and yeast cells), clathrin-coated vesicles (CCVs) for delivery to the lytic vacuole. This hypothesis was tested using a biochemical approach. An affinity column, to which a synthetic peptide representing the proaleurain vacuolar sorting determinant was coupled, was used to select proteins from lysates of CCV membranes, where the CCVs were purified from developing pea seeds (Kirsch *et al.*, 1994). Using such an approach, an 80 kD binding protein, termed BP-80, was purified. BP-80 is a type I integral membrane protein that contains a single transmembrane domain and a short cytoplasmic tail. It belongs to a gene family of proteins termed vacuolar sorting receptor (VSR) proteins (Kirsch *et al.*, 1994; Paris *et al.*, 1996, 1997). Several VSR homologs have been cloned from other plant species, including *Arabidopsis* and pumpkin (Ahmed *et al.*, 1997b; Shimada *et al.*, 1997).

Results from protease digestion assays on CCV membranes indicated that the ligand-binding domain was within the lumen of the vesicles (Kirsch *et al.*, 1994), and, from its cDNA sequence, BP-80 was predicted to be a type I integral membrane protein. Consistent with this prediction, when truncated BP-80 lacking transmembrane domain and cytoplasmic tail sequences was expressed in transient tobacco suspension culture cells, it was secreted outside of the cells (Paris *et al.*, 1997).

The short cytoplasmic tail of BP-80 and all homologs characterized to date contain a conserved tyrosine motif, YMPL, that is thought to be involved in traffic of the proteins in CCVs (Ahmed *et al.*, 1997b; Paris *et al.*, 1997; Shimada *et al.*, 1997). Pea BP-80 is abundant in highly purified CCVs (Hinz *et al.*, 1999). The cytoplasmic tail sequence of an *Arabidopsis* homolog of BP-80 was found to interact *in vitro* with mammalian AP-1 clathrin adaptor protein complex that participates in the formation of CCVs at a location previously thought to be the TGN (Sanderfoot *et al.*, 1998). These results are difficult to interpret in light of recent studies in mammalian cells and yeast which demonstrate clearly that formation of CCVs at the TGN is mediated by GGA (Golgi-localized, γ-ear-containing, ARF-binding protein) proteins that interact specifically with motifs in the cytoplasmic tails of sorting receptors (Tooze, 2001). It is likely that AP-1 participates in CCV formation at other locations, perhaps to recycle receptors back from late endosomes. AP-1 adaptor proteins are well characterized in plants (Robinson *et al*., 1998), while plant GGA proteins have not yet been identified. A role for the cytoplasmic tail of BP-80 in recycling from a prevacuolar compartment was indicated by the finding that the full length protein had a substantially longer half-life than a truncated form lacking only its cytoplasmic tail, when each was transiently expressed in tobacco suspension culture protoplasts (Jiang & Rogers, 1998).

8.3.2 Ligand specificity of binding by BP-80 and homologs

An *in vitro* binding assay using BP-80 from solubilized membranes and [^{125}I]-labeled proaleurain peptide sequence SSSFADSNPIRPVTDRAASTYC, was developed to study the affinity and specificity of the interaction (Kirsch *et al*., 1994). The k_d for the proaleurain peptide was estimated to be 37 nM. A peptide representing the prosporamin vacuolar sorting determinant, SRFNPIRLPT, competed weakly for binding, while a peptide representing the mutated inactive prosporamin determinant, SRFNPGRLPT, and a peptide representing the barley lectin C-terminal sorting determinant for the protein storage vacuole pathway, VFAEAIAANSTLVAE, showed no competition for binding. Thus, the ability of the small number of different peptides that were tested to interact with BP-80 seemed to parallel their activity as sorting determinants for the lytic vacuole pathway. A second important result that came from these studies was an estimate of the pH dependence of ligand binding (Kirsch *et al*., 1994). The BP-80–proaleurain peptide interaction had a narrow pH optimum of 6.0–6.5, and binding efficiency rapidly dropped off above and below that range such that it was only about half maximal at either pH 5 or pH 7.5; binding was essentially completely abolished at pH 4.0. This pattern would be consistent with what would be expected for a sorting receptor that would bind proteins at the mildly acidic pH of the *trans*-Golgi and release them into the acidic environment of a prevacuolar compartment (Kirsch *et al*., 1994).

A second approach utilized affinity columns coupled with different synthetic peptides, where their ability to select pea BP-80 from pea membrane lysates was tested (Kirsch *et al*., 1996). These studies demonstrated that BP-80 was able to bind to the vacuolar sorting determinant of prosporamin and to a peptide representing the C-terminus of Brazil nut 2S albumin but not to the C-terminal propeptide of barley lectin. Similar studies used lysates of *Arabidopsis* microsomes to study binding of VSR proteins that were recognized by antiserum raised to the lumenal domain of an *Arabidopsis* BP-80 homolog (Ahmed *et al*., 2000). There are seven genes for this protein family in *Arabidopsis* (Hadlington & Denecke, 2000), and the antiserum presumably would have detected any or all of the family members. Binding at a neutral pH was observed for peptides representing the native prosporamin and proaleurain sequences, with bound protein eluted at pH 4. No binding was observed to the prosporamin and proaleurain sequences where NPIR was mutated to NPGR, and no binding was observed for the barley lectin and tobacco chitinase C-terminal propeptides (Ahmed *et al.*, 2000). Interestingly, in each instance binding to the affinity column was incomplete, with a more than half of the immunologically detected VSR proteins remaining in the flow through fraction (Ahmed *et al.*, 2000). This observation might indicate that different VSR protein family members have different ligand binding specificities, although technical reasons for inefficient binding cannot be excluded.

A particularly interesting investigation into VSR protein family members has evolved from isolation of two of the proteins from precursor accumulating (PAC) vesicles that were purified from developing pumpkin cotyledons (Shimada *et al.*, 1997). PAC vesicles represent prevacuolar compartments for protein storage vacuoles in seed embryos (Hara-Nishimura *et al.*, 1998). These organelles receive protein traffic directly from the ER, as well as from the Golgi complex (Hara-Nishimura *et al.*, 1998), and then deliver their contents to the developing PSVs. Storage proteins are the most abundant type of proteins in PAC vesicles (Hara-Nishimura *et al.*, 1998; Shimada *et al.*, 1997), and the authors' hypothesis was that the VSR proteins, termed PV82 and PV72, represent receptors responsible for sorting storage proteins into the PSV pathway. This hypothesis would seem to contradict the observations that PV82 and PV72 were purified from PAC vesicle membrane lysates by chromatography on a proaleurain peptide affinity column (Shimada *et al.*, 1997), and the sequences of PV72 and the *Arabidopsis* BP-80 homolog VSR$_{At1}$ (Paris *et al.*, 1997), otherwise known as AtELP (Ahmed *et al.*, 1997b, 2000) or atbp80b (Hadlington & Denecke, 2000), are essentially identical.

The abilities of PV72 and PV82 to bind to synthetic peptides representing portions of the pumpkin storage protein 2S albumin were studied using an affinity chromatography assay (Shimada *et al.*, 1997). Both proteins bound efficiently to peptide 2S-I, sequence SRDVLQMRGIENPWRREG, but only weakly to peptide 2S-C, sequence KARNLPSMCGIRPQRCDF. The motif within the 2S-I peptide, RREG, necessary for efficient binding, was identified by deletion and mutation analyses. Similar studies indicated that NLPS within the 2S-C sequence was functionally important (Shimada *et al.*, 1997). No distinction between PV72 and PV82 in their ability to interact with these sequences was noted.

The abilities of PV72 to interact with ligands were studied further using a truncated form of the protein representing the full length lumenal domain but lacking transmembrane domain and cytoplasmic tail. This lumenal domain was expressed in insect cells and was purified (Watanabe *et al.*, 2002); its structure would be essentially the same as shown for tBP-80 in Fig. 8.2. The method of expression ensured that the truncated form had undergone proper folding and intramolecular disulfide bond formation in the ER. Binding of the purified proteins to synthetic peptides was assayed by affinity chromatography and by surface plasmon resonance. In the latter approach, peptides were immobilized on chips and the protein was presented in solution. The lumenal domain protein bound the 2S-I peptide with a K_d value of 0.2 μM, but binding was found to be Ca^{2+}-dependent and pH-independent. Thus binding was stable at pH 4.0 in the presence of Ca^{2+}, while the ligand was released at either pH 7.0 or 4.0 by application of EGTA. Further truncation

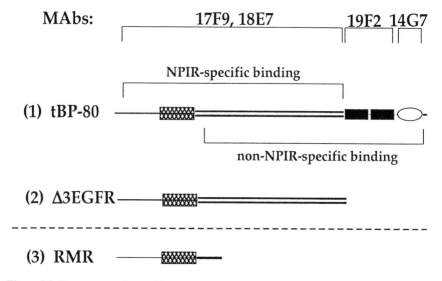

Figure 8.2 The structure of recombinant proteins used in ligand binding studies. Shown are linear diagrams drawn approximately to scale for (1) the lumenal domain of BP-80, (2) a truncation lacking the three EGF repeats, and (3) the lumenal domain of RMR protein JR702. The cross-hatched box represents the domain that is conserved in both VSR and RMR proteins. The horizontal stripes represent the central unique domain of BP-80, and the EGF repeats are represented by two black rectangles and a single open oval to indicate the latter is likely to co-ordinate Ca^{2+}. Regions in BP-80 that have been associated with NPIR-specific and non-NPIR-specific ligand binding are indicated by brackets. MAbs refers to monoclonal antibodies that have their epitopes on regions of BP-80 as indicated.

of the lumenal domain, to remove one, two, or three of the epidermal growth factor (EGF) repeats, greatly increased the K_d to 2–4 µM. Thus, the EGF repeats, in particular the third, most C-terminal repeat, appeared to mediate the Ca^{2+}-dependent binding, and all repeats played a crucial role in interaction with the ligands (Watanabe et al., 2002).

Based on these studies, the authors proposed that "Ca^{2+} regulates the vacuolar sorting mechanism in higher plants" (Watanabe et al., 2002). They point out that previous studies of interactions of VSR proteins with peptide ligands in affinity chromatography assays (Ahmed et al., 2000; Kirsch et al., 1994, 1996) utilized Ca^{2+}-containing buffers, and that elution from these affinity resins at pH 4.0 utilized buffers containing EGTA. Thus elution could be explained by chelation of Ca^{2+} (Watanabe et al., 2002). This explanation, however, would not apply to the soluble binding assay developed for BP-80, where all buffers contained 1 mM Ca^{2+} but the buffer pH was varied. Those experiments documented a clear pH-dependence on binding of the proaleurain peptide (Kirsch et al., 1994). The results also appear to be at odds with similar studies utilizing truncated recombinant forms of BP-80, where deletion of all three EGF repeats generated a protein that retained the ability to bind the proaleurain peptide in a manner dependent upon the NPIR motif, and with an affinity that was indistinguishable from that of the full length lumenal domain (Cao *et al.*, 2000).

It is possible that the discrepancy between the two studies may be explained by the nature of the ligands that were studied and the way in which they were presented. The proaleurain peptide, when studied in solution binding assays, appears to interact with BP-80 in two separate ways, one that is NPIR-specific, and the other that does not depend on NPIR (Cao *et al.*, 2000) (see Section 8.5, below). In contrast, assays using the immobilized 2S-I peptide (Watanabe *et al.*, 2002) probably studied binding by PV72 that did not involve the NPIR-specific site.

8.3.3   Studies to address functions of BP-80 and homologs in plant cells

The evidence that BP-80 and homologs, the VSR proteins, function as sorting receptors in plant cells can be described as *guilt by association*, meaning that all of the experimental results were consistent with that function but none were conclusive. However, a direct test of the hypothesis has recently been successfully accomplished in a heterologous system, *Saccharomyces cerevisiae* (see Section 8.4 below). There are two related questions with respect to function, namely do the VSR proteins function as sorting receptors and, if so, which pathway do they follow?

Strong evidence that BP-80 (Cao *et al.*, 2000; Kirsch *et al.*, 1994, 1996) and the *Arabidopsis* VSR proteins (Ahmed *et al.*, 2000) function as sorting receptors for the lytic vacuole pathway was derived from the peptide-binding studies, summarized above. These studies demonstrated that the proteins bound to peptides that contained functionally defined vacuolar sorting determinants for the lytic pathway, but did not bind when the central NPIR in the sorting determinants was mutated to a non-functional NPGR. Additionally, the proteins did not bind to peptides representing the C-terminal vacuolar sorting determinants from tobacco chitinase and barley lectin that direct proteins into the protein storage vacuole pathway.

In a different approach, the distribution of BP-80 and pea storage proteins was determined in highly purified CCVs, specific for lytic vacuole pathway traffic, and in dense vesicles, specific for the protein storage vacuole pathway. Both vesicle types were prepared from developing pea cotyledons. The purified CCVs contained abundant BP-80 but little of the pea storage proteins, vicilin and legumin (Hohl *et al.*, 1996). In contrast, the storage proteins were highly enriched in purified dense vesicles which contained no detectable BP-80 (Hinz *et al.*, 1999). Thus BP-80 was tightly associated with traffic in CCVs. An important point was raised from observations on the developing pea cotyledon system that dense vesicles derived from the *trans*-Golgi frequently appeared to have clathrin buds forming from them (Robinson *et al.*, 1997; Robinson & Hinz, 1999). This phenomenon appeared to parallel the known mechanisms in mammalian cells whereby missorted lysosomal proteins were retrieved from regulated secretory vesicles by budding of the mannose 6-phosphate receptor into CCVs that formed on the surface of the secretory vesicles (Robinson *et al.*, 1998).

An elegant distinction between sorting of pea storage proteins and the function of BP-80 was defined by studies using quantitative immunoEM to analyze the distribution of the proteins in Golgi stacks (Hillmer *et al.*, 2001). The storage proteins were observed to form aggregates in the *cis*-Golgi where intense labeling with anti-storage protein antibodies was measured, with little labeling in the *trans*-Golgi. In contrast, the distribution of BP-80 was the reverse where maximal labeling occurred in the *trans*-Golgi and little labeling was observed in the *cis*-Golgi (Hillmer *et al.*, 2001). A similar approach utilized transgenic *Arabidopsis* plants expressing barley lectin as a marker for the PSV pathway, and prosporamin as a marker for the lytic vacuole pathway. Labeling patterns within Golgi stacks obtained with antiserum to the *Arabidopsis* VSR proteins predominantly over-lapped labeling for prosporamin, but showed no co-localization with barley lectin (Ahmed *et al.*, 2000).

In contrast to these results, two other studies were thought to support the concept that VSR proteins from pumpkin (Shimada *et al.*, 1997) and tobacco (Miller *et al.*, 1999) functioned in the protein storage vacuole pathway. In the first study, as noted above, the pumpkin BP-80 homologs PV72 and PV82 were found to be the abundant proteins in purified PAC vesicles. PAC vesicles represent pre-vacuolar organelles for PSVs and are filled with storage proteins (Hara-Nishimura *et al.*, 1998). Thus the circumstances suggested that PV72 and PV82 were present in PAC vesicles because they were storage protein sorting receptors (Shimada *et al.*, 1997). One serious limitation to this hypothesis, however, is that binding of either proteins to the 2S albumin precursor protein in its native conformation has not been demonstrated.

This argument assumes that PAC vesicles have only one function, and that is to package storage proteins. However, the original characterization of PAC vesicles demonstrated that they received proteins both directly from the ER as well as from the Golgi apparatus, and that they contained not only an electron dense central core, where storage proteins were present but also a peripheral lucent zone that appeared to be comprised of internal vesicles and contained glycoproteins (Hara-Nishimura *et al.*, 1998). Thus they are multivesicular bodies, and such organelles are best characterized as prevacuolar compartments/endosomes where vesicular pathways both deliver and retrieve proteins (Lemmon & Traub, 2000). Therefore, it is possible that PV72 and PV82 are present in PAC vesicles because they have a role in recycling missorted proteins back to the Golgi apparatus.

A second possible explanation for the presence of the BP-80 homologs in PAC vesicles developed out of the studies on the organization of PSVs (Jiang *et al.*, 2000, 2001). These studies demonstrated that PSVs are themselves multivesicular bodies, where the storage compartment containing storage proteins surrounds internal, membrane-defined vesicles. The membrane of these internal vesicles contains markers, γ-TIP and V-PPase, that are characteristic of lytic vacuoles, and within the vesicles are proteins that specifically traffic to lytic vacuoles (Jiang *et al.*, 2001). Additionally, in transgenic tobacco plants, the traffic of integral membrane reporter proteins that are specific for either the storage compartment or the lytic vacuole was studied. One such reporter protein, termed Re-F-B-B (see Section 8.6, below), traffics specifically to the lytic vacuole prevacuolar compartment where it is proteolytically processed to a smaller form. In transgenic plants, Re-F-B-B was directed to the internal vesicles of the PSV (Jiang *et al.*, 2001). In contrast, a second chimeric reporter protein, termed Re-F-B-α, traffics directly from the ER to the PSV or its equivalent where it remains in its full length form (Jiang & Rogers, 1998). In transgenic plants, Re-F-B-α was directed to the PSV storage compartment (Jiang *et al.*, 2000). The relevant point here is that, in developing seeds and in certain root tip cells, both the reporter proteins were localized to organelles that probably represent tobacco PAC vesicles, and comparisons to an internal marker for the PAC vesicles indicated that essentially all would contain a given reporter protein. However, in the PAC vesicles Re-F-B-α remained in its full length, unprocessed form (Jiang *et al.*, 2000), while Re-F-B-B had been proteolytically processed in a manner similar to what occurs in the lytic vacuole prevacuolar compartment (Jiang *et al.*, 2001). In other words, the results indicated that a given PAC vesicle could contain both reporter proteins, but one would stay intact while the other would be proteolytically processed. Thus, it remains a possibility that PAC vesicles have two compartments, similar to what occurs in PSVs, and that they transport to one, the storage compartment, proteins that will be protected, while by a separate pathway they transport proteins to the other, the lytic compartment, proteins that will be exposed to a degradative environment (Jiang & Rogers, 2001). The topology of membrane organization that would permit such an arrangement remains to be elucidated. However, this hypothesis would allow PV72 and PV82 to be present in pumpkin PAC vesicles because they were delivering lytic enzymes to one of the two compartments, separate from the compartment containing storage proteins.

The second set of experiments that argued for a role by VSR proteins in sorting to the PSV came from the studies of a 46 kDa protease inhibitor precursor, Na-PI, which is expressed in tobacco flower stigmas. The mature forms of the inhibitor represent 6 kDa products that accumulate in vacuoles (Miller *et al.*, 1999). Targeting to vacuoles is at least partially mediated by a C-terminal propeptide because deletion of the propeptide resulted in secretion of some of the proteins when expressed in tobacco BY-2 cells, and therefore the protein's destination was thought to be a PSV. Crosslinking experiments followed by immunoprecipitation with anti-BP-80 monoclonal antibodies demonstrated that the Na-PI precursor interacted with tobacco BP-80 homologs. The authors concluded that the tobacco VSR proteins bound the C-terminal propeptide and thereby directed it to a vacuole (Miller *et al.*, 1999). In response (Jiang & Rogers, 1999b), it was pointed out that the Na-PI by definition is a vegetative storage protein because it accumulated in vacuoles in vegetative tissues. In this location, it is exposed to proteases that cut it into small fragments, consistent with the concept that vegetative storage proteins are predominantly accumulated in vacuoles associated with the lytic vacuole pathway. The experiment where the C-terminal propeptide was deleted from the precursor could not be interpreted because the fraction of the precursor that was probably directed to the lytic vacuole and degraded to small fragments was not measured. Very likely the tobacco BP-80 homologs would function in directing a substantial portion, perhaps most, of the precursor to the lytic vacuole pathway and would have no association with targeting mediated by the C-terminal propeptide. The presence of the C-terminal propeptide and its proposed function in sorting to the PSV pathway might simply represent a second determinant that was independent of VSR protein functions (Jiang & Rogers, 1999b).

8.4   BP-80 as a sorting receptor: proof of function in yeast

Although substantial evidence, discussed above, supports the concept that the VSR proteins function as sorting receptors, it has been difficult to prove this functional role *in vivo*. However, in recent studies the function of BP-80 as a vacuolar sorting receptor has been proved using an *in vivo* test in yeast (Humair *et al.*, 2001).

Green fluorescent protein (GFP) to which a functional signal peptide was attached has been used as a marker for studying vacuolar targeting in plant cells (Neuhaus, 2000). Both C-terminal and N-terminal, sequence-specific vacuolar sorting determinants have been fused at appropriate positions to signal peptide-GFP and traffic of the chimeric proteins were studied by expression in transgenic tobacco mesophyll protoplasts. It was demonstrated that the sequence-specific and the C-terminal sorting determinants directed the GFP fusion proteins to two different vacuolar compartments that coexisted in the same single cell (Di Sansebastiano *et al.*, 1998, 2001). For example, when the whole propeptide from barley aleurain was fused to GFP (termed Aleu-GFP) and the resulting Aleu-GFP fusion expressed in tobacco leaf protoplasts, the Aleu-GFP fusion was found to accumulate in an acidic vacuole. In contrast, when the chitinase propeptide was fused to the C-terminus of GFP (termed GFP-Chi) and expressed in tobacco mesophyll protoplasts, the GFP-Chi fusion was excluded from acidic vacuoles but accumulated in neutral vacuoles in most of the cell types (Di Sansebastiano *et al.*, 1998, 2001). These results demonstrated the usefulness of GFP fusion proteins as probes for the two pathways to distinct vacuoles in plant cells.

GFP fusions containing the two types of vacuolar sorting determinants were subsequently used as probes to study the functional roles of BP-80 *in vivo* using a yeast system (Humair *et al.*, 2001). These studies used a modified aleurain vacuolar sorting determinant. The sequence RTANFADENPIRQVVSDSFHELES from *Petunia* aleurain was placed between the signal peptide and GFP sequences to give the modified Aleu-GFP. When expressed alone in a mutant yeast strain having a deletion of the yeast vacuolar sorting receptor Vps10p (Δvps10p), signal peptide-GFP, modified Aleu-GFP, and GFP-Chi were all secreted from the cells and did not accumulate in the vacuole. However, in Δvps10p yeast cells co-expressing BP-80 and modified Aleu-GFP, the receptor directed the Aleu-GFP fusion to the yeast vacuole via the prevacuolar compartment. In contrast, when co-expressed with BP-80, both signal peptide-GFP and GFP-Chi were secreted from the cells and did not accumulate in the vacuole. The most direct interpretation of these results is that BP-80 physically interacted with the *Petunia* aleurain vacuolar sorting determinant within the yeast secretory pathway and caused the Aleu-GFP fusion protein to be sorted to the yeast vacuole (Humair *et al.*, 2001). The study thus confirms the hypothesis, based on *in vitro* binding studies, that BP-80 could interact with ligands within the lumenal environment of the ER or the Golgi apparatus, and that the interaction would be specific for the sequence-specific vacuolar sorting determinants known to direct proteins to the plant lytic vacuole. In future, this system should be useful for testing the interactions between various receptors and their potential ligands *in vivo*, experiments that would be difficult to accomplish using a plant cell system. For example, it is not clear why there should be genes for seven BP-80 homologs in *Arabidopsis*. One possibility is that different VSR proteins might recognize different ligand(s) for vacuolar targeting, as discussed above.

8.5   Structural requirements for ligand binding by VSR proteins

In order to study the ligand binding characteristics of BP-80 in more detail, a truncated form of BP-80 lacking transmembrane domain and cytoplasmic tail sequences (tBP-80) was expressed as a soluble protein in *Drosophila* S2 cells from which it was secreted into the medium and could be purified (Cao *et al.*, 2000) (Fig. 8.2). tBP-80 contains three structurally defined regions: an N-terminal RMR homology domain (Jiang *et al.*, 2000) (Fig. 8.2, cross-hatched area), a unique central region (Fig. 8.2, horizontal stripes), and three C-terminal epidermal growth factor (EGF) repeats (Fig. 8.2, black boxes and open oval). Three further truncations lacking one, two, or all three EGF repeat motifs were also expressed; these were termed Δ1EGFR, Δ2EGFR, and Δ3EGFR. Four monoclonal antibodies raised against BP-80 were used to identify different portions of these recombinant proteins (Fig. 8.2). One, 14G7, recognized only tBP-80 and therefore required the C-terminal-most (third) EGF repeat for its binding epitope. Another, 19F2, did not bind to Δ3EGFR but recognized the other three recombinant proteins; therefore, its epitope for binding required the presence of the first and second EGF repeats. Two, 17F9 and 18E7, recognized all four recombinant proteins and therefore interacted with epitopes on the N-terminal 43 kDa portion containing the RMR homology and unique domains that were expressed in recombinant form as Δ3EGFR (Cao *et al.*, 2000).

A fluorescent tag was coupled to the proaleurain peptide and used in a binding assay, where bound peptide was separated from free by gel filtration chromatography (Cao *et al.*, 2000). tBP-80 bound the peptide as a monomer with high affinity, and ~80% of binding was competed by sporamin peptide (sequence SRFNPIRLPT) in 10^4-fold excess. This result indicated that ~20% of binding was due to non-NPIR determinants. In contrast, Δ3EGFR did not bind proaleurain peptide alone, but when mixed with either monoclonal antibody 17F9 or 18E7 the Δ3EGFR-antibody complex bound the peptide in a manner that was indistinguishable from tBP-80. Importantly, however, this binding was completely competed by the sporamin peptide. We therefore conclude that the NPIR-specific binding site was contained completely within the N-terminal ~2/3 portion comprised of the RMR and unique domains of the protein (Fig. 8.2), that the EGF repeats affected the conformation of the NPIR-specific binding site such that it was accessible to ligand, and that the monoclonal antibodies, by interacting at two different sites on the Δ3EGFR protein, substituted for the EGF repeats and thereby altered the conformation of Δ3EGFR to permit ligand binding.

These studies did not separate the RMR homology domain from the central unique domain to address their individual roles in NPIR-specific binding. Recent studies provided some insight into this question by addressing the ability of RMR proteins to interact with peptide ligands. The lumenal domain of the RMR protein encoded by JR702 (Jiang *et al.*, 2000) (Fig. 8.2) was expressed and secreted from *Drosophila* S2 cells and purified to about 50% homogeneity (Park *et al.*, 2002). The ability of the protein to bind to different peptide ligands was tested using peptide affinity columns. There was no detectable binding to the proaleurain peptide column. In contrast, there was a strong binding to columns carrying the C-terminal propeptides of tobacco chitinase (Neuhaus *et al.*, 1994) and barley lectin (Dombrowski *et al.*, 1993), and binding to the barley lectin peptide was abolished by the addition of two Gly residues to the C-terminus of the barley lectin sequence (Park *et al.*, 2002). Application of these results to postulating a functional role for the RMR homology domain in VSR proteins such as BP-80 is limited because RMR proteins have additional peptide sequence in their lumenal domains (Jiang *et al.*, 2000). Nevertheless, the results argue that the RMR homology domain in BP-80 would not by itself form the NPIR-specific binding site, and that its interaction with the central unique domain would probably be necessary for that function. Recently the lumenal domain of the BP-80 homolog, VSR_{At1} (Paris *et al.*, 1997), was expressed in *Drosophila* S2 cells, purified to homogeneity from the medium, crystallized, and complete X-ray diffraction data at a level of 3 Å was obtained (S. W. Rogers, J. C. Rogers, and C. H. Kang, unpublished data). It is most likely that the three-dimensional structure of this molecule will be solved in the near future to answer this and other questions about structure/function relationships.

Results described above indicated that the non-NPIR-specific binding (i.e. that was not competed by the sporamin peptide) site required the presence of the EGF repeats (Cao *et al.*, 2000). This concept would be consistent with the observation by Watanabe *et al.* (2002) that Ca^{2+}-dependent binding to the proaleurain peptide required the presence of the EGF repeats on the pumpkin BP-80 homolog, PV72. In contrast, pH-dependent, NPIR-specific binding appears to be defined, exclusively, by the structure of the *unique* domain lacking EGF repeats. It is possible that the use of a soluble peptide for binding studies by Cao *et al.* (2000) allowed access for the peptide ligand to an NPIR-specific binding site that might be sterically less favorably positioned in the soluble tBP-80 as compared to the site in BP-80 when attached to a membrane. In contrast, Watanabe *et al.* (2002) utilized the peptide attached to a rigid matrix, and it is possible that this approach prevented the NPIR motif on the peptide from reaching the NPIR-specific binding site of PV72.

Further, mapping of the NPIR-specific ligand binding site of tBP-80 utilized antibodies that recognized either the N-terminus or the C-terminus of the molecule. Polyclonal rabbit antibodies, termed RA3, raised to a peptide representing the N-terminus of BP-80, identified denatured molecules containing the N-terminus, while monoclonal antibody 14G7 recognized the C-terminus. Infact, properly folded tBP-80 was not recognized by RA3, indicating that the N-terminus is normally not accessible to the surface. Limited proteolysis of purified tBP-80 with endo-proteinase Asp-N was used to define regions of the molecule involved in the two different types of ligand binding, NPIR-specific, and non-NPIR-specific (Cao *et al.*, 2000). Two predominant protease-resistant forms of tBP-80 were obtained. The first represented the N-terminal RMR homology domain plus the unique central domain, a fragment similar to the recombinant Δ3EGFR; the second ~43 kDa fragment represented the central unique domain connected to the C-terminal EGF repeat-containing domain. The proteolytic cleavage to yield the second form removed a unit that was essentially equivalent to the RMR homology domain. As noted before, NPIR-specific binding was associated with the first, Δ3EGFR form. Analysis of the second form was accomplished by allowing the partial proteolysis products to bind to a proaleurain peptide column alone, in the presence of sporamin peptide to prevent NPIR-specific binding, or in the presence of the proaleurain peptide to prevent all but non-specific binding. Protein that bound to the column was then eluted and analyzed on a western blot probed with the C-terminal-specific monoclonal antibody 14G7; thus fragments could be mapped to the C-terminus of tBP-80. In the absence of peptide competitor, full-length undigested tBP-80 and a lesser amount of the C-terminal ~43 kDa protein was bound to the proaleurain peptide column. Sporamin peptide prevented the full-length form from binding but did not prevent the C-terminal ~43 kDa form from binding. This indicates that binding for the latter does not involve NPIR. Proaleurain peptide prevented both the full-length and C-terminal ~43 kDa forms from binding. This would support the concept that binding of the C-terminal ~43 kDa form involves sequence deter-minants that are on the proaleurain peptide but are separate from NPIR, so-called non-NPIR-specific binding determinants (Cao *et al.*, 2000) (Fig. 8.2).

8.6 Chimeric integral membrane reporter proteins as probes of the Golgi to lytic vacuole pathway

The different mechanisms by which soluble proteins and integral membrane proteins are sorted from the Golgi into pathways to vacuoles interact in an obvious manner. If sorting of a soluble protein requires binding by an integral membrane receptor protein, it is the mechanisms that direct the receptor into the appropriate transport vesicles that ultimately govern the sorting process for both. At present, there is only one pathway known for transport between the Golgi apparatus and the lytic prevacuolar compartment in plant cells, and that has largely been defined by studying the traffic of intact BP-80 and of chimeric reporter proteins with BP-80 transmembrane domain and cytoplasmic tail sequences (Brandizzi *et al.*, 2002; Jiang & Rogers, 1998).

The mechanism by which BP-80 reaches its final destination was addressed using a chimeric integral membrane reporter protein (Jiang & Rogers, 1998). The hypothesis that the transmembrane domain and cytoplasmic tail sequences of BP-80 were specific and sufficient for correct targeting of the receptor was tested. A chimeric integral membrane reporter protein was designed and tested using transient expression in tobacco suspension culture cells (Jiang & Rogers, 1998). The chimeric reporter protein contained a lumenal domain constructed from barley proaleurain with a mutated vacuolar sorting determinant connected to linker sequences that included a FLAG epitope tag and a short Ser/Thr-rich sequence from BP-80. The linker was, in turn, connected to the BP-80 transmembrane and cytoplasmic tail sequences (Jiang & Rogers, 1998). The strategy behind the experiments was that the reporter would be processed into a mature form if it trafficked to the correct destination, which in turn would demonstrate the role of transmembrane domain and cytoplasmic tail sequences in correct targeting. Indeed, when this reporter protein was expressed in tobacco suspension culture cells, it reached small organelles of ≤1 μm size, prevacuolar compartments, where the proaleurain moiety was proteolytically processed into mature aleurain and released from membrane attachment.

Two different experimental approaches were used to demonstrate that this reporter protein transited the Golgi apparatus. First, it acquired Golgi-specific complex modifications (Lerouge *et al.*, 1998) to Asn-linked glycans (Jiang & Rogers, 1998). Second, when the reporter protein was modified by the insertion of three tandemly repeated Kex2p substrate sequences in the linker region, it was cleaved by a Golgi-localized Kex2p-like protease that released intact proaleurain from membrane attachment (Jiang & Rogers, 1999a). As a result, the released proaleurain was secreted.

Both subcellular fractionation and confocal immunofluorescence approaches were also used to study the fate of the reporter protein in transgenic tobacco culture cells. When double-labeling experiments using antibodies specific for the reporter protein and for the endogenous tobacco VSR proteins were performed, the reporter protein co-localized with the tobacco VSR proteins within cytosolic organelles of ≤1 μm in size (Jiang & Rogers, 1998). In addition, these organelles floated on ficoll step gradients with the fraction enriched in vacuoles (Jiang & Rogers, 1998). Furthermore, recent results have demonstrated that VSR proteins are predominantly concentrated in lytic prevacuolar compartments in different cell types and species (Li *et al.*, 2002). Thus, the localization of VSR proteins can be used to identify prevacuolar organelles on the lytic vacuole pathway (see Section 8.7 below).

Localization of the reporter protein and of VSR proteins at a steady state to prevacuolar organelles indicates that the transmembrane/cytoplasmic tail sequences of the proteins function to concentrate them there, as opposed to the possibility that the proteins might be concentrated in the Golgi apparatus and only transiently exit. Could separate functions for transmembrane or for cytoplasmic tail sequences be defined? In the case of the chimeric proaleurain/BP-80 reporter protein, substitution of 12 amino acids representing the C-terminal cytoplasmic tail of γ-TIP, SRTHEQLPTTDY, for the 31 C-terminal residues of the BP-80 cytoplasmic tail resulted in proteolytic processing of the reporter to an extent that was similar to the unmodified protein (Jiang & Rogers, 1998). Additionally, simple deletion of the 31 C-terminal residues gave the same result; processing of the reporter occurred in a manner similar to that observed with the intact protein. Pulse-chase assays with these two modified reporters were not performed and therefore small differences in the efficiency of export from the Golgi apparatus could not be excluded. However, both these results strongly indicated that the deleted sequences, which are highly conserved in other VSR proteins and included the YMPL motif that is thought to participate in interactions with clathrin adaptor proteins, were not necessary for traffic out of the Golgi apparatus (Jiang & Rogers, 1998). Both constructs retained five residues immediately adjacent to the transmembrane domain, KYRIR, and it remains a possibility that these residues could direct in an unexpected, positive way export from the Golgi apparatus. These results have similarities to the studies in yeast, where the transmembrane domain alone of some single pass proteins is sufficient to direct the protein to the prevacuolar compartment (Roberts *et al.*, 1992). Subsequent studies hae emphasized, however, that both the length of the transmembrane domain and its amino acid composition greatly affect localization of a protein within the secretory pathway (Lewis *et al.*, 2000; Reggiori *et al.*, 2000).

In a different approach, Brandizzi *et al.* (2002) fused the transmembrane domain and first five residues of the cytoplasmic tail of BP-80 to signal peptide-green fluorescent protein (GFP) and expressed the chimeric protein in tobacco leaf epidermal cells. This GFP reporter protein co-localized completely with a yellow fluorescent protein-tagged Golgi marker (Brandizzi *et al.*, 2002). The authors concluded that the BP-80 transmembrane domain limited their protein to membranes in the Golgi apparatus, and that positive information is needed to transit out of the Golgi apparatus to the lytic prevacuolar compartment. Thus, on the surface, the two studies (Brandizzi *et al.*, 2002; Jiang & Rogers, 1998) appeared to yield contradictory results.

In reality, however, it is likely that the results from the two studies are fully compatible. Brandizzi *et al.* (2002) assessed the location of the most abundant concentrations of their GFP fusion proteins but their data emphasize the limitations of that approach. For example, when analyzing a second set of fusion proteins that incorporated a mammalian transmembrane domain, the authors performed western blots to assess the size of the GFP-containing molecules. For the two fusion proteins that were largely localized to plasma membrane or ER respectively, ~95% of the GFP remained in the intact fusion proteins and only a trace amount was the size of GFP alone, the result of proteolytic cleavage from membrane attachment. In contrast, for the fusion protein that localized to the Golgi, TM20, 28% of the GFP had been cleaved from membrane attachment, probably because it had exited the Golgi and had been delivered to the lytic vacuole (Brandizzi *et al.*, 2002). Similar studies of the BP-80 transmembrane domain fusion protein were not performed, but it is likely that the confocal fluorescence assay could not have detected 20% of the GFP in structures other than the Golgi, for example the lytic prevacuolar compartment and vacuole. In contrast, the [^{35}S]-labeling/immuno-precipitation assay was quite sensitive and could detect small amounts of the reporter protein that had been proteolytically processed (Jiang & Rogers, 1998), but the location of the abundant amount of the intact reporter protein, lacking the cytoplasmic tail, was not determined. Thus in both the studies it is likely that most of each reporter protein at a steady state was localized to the Golgi apparatus, but also that a measurable amount of each could exit the Golgi and be transported to the lytic PVC.

Other points should also be considered. It is important to note that the two reporter constructs had important differences localized to the lumenal amino acid sequences immediately adjacent to the transmembrane domain. In the proaleurain reporter construct, these amino acids came from BP-80 and defined a Ser/Thr-rich motif, SKTASQAKST. In contrast, in the GFP reporter construct, the corresponding amino acids were THGMDELYKST. Thus, the overall charge of the sequence was changed from basic to acidic, and the Ser/Thr cluster which may be the target for O-linked glycosylation in BP-80 was disrupted. It is well established that the stalk regions of Golgi glycosyltransferases adjacent to the transmembrane sequences can play important roles in protein–protein interactions that, in turn, participate in retention of the proteins to specific Golgi cisternae (Nilsson *et al.*, 1993). This consideration raises two possibilities that could have affected the results. One is that the GFP reporter sequence might cause the protein to interact with a resident Golgi protein and thereby to be retained. The fact that the insertion of three additional amino acids into the BP-80 transmembrane sequence caused the GFP reporter to move to the plasma membrane (Brandizzi *et al.*, 2002) would not exclude this possibility, because the specific sequence arrangement and amino acid composition of transmembrane sequences can be a separate factor that contributes to Golgi localization (Nilsson *et al.*, 1993). A second possibility is that the native BP-80 sequence could have interacted with other proteins, e.g. endogenous tobacco VSR proteins, and caused the proaleurain reporter (Jiang & Rogers, 1998) to be passively drawn into vesicles destined for the lytic prevacuolar compartment. Until the influences of these two different lumenal flanking sequences are tesed directly it will not be possible to exclude either possibilities.

A third study fused signal peptide-GFP to the C-terminal 68 amino acids of pumpkin PV72 to yield a construct that included the PV72 Ser/Thr-rich lumenal sequence, GNIGSTVTSWSVVK adjacent to the transmembrane domain (Mitsuhashi *et al.*, 2000). In contrast to the two reporter proteins discussed above, this reporter protein included an intact cytoplasmic tail. When this GFP-PV72 fusion protein was expressed in tobacco BY-2 suspension culture cells, its localization was dependent upon the state of growth and differentiation of the cells. In three day-old suspension culture cells, the GFP localized to small, ≤1 μm punctate structures whose appearance would be consistent with either Golgi or prevacuolar organelles. Other markers for these organelles were not used to allow a specific identification and no note was made about whether they were mobile (a defining feature of plant Golgi stacks), but the authors concluded that the protein was localized to Golgi (Mitsuhashi *et al.*, 2000). In striking contrast, in cells from ten day-old callus tissue, the GFP strongly and diffusely labeled large vacuoles. Western blots demonstrated that in the former instance the GFP was exclusively in the intact fusion protein, while in the latter instance the GFP had been reduced in size to that of recombinant GFP alone. Thus, in the latter instance, the fusion protein had trafficked to a place where GFP had been separated from membrane attachment and accumulated in the vacuole lumen. These observations raise an important point that should be considered with respect to the interpretation of the BP-80 fusion protein studies (Brandizzi *et al.*, 2002; Jiang & Rogers, 1998). Traffic of a membrane protein may be affected by the state of cell growth and differentiation. For example, perhaps the lipid composition and thickness of membranes in the different compartments in the secretory pathway can vary, with resultant effects on localization of integral membrane proteins (Brandizzi *et al.*, 2002).

8.7   The lytic prevacuolar compartment: identity and function

Even though this volume deals with Golgi functions, it is important to include a discussion of the organelle downstream from the Golgi complex on the lytic vacuole pathway, the prevacuolar compartment, because protein traffic between the two organelles interacts in important ways. Prevacuolar compartments (PVCs) are defined as organelles that receive cargo from transport vesicles and subsequently deliver that cargo to the vacuole by fusion with the tonoplast (Bethke & Jones, 2000). Based on precedents from mammalian and yeast systems (Lemmon & Traub, 2000), they are intermediate organelles on the pathways to vacuoles from both Golgi and from endocytosis from the plasma membrane. Pathways for endocytosis in plant cells are not well defined (Battey *et al.*, 1999) (although recent studies (Emans *et al.*, 2002; Ueda *et al.*, 2001) provide important new approaches), and hence our attention will focus primarily on Golgi to PVC traffic.

PVCs exist to provide a place from which missorted proteins can be retrieved and sent back to the Golgi apparatus (Lemmon & Traub, 2000). Inherent in this model is the concept that delivery of proteins to vacuoles is, in most cases, an irreversible destination. Thus, PVCs must be defined both functionally (a site for retrieval) and morphologically (on the pathways to vacuoles but separate from them) (Robinson *et al.*, 2000).

In yeast and mammalian systems, two different types of proteins have served as markers for the PVC/endosome. First, sorting receptors that directed soluble proteins at the TGN into transport vesicles destined for the PVC/endosome were defined. These sorting receptors cycle between the TGN and PVC (Robinson & Hinz, 1997). In mammalian cells, the mannose 6-phosphate receptor has this function but, interestingly, is predominantly concentrated in late endosomes (Griffiths *et al.*, 1988). Second, individual Rab GTPases and syntaxins are associated with each organelle. In yeast, for example, Pep12p is a syntaxin specifically associated with the PVC/endosome (Pelham, 2000).

In contrast to yeast and mammalian cells, plant cells contain two functionally distinct vacuoles: the protein storage vacuole (PSV) and lytic vacuole (Hoh *et al.*, 1995; Jauh *et al.*, 1999; Paris *et al.*, 1996). Additionally, multiple pathways using distinct transport vesicles are responsible for transporting proteins to the PSV and lytic vacuole in plant cells (Hara-Nishimura *et al.*, 1998; Hinz *et al.*, 1999; Jiang & Rogers, 2001). There is increasing evidence suggesting that PVCs exist in plant cells and that they play a similar role in protein trafficking in the plant secretory pathway (Bethke & Jones, 2000; Robinson *et al.*, 2000). Identification of the separate PVCs for lytic and protein storage vacuoles will enable functional definitions of their roles in the complex plant vacuolar system and the multiple vesicular pathways leading to vacuoles.

In spite of the important role of PVCs in mediating protein traffic to vacuoles in the plant secretory pathway, identification and characterization of plant PVCs, either functionally or morphologically, have been challenging due to the complexity of the plant vacuolar systems and the existence of multiple pathways of vacuolar targeting (Jiang & Rogers, 2001; Robinson *et al.*, 2000).

Several approaches have been used to identify and characterize PVCs in plant cells. In *Arabidopsis*, AtPep12p or AtSYP21 (Sanderfoot *et al.*, 2000), a yeast Pep12p homolog that can functionally complement the yeast *pep12* mutant, has been localized to what was termed a PVC (a late post-Golgi compartment) by immunoEM and subcellular fractionation in *Arabidopsis* root tip cells (Bassham *et al.*, 1995; Conceicao *et al.*, 1997). The morphological characterization of these structures has been questioned (Robinson *et al.*, 2000), and no evidence was provided to document a function for the structures that were identified.

Another approach focused on BP-80 and other members of the VSR protein family. Studies on the subcellular localization of VSR proteins demonstrated the presence of putative PVCs for the lytic vacuolar pathway. The subcellular localization of VSR proteins has been carried out mainly using two techniques: immunoEM and subcellular fractionation. Studies using these two approaches have shown that VSR proteins are not present in the tonoplast (Ahmed *et al.*, 1997b; Hinz *et al.*, 1999; Paris *et al.*, 1997; Sanderfoot *et al.*, 1998). BP-80 has been localized by immunoEM to the Golgi apparatus and to what was considered to be a lytic PVC. The latter was characterized by clear, approximately 250 nm diameter vesicles that appeared to fuse with adjacent vacuoles. Both immunoEM and confocal immunofluorescence studies indicated that BP-80 was not present in tonoplasts in the root tip and in the cotyledon cells of pea (Paris *et al.*, 1997). *AtELP* (Ahmed *et al.*, 1997a,b), a BP-80 homolog from *Arabidopsis* equivalent to VSR_{At1} (Paris *et al.*, 1997) has been located to the Golgi apparatus and to a putative PVC characterized by ~100 nm diameter tubules in *Arabidopsis* root tip cells (Sanderfoot *et al.*, 1998). It was, however, suggested that the tubular structures identified were most consistent with the appearance of *trans*-Golgi network (Robinson *et al.*, 2000).

In subcellular fractionation studies, BP-80 was found to co-fractionate partially with Golgi membranes in developing pea cotyledon cells (Hinz *et al.*, 1999). Similar results were obtained with VSR$_{At1}$/AtELP in fractionations of *Arabidopsis* tissue (Ahmed *et al.*, 1997b; Sanderfoot *et al.*, 1998). Additionally, AtELP co-fractionated in sucrose density gradient fractionations of *Arabidopsis* root membrane with AtPep12p (AtSYP21p) (Sanderfoot *et al.*, 1998). These results together suggest that VSR proteins are localized in both the Golgi apparatus and a putative post-Golgi PVC in these cells. The limitations to sucrose gradient fractionation, however, focus on the fact that the number and identity of organelles in pathways to vacuoles in plant cells are largely not known. If one does not know how many different organelles might be present in a plant cell, and one does not have antibodies to identify proteins that are specific for each organelle, how would one interpret and tie to a specific organelle the position of a SNARE or VSR protein within the gradient? This consideration would argue that microscopy coupled with a functional assay to tie a particular protein to a particular organelle will continue to be the means by which a PVC can be identified.

More recently, PVCs in pea and tobacco cells have been identified and characterized using confocal immunofluorescence with antibodies specific for proteins resident in different organelles in the plant secretory pathway (Li *et al.*, 2002). The localization of BP-80 in pea root tip cells was first determined using various anti-BP-80 antibodies for confocal immunofluorescence. These antibodies are described in Section 8.5 and include the polyclonal anti-N-terminal peptide RA3 antibodies that are specific for pea BP-80 (Paris *et al.*, 1997), and the four monoclonal antibodies whose epitopes are mapped in Fig. 8.2. The latter cross-react with VSR proteins from other species (Jiang & Rogers, 1998; Miller *et al.*, 1999). These five antibodies were tested in various combinations in double-labeling experiments in pea root tip cells and results demonstrated that all these antibodies co-localized (Li *et al.*, 2002). For identifying Golgi organelles, three different markers were used: JIM84, a rat monoclonal antibody directed against Lewis α-containing *N*-glycans that are localized in the *trans*-Golgi (Fitchette *et al.*, 1999; Satiat-Jeunemaitre & Hawes, 1992); a Man1-GFP fusion protein that has been shown to localize to *cis*-Golgi in transgenic tobacco BY-2 cells (Nebenführ *et al.*, 1999), and polyclonal antibodies raised to a recombinant form of the *cis*-Golgi enzyme, mannosidase I (Nebenführ *et al.*, 1999).

The use of confocal immunofluorescence has allowed a direct comparison of the localization of VSR proteins to that of established Golgi markers, and to a quantitative analysis of the distribution of VSR proteins between Golgi and PVCs in hundreds of cells. Using this approach, it was, thus, demonstrated that VSR-labeled organelles are largely separate from ER, Golgi apparatus and vacuolar compartments, a finding that indicates VSR proteins must be concentrated on PVCs (Li *et al.*, 2002). Labeling with the anti-BP-80 antibodies largely co-localized with anti-AtPep12p (AtSYP21p) labeling, and labeling with anti-AtPep12p was largely separate from that of the Golgi markers. Similar results were obtained in two different systems, pea root tip cells and BY-2 tobacco suspension culture cells

(Li *et al.*, 2002). The results are consistent with the hypothesis that VSR proteins are predominantly concentrated on post-Golgi, lytic prevacuolar compartments (Li *et al.*, 2002). Thus, anti-VSR antibodies can be used as markers for defining PVCs in various plant cells. Additionally, these observations lead to the hypothesis that VSR proteins only recycle back to Golgi briefly for selection of transit cargo molecules and then return to the PVCs for cargo delivery.

Acknowledgements

This work was partially supported by grants from the Research Grants Council of Hong Kong (project CUHK4156/01M and CUHK4260/02M), Chinese University of Hong Kong (project 2030262) and Germany/Hong Kong Joint Research Scheme (project 2900102) to L. Jiang. Research in the Rogers Laboratory is supported by grants from the US Department of Energy, the National Science Foundation, and the Human Frontier Science Program.

References

Ahmed, S.U., Bar-Peled, M. & Raikhel, N.V. (1997a) Addendum: cloning and subcellular location of an *Arabidopsis* receptor-like protein that shares common features with protein-sorting receptors of eukaryotic cells. *Plant Physiol.*, **115**, 311–312.

Ahmed, S.U., Bar-Peled, M. & Raikhel, N.V. (1997b) Cloning and subcellular location of an *Arabidopsis* receptor-like protein that shares common features with protein-sorting receptors of eukaryotic cells. *Plant Physiol.*, **114**, 325–336.

Ahmed, S.U., Rojo, E., Kovaleva, V., Venkataraman, S., Dombrowski, J.E., Matsuoka, K. & Raikhel, N.V. (2000) The plant vacuolar sorting receptor AtELP is involved in transport of NH_2-terminal propeptide-containing vacuolar proteins in *Arabidopsis thaliana. J. Cell Biol.*, **149**, 1335–1344.

Ahn, K., Yeyeodu, S., Collette, J., Madden, V., Arthur, J., Li, L. & Erickson, A.H. (2002) An alternate targeting pathway for procathepsin L in mouse fibroblasts. *Traffic*, **3**, 147–159.

Bassham, D.C., Gal, S., Conceicao, A.S. & Raikhel, N.V. (1995) An *Arobidopsis* syntaxin homologue isolated by functional complementation of a yeast pep12p mutant. *Proc. Natl. Acad. Sci. USA*, **92**, 7262–7266.

Battey, N.H., James, N.C., Greenland, A.J. & Brownlee, C. (1999) Exocytosis and endocytosis. *Plant Cell*, **11**, 643–660.

Bethke, P.C. & Jones, R.L. (2000) Vacuoles and prevacuolar compartments. *Curr. Opin. Cell Biol.*, **3**, 469–475.

Boevink, P., Oparka, K., Santa Cruz, S., Martin, B., Betteridge, A. & Hawes, C. (1998) Stacks on tracks: the plant Golgi apparatus trafficks on an actin/ER network. *Plant J.*, **15**, 441–447.

Brandizzi, F., Frangne, N., Marc-Martin, S., Hawes, C., Neuhaus, J.-M. & Paris, N. (2002) In plants the destination for single pass membrane proteins is markedly influenced by the length of the hydrophobic domain. *Plant Cell*, **14**, 1077–1092.

Cao, X., Rogers, S.W., Butler, J., Beevers, L. & Rogers, J.C. (2000) Structural requirements for ligand binding by a probable plant vacuolar sorting receptor. *Plant Cell*, **12**, 439–506.

Conceicao, A., Marty-Mazars, D., Bassham, D.C., Sanderfoot, A.A., Marty, F. & Raikhel, N.V. (1997) The syntaxin homologue AtPEP12p resides on a late post-Golgi compartment in plants. *Plant J.*, **9**, 571–582.

Di Sansebastiano, G.P., Paris, N., Marc-Martin, S. & Neuhaus, J.-M. (1998) Specific accumulation of GFP in a non-acidic vacuolar compartment via a C-terminal propeptide-mediated sorting pathway. *Plant J.*, **15**, 449–457.

Di Sansebastiano, G.P., Paris, N., Marc-Martin, S. & Neuhaus, J.-M. (2001) Regeneration of a lytic central vacuole and of neutral peripheral vacuoles can be visualized by green fluorescent proteins targeted to either type of vacuoles. *Plant Physiol.*, **126**, 78–86.

Dombrowski, J.E., Schroeder, M.R., Bednarek, S.Y. & Raikhel, N.V. (1993) Determination of the functional elements within the vacuolar targeting signal of barley lectin. *Plant Cell*, **5**, 587–596.

Emans, N., Zimmermann, S. & Fischer, R. (2002) Uptake of a fluorescent marker in plant cells is sensitive to brefeldin A and wortmannin. *Plant Cell*, **14**, 71–86.

Fitchette, A.C., Cabanes-Macheteau, M., Marvin, L., Martin, B., Satiat-Jeunemaitre, B., Gomord, V., Lerouge, K., Faye, L. & Hawes, C. (1999) Biosynthesis and immunolocalization of Lewis α-containing N-glycans in the plant cell. *Plant Physiol.*, **121**, 333–343.

Frigerio, L., de Virgilio, M., Prada, A., Faoro, F. & Vitale, A. (1998) Sorting of phaseolin to the vacuole is saturable and requires a short C-terminal peptide. *Plant Cell*, **10**, 1031–1042.

Griffiths, G., Hofack, B., Simons, K., Mellman, I. & Kornfeld, S. (1988) The mannose 6-phosphate receptor and the biogenesis of lysosome. *Cell*, **52**, 329–341.

Guruprasad, K., Törmäkangas, K., Kervinen, J. & Blundell, T.L. (1994) Comparative modelling of barley-grain aspartic proteinase: a structural rationale for observed hydrolytic specificity. *FEBS Lett.*, **352**, 131–136.

Hadlington, J.L. & Denecke, J. (2000) Sorting of soluble proteins in the secretory pathway of plants. *Curr. Opin. Plant Biol.*, **3**, 461–468.

Hara-Nishimura, I. & Maeshima, M. (2000) Vacuolar processing enzymes and aquaporins. *Annu. Plant Rev.*, **5**, 20–42.

Hara-Nishimura, I., Shimada, T., Hatano, K., Takeuchi, Y. & Nishimura, M. (1998) Transport of storage proteins to protein storage vacuoles is mediated by large precursor-accumulating vesicles. *Plant Cell*, **10**, 825–836.

Hillmer, S., Movafeghi, A., Robinson, D.G. & Hinz, G. (2001) Vacuolar storage proteins are sorted in the cis-cisternae of the pea cotyledon Golgi apparatus. *J. Cell Biol.*, **152**, 41–50.

Hinz, G., Hillmer, S., Baumer, M. & Hohl, I. (1999) Vacuolar storage proteins and the putative vacuolar sorting receptor, BP-80, exit the Golgi apparatus of developing pea cotyledons in different transport vesicles. *Plant Cell*, **11**, 1509–1524.

Hoh, B., Hinz, G., Jeong, B.-K. & Robinson, D.G. (1995) Protein storage vacuoles form de novo during pea cotyledon development. *J. Cell Sci.*, **108**, 299–310.

Hohl, I., Robinson, D.G., Chrispeels, M.C. & Hinz, G. (1996) Transport of storage proteins to the vacuole is mediated by vesicles without a clathrin coat. *J. Cell Sci.*, **109**, 2539–2550.

Holwerda, B.C., Galvin, N.J., Baranski, T.J. & Rogers, J.C. (1990) In vitro processing of aleurain, a barley vacuolar thiol protease. *Plant Cell*, **2**, 1091–1106.

Holwerda, B.C., Padgett, H.S. & Rogers, J.C. (1992) Proaleurain vacuolar targeting is mediated by short contiguous peptide interactions. *Plant Cell*, **4**, 307–318.

Humair, D., Hernández Felipe, D., Neuhaus, J.-M. & Paris, N. (2001) Demonstration in yeast of the function of BP-80, a putative plant vacuolar sorting receptor. *Plant Cell*, **13**, 781–792.

Jauh, G.-Y., Fischer, A.M., Grimes, H.D., Ryan, C.A. & Rogers, J.C. (1998) δ-Tonoplast intrinsic protein defines unique plant vacuole functions. *Proc. Natl. Acad. Sci. USA*, **95**, 12995–12999.

Jauh, G.Y., Philips, T.E. & Rogers, J.C. (1999) Tonoplast intrinsic protein isoforms as markers for vacuolar functions. *Plant Cell*, **11**, 1867–1882.

Jiang, L., Phillips, T.E., Rogers, S.W., Hamm, C.A., Drozdowicz, Y.M., Rea, P.A., Maeshima, M. & Rogers, J.C. (2001) The protein storage vacuole: a compound organelle. *J. Cell Biol.*, **155**, 991–1002.

Jiang, L., Phillips, T.E., Rogers, S.W. & Rogers, J.C. (2000) Biogenesis of the protein storage vacuole crystalloid. *J. Cell Biol.*, **150**, 755–769.

Jiang, L. & Rogers, J.C. (1998) Integral membrane protein sorting to vacuoles in plant cells: evidence for two pathways. *J. Cell Biol.*, **143**, 1183–1199.

Jiang, L. & Rogers, J.C. (1999a) Functional analysis of a Golgi-localized Kex2p-like protease in tobacco suspension culture cells. *Plant J.*, **18**, 23–32.

Jiang, L. & Rogers, J.C. (1999b) The role of BP-80 and homologs in sorting proteins to vacuoles. *Plant Cell*, **11**, 2069–2071.

Jiang, L. & Rogers, J.C. (1999c) Sorting of membrane proteins to vacuoles in plant cells. *Plant Sci.*, **146**, 55–67.

Jiang, L. & Rogers, J.C. (2001) Compartmentation of proteins in the protein storage vacuole, a compound organelle in plant cells. *Adv. Bot. Res.*, **35**, 139–170.

Kim, D.H., Eu, Y.-J., Yoo, C.M., Kim, Y.-W., Pih, K.T., Jin, J.B., Kim, S.J., Stenmark, H. & Hwang, I. (2001) Trafficking of phosphatidylinositol 3-phosphate from the *trans*-Golgi network to the lumen of the central vacuole in plant cells. *Plant Cell*, **13**, 287–301.

Kirsch, T., Paris, N., Butler, J.M., Beevers, L. & Rogers, J.C. (1994) Purification and initial characterization of a potential plant vacuolar targeting receptor. *Proc. Natl. Acad. Sci. USA*, **91**, 3403–3407.

Kirsch, T., Saalbach, G., Raikhel, N.V. & Beevers, L. (1996) Interaction of a potential vacuolar targeting receptor with amino- and carboxyl-terminal targeting determinants. *Plant Physiol.*, **111**, 469–474.

Lemmon, S.K. & Traub, L.M. (2000) Sorting in the endosomal system in yeast and animal cells. *Curr. Opin. Cell Biol.*, **12**, 457–466.

Lerouge, P., Cabanes-Macheteau, M., Rayon, C., Fischette-Lainé, A.-C., Gomord, V. & Faye, L. (1998) N-glycoprotein biosynthesis in plants: recent developments and future trends. *Plant Mol. Biol.*, **38**, 31–48.

Lewis, M.J., Nichols, B.J., Prescianotto-Baschong, C., Riezman, H. & Pelham, H.R. (2000) Specific retrieval of the exocytic SNARE snc1p from early yeast endosomes. *Mol. Biol. Cell*, **11**, 23–38.

Li, Y.B., Rogers, S.W., Tse, Y.C., Lo, S.W., Sun, S.S.M., Jauh, G.-Y. & Jiang, L. (2002) BP-80 and homologs are concentrated on post-Golgi, lytic prevacuolar compartments. *Plant Cell Physiol.*, **43**, 726–742.

Maeshima, M., Sasaki, T. & Asahi, T. (1985) Characterization of major proteins in sweet potato tuberous roots. *Phytochemistry*, **24**, 1899–1902.

Maeshima, M. & Yoshida, S. (1989) Purification and properties of vacuolar membrane proton-translocating inorganic pyrophosphatase from mung bean. *J. Biol. Chem.*, **264**, 20068–20073.

Matsuoka, K., Matsumoto, S., Hattori, T., Machida, Y. & Nakamura, K. (1990) Vacuolar targeting and posttranslational processing of the precursor to the sweet potato tuberous root storage protein in heterologous plant cells. *J. Biol. Chem.*, **265**, 19750–19757.

Matsuoka, K. & Nakamura, K. (1991) Propeptide of a precursor to a plant vacuolar protein required for vacuolar targetting. *Proc. Natl. Acad. Sci. USA*, **88**, 834–838.

Matsuoka, K. & Nakamura, K. (1999) Large alkyl side chains of isoleucine and leucine in the NPIRL region constitute the core of the vacuolar sorting determinant of sporamin precursor. *Plant Mol. Biol.*, **41**, 825–835.

Matsuoka, K. & Neuhaus, J.-M. (1999) *Cis*-elements of protein transport to the plant vacuoles. *J. Exp. Bot.*, **50**, 165–174.

Matsuoka, K., Watanabe, N. & Nakamura, K. (1995) O-glycosylation of a precursor to a sweet potato vacuolar protein, sporamin, expressed in tobacco cells. *Plant J.*, **8**, 877–889.

Maurel, C. (1997) Aquaporins and water permeability of plant membranes. *Annu. Rev. Plant Physiol. Plant Mol. Biol.*, **48**, 399–429.

Miller, E.A., Lee, M.C.S. & Anderson, M.A. (1999) Identification and characterization of a prevacuolar compartment in stigmas of *Nicotiana alata*. *Plant Cell*, **11**, 1499–1508.

Mitsuhashi, N., Shimada, T., Mano, S., Nishimura, M. & Hara-Nishimura, I. (2000) Characterization of organelles in the vacuolar sorting pathway by visualization with GFP in tobacco BY-2 cells. *Plant Cell Physiol.*, **41**, 993–1001.

Nakamura, K., Matsuoka, K., Mukumoto, F. & Watanabe, N. (1993) Processing and transport to the vacuole of a precursor to sweet potato sporamin in transformed tobacco cell line BY-2. *J. Exp. Bot.*, **44** (Suppl.), 331–338.

Nebenführ, A., Gallagher, L.A., Dunahay, T.G., Frohlick, J.A., Mazurkiewicz, A.M., Meehl, J.B. & Staehelin, L.A. (1999) Stop-and-go movements of plant Golgi stacks are mediated by the acto-myosin system. *Plant Physiol.*, **121**, 1127–1141.

Neuhaus, J.-M., Pietrzak, M. & Boller, T. (1994) Mutation analysis of the C-terminal vacuolar targeting peptide of tobacco chitinase: low specificity of the sorting system, and gradual transition between intracellular retention and secretion into the extracellular space. *Plant J.*, **5**, 45–54.

Neuhaus, J.-M. & Rogers, J.C. (1998) Sorting of proteins to vacuoles in plant cells. *Plant Mol. Biol.*, **38**, 127–144.

Neuhaus, J.M. (2000) GFP as a marker for vacuoles in plants. *Annu. Plant Rev.*, **5**, 254–269.

Nilsson, T., Slusarewica, P., Hoe, M.H. & Warren, G. (1993) Kin recognition. A model for the retention of Golgi enzymes. *FEBS Lett.*, **330**, 1–4.

Okita, T.W. & Rogers, J.C. (1996) Compartmentation of proteins in the endomembrane system of plant cells. *Annu. Rev. Plant Physiol. Plant Mol. Biol.*, **47**, 327–350.

Paris, N., Rogers, S.W., Jiang, L., Kirsch, T., Beevers, L., Phillips, T.E. & Rogers, J.C. (1997) Molecular cloning and further characterization of a probable plant vacuolar sorting receptor. *Plant Physiol.*, **115**, 29–39.

Paris, N., Stanley, C.M., Jones, R.L. & Rogers, J.C. (1996) Plant cells contain two functionally distinct vacuolar compartments. *Cell*, **85**, 563–572.

Park, J.H., Rogers, S.W., Paris, N. & Rogers, J.C. (2002) The ligand binding specificity of a plant RMR protein is consistent with the function of a sorting receptor for the protein storage vacuole pathway. *Submitted for publication.*

Pelham, H.R. (2000) SNAREs and the secretory pathway – lessons from yeast. *Exp. Cell Res.*, **247**, 1–8.

Reggiori, F., Black, M.W. & Pelham, H.R.B. (2000) Polar transmembrane domains target proteins to the interior of the yeast vacuole. *Mol. Biol. Cell.*, **11**, 3737–3749.

Roberts, C.J., Nothwehr, S.F. & Stevens, T.H. (1992) Membrane protein sorting in the yeast secretory pathway: evidence that the vacuole may be the default compartment. *J. Cell Biol.*, **119**, 63–83.

Robinson, D.G., Baumer, M., Hinz, G. & Hohl, I. (1997) Ultrastructure of the pea cotyledon Golgi apparatus: origin of dense vesicles and the action of brefeldin A. *Protoplasma*, **200**, 198–209.

Robinson, D.G. & Hinz, G. (1997) Vacuole biogenesis and protein transport to the plant vacuole: a comparison with the yeast vacuole and the mammalian lysosome. *Protoplasma*, **197**, 1–25.

Robinson, D.G. & Hinz, G. (1999) Golgi-mediated transport of seed storage proteins. *Seed Sci. Res.*, **9**, 267–283.

Robinson, D.G., Hinz, G. & Holstein, S.E.H. (1998) The molecular characterization of transport vesicles. *Plant Mol. Bio.*, **38**, 49–76.

Robinson, D.G., Rogers, J.C. & Hinz, G. (2000) Post-Golgi, prevacuolar compartments. *Annu. Plant Rev.*, **5**, 270–298.

Runeberg-Roos, P., Törmäkangas, K. & Östman, A. (1991) Primary structure of a barley-grain aspartic proteinase: a plant aspartic proteinase resembling mammalian cathepsin D. *Eur. J. Biochem.*, **202**, 1021–1027.

Sanderfoot, A.A., Ahmed, S.U., Marty-Mazars, D., Rapoport, I., Kirchhausen, T., Marty, F. & Raikhel, N.V. (1998) A putative vacuolar cargo receptor partially colocalizes with AtPEP12p on a pre-vacuolar compartment in *Arabidopsis* roots. *Proc. Natl. Acad. Sci. USA*, **95**, 9920–9925.

Sanderfoot, A.A., Assaad, F.F. & Raikhel, N.V. (2000) The *Arabidopsis* genome. An abundance of soluble N-ethylmaleimide-sensitive factor adaptor protein receptors. *Plant Physiol.*, **124**, 1558–1569.

Sarafian, V., Kim, Y., Poole, R.J. & Rea, P.A. (1992) Molecular cloning and sequence of cDNA encoding the pyrophosphate-energized vacuolar membrane proton pump of *Arabidopsis thaliana*. *Proc. Natl. Acad. Sci. USA*, **89**, 1775–1779.

Satiat-Jeunemaitre, B. & Hawes, C. (1992) Redistribution of a Golgi glycoprotein in plant cells treated with Brefeldin A. *J. Cell Sci.*, **103**, 1153–1166.

Schekman, R. & Orci, L. (1996) Coat proteins and vesicle budding. *Science*, **271**, 1526–1533.

Shimada, T., Kuroyanagi, M., Nishimura, M. & Hara-Nishimura, I. (1997) A pumpkin 72-kDa membrane protein of precursor-accumulating vesicles has characteristics of a vacuolar sorting receptor. *Plant Cell Physiol.*, **38**, 1414–1420.

Staehelin, L.A. (1997) The plant ER: a dynamic organelle composed of a large number of discrete functional domains. *Plant J.*, **11**, 1151–1165.

Tooze, S.A. (2001) GGAs tie up the loose ends. *Science*, **292**, 1663–1665.

Törmäkangas, K., Hadlington, J.L., Pimpl, P., Hillmer, S., Brandizzi, F., Teeri, T.H. & Denecke, J. (2001) A vacuolar sorting domain may also influence the way in which proteins leave the endoplasmic reticulum. *Plant Cell.*, **13**, 2021–2032.

Ueda, T., Yamaguchi, M., Uchimiya, H. & Nakano, A. (2001) Ara6, a plant-unique novel type Rab GTPase, functions in the endocytic pathway of *Arabidopsis thaliana. EMBO J.*, **20**, 4730–4741.

Watanabe, E., Shimada, T., Kuroyanagi, M., Nishimura, M. & Hara-Nishimura, I. (2002) Calcium-mediated association of a putative vacuolar sorting receptor PV72 with a propeptide of 2S albumin. *J. Biol. Chem.*, **277**, 8708–8715.

Wee, E.G.T., Sherrier, D.J., Prime, T.A. & Dupree, P. (1998) Targeting of active sialytransferase to the plant Golgi apparatus. *Plant Cell*, **10**, 1759–1768.

Yeh, K.-W., Chen, J.-C., Lin, M.-I., Chen, Y.-M. & Lin, C.-Y. (1997) Functional activity of sporamin from sweet potato (*Ipomoea batatas* Lam.): A tuber storage protein with trypsin inhibitory activity. *Plant Mol. Biol.*, **33**, 565–570.

Yeyeodu, S., Ahn, K., Madden, V., Chapman, R., Song, L. & Erickson, A.H. (2000) Procathepsin L self-association as a mechanism for selective secretion. *Traffic*, **1**, 724–737.

9 Sorting of storage proteins in the plant Golgi apparatus

Giselbert Hinz and Eliot M. Herman

9.1 Introduction

Plants store proteins in embryonic and vegetative cells to provide metabolic resources for subsequent stages of growth and development (Shewry, 1995; Shewry *et al.*, 1995; Müntz, 1998, for review). Breaking dormancy, whether from a storage state such as in seeds or renewal of growth after a seasonal dormant state, is a critical stage in the life cycle of plants. Among stored reserve systems, seeds are the best studied example of this phenomena. The proteins synthesized and stored during seed maturation become mobilized during germination to support the rapid growth of the new seedling until it is capable of independent autotrophic growth. Other parallel examples of this developmental pattern are the storage of proteins in bark, tubers, and other organs that are required to support spring regrowth after the winter dormancy. Similarly, many vegetative organs, for example soybeans, accumulate vegetative storage proteins in leaves during vegetative growth. These are mobilized following seed set with the recycled nitrogen and carbon being made available to support the formation of storage substances accumulated in seeds. In human affairs, it is the seeds of rice, wheat, maize, and soybeans that are the great agricultural commodities, and the storage of protein in these seeds accounts for a large fraction of the available amino acids for both humans and animals throughout the world. With an increasing world population and the critical need for balanced nutrition, the molecular biology and biochemistry of storage proteins as well as the cellular and physiological mechanisms regulating their synthesis are of practical as well as academic interest. As biotechnological approaches are adopted to modify seed components to increase the quality of seed proteins and to produce specialty products, understanding the biological mechanisms of seed protein accumulation is essential to assure that these products are correctly processed and accumulated in seeds.

9.2 Seed storage proteins and ancillary proteins are stored in two different types of storage organelles, protein bodies and protein storage vacuoles

Seed protein storage organelles are termed protein bodies (PBs) and protein storage vacuoles (PSV) with both terms often used in the literature to specify the same organelles. This confusion resulted from cytological definition of these organelles prior to the elucidation of the molecular and biochemical details of the included proteins that resulted in a clear differentiation between PBs and PSV. PBs are now defined as organelles that are derived by direct budding from the ER and possess a limiting membrane derived from the ER that retains bound ribosomes. In maize endosperm, the PBs are not budded from the ER and remain connected at the terminal end of the ER. PBs are osmotically inactive and do not appear to have any active transport processes. PBs appear to contain only storage proteins and lack any other type of proteins except for a small quantity of ER lumen proteins that are likely incidentally captured in the PB matrix during their formation. The matrix contained within the PBs can be either amorphous or contains multiple domains that sequester aggregated storage proteins (Lending *et al.*, 1988; Lending & Larkins, 1989; Coleman *et al.*, 1996). PBs until recently appeared to be restricted to cereals as organelles that contain hydrophobic storage proteins but are also found in dicotyledonous seeds. The zibethins, unglycosylated storage albumins in the cotyledons of the durian seeds (*Durio zibethicus* L.), are retained in swollen regions of the ER which later form the PBs (Brown *et al.*, 2001). Shortly before the onset of desiccation, when the storage parenchyma cells are completely filled with PSVs, the 11S storage globulin legumin of broad bean (Adler & Müntz, 1983) and pea (Robinson *et al.*, 1995) is retained in the ER lumen forming aggregates in the dilated cisternae of smooth ER.

9.3   Origin of PSVs and relationship to lytic vacuole

PSVs are a different type of organelle and are analogous to the vegetative or lytic vacuoles found in most plant cells (Wink, 1993). Although this was not initially appreciated because the PSV is filled with protein, the PSVs are bound by a tonoplast that contains active transporters that provide an ATP and pyrophosphate-driven proton transport that maintains an acid lytic environment within the PSV. The presence of a distinct set of water channels or aquaporins (tonoplast intrinsic protein, TIP) in their membrane distinguishes PSV from the vegetative or lytic vacuoles: α- and δ-TIP isoforms characterize the tonoplast of the PSV, whereas the tonoplast of the lytic vacuole contains the aquaporin isoform γ-TIP (Jauh *et al.*, 1999). While the TIP family is largely being characterized with the three distinct general isoforms, analysis of the sequences within GenBank show other intermediate sequences that may have roles in specialized vacuoles of other organs or in transitional states from one type of vacuole to another. For instance, vacuoles containing a protein cross-reactive with an anti-α-TIP antibody are not only present in seed tissue but have also been detected in barley and pea roots storing a seed-lectin (Paris *et al.*, 1996). Moreover, vacuoles containing proteins cross-reactive with anti-α- and γ-TIP antibodies can be present in a single cell at a given time (Hoh *et al.*, 1995; Paris *et al.*, 1996). In roots, these two vacuoles later fuse to form the central vacuole. In developing pea cotyledons, the α-TIP vacuole seems to develop *de novo* after the onset of the maturation phase whereas the γ-TIP vacuole becomes degraded, probably by autophagy (Hoh *et al.*, 1995). A tubular-like, smooth membrane system filled with storage proteins first surrounds the vegetative vacuole. The tubular membrane then dilates and the surrounded vegetative vacuole becomes degraded. α-TIP is present in the tonoplast of the newly formed PSV, but not in the tonoplast of the degraded vegetative vacuole, and, *vice versa*, γ-TIP is only detectable in the tonoplast of the vegetative vacuole. Later, the α-TIP PSV fragments into several hundred μm sized small PSVs that are completely filled with storage protein (Craig *et al.*, 1979, 1980). This fragmentation leads to an increase of about hundred times in surface area as compared to the central PSV from which the small ones originate (Craig *et al.*, 1979, 1980). This significant increase in surface area indicates that, in addition to the transport of the soluble storage proteins, there must be a net flux of membrane into the PSV. This is by no means self-evident because the lytic or vegetative vacuole shows a steady state in membrane transport. This steady state is maintained by the presence of a *filter-organelle*, the so-called prevacuolar compartment (PVC), probably homologous to the mammalian or yeast late-endosome (Bassham *et al.*, 1995; Sanderfoot *et al.*, 1998, as reviewed in Robinson *et al.*, 2000). Because vacuolar sorting vesicles arriving from the Golgi release their cargo into these PVC and the receptors are recycled back to the Golgi by vesicles, no net loss of membrane from the Golgi to the vacuole or the PVC can occur. This probably does not happen in seed cells. Accordingly, in pea seeds α-TIP is sorted via the same vacuolar tranport vesicles as the storage globulins are (Hinz *et al.*, 1999). Whether there is a specialized PVC on the pathway to the PSV is still a matter of debate. Evidence for this proposal comes from morphological observations demonstrating the presence of so-called multivesicular bodies filled with storage protein and which appear to fuse with PSV (Robinson *et al.*, 1998a).

Cotyledons and roots are not the only tissues where two vacuoles have been observed simultaneously. In tobacco leaves, two distinct vacuoles can be distinguished by the different acidity of their lumen (Di Sansebastiano *et al.*, 1998) which later fuse during cell development forming the large central vacuole (Di Sansebastiano *et al.*, 2001). Whether the tonoplast of these two vacuoles also contain different TIP-isoforms has not been examined. An additional vacuole, a so-called secondary vacuole, besides the PSV has been described in barley aleurone (Swanson *et al.*, 1998) which, however, also contained α-TIP like the PSV.

Although vacuolar proteins are sorted via the Golgi apparatus the origin of the new vacuoles after mitosis, and the extent to which they are Golgi or ER derived, is still a matter of debate (as reviewed by Robinson & Hinz, 1997; Marty, 1999) as is the origin of the newly formed PSV in the pea seeds. Based on the immunological observation that the ER chaperone BiP is present in the PSV of pea cotyledons, it has been proposed that the PSV originates from the ER rather than from the Golgi (Robinson *et al.*, 1995). BiP carries the carboxyterminal ER-retention signal KDEL and is a major ER-lumen constituent (Pelham, 1988; Hadlington & Denecke, 2000; see Chapter 6), and hence the presence of BiP may be due to the PSV having an ER-derived ontogeny. This is supported by the observation that the glutelin-containing vacuole of rice endosperm, which is analogous to the PSV (Krishnan *et al.*, 1986), also contains other KDEL-bearing reticuloplasmins like calreticulin (Torres *et al.*, 2001).

The primary storage proteins of PSVs are the 7S and 11S seed storage globulins which are members of large gene families and are the most prominent PSV constituents (see Shewry, 1995; Shewry *et al.*, 1995 for review). The storage globulins are coded by multigene families (Casey, 1979; Müntz, 1998; Casey & Domoney, 1999) and are further classified into two groups with respect to their sedimentation coefficients: the 11S globulins of the legumin type (like pea legumin and soybean glycinin) and the 7S globulins of the vicilin type (like pea vicilin and bean phaseolin). They are found in diverse species of plants. Storage proteins have an ancient origin as indicated by their presence in *Ginkgo biloba* seeds, conifers as well as in extensively characterized dicotyledonous and monocotyledonous seeds (Higuchi & Fukazawa, 1987; Häger *et al.*, 1995). These proteins appear to have evolved from a common ancestor (Shutov *et al.*, 1995). In PSVs that possess crystalloids, the 7S proteins are in the peripheral matrix, while the 11S proteins are the primary constituent of the crystalloid (Hara-Nishimura *et al.*, 1985; Jiang *et al.*, 2001). In other species, the 7S and 11S proteins are uniformly codistributed throughout the PSV matrix. In the ER, the precursor polypeptides of the 11S globulins form trimers with a sedimentation coefficient of about 9S and a molecular mass of 180 kDa. After import into the PSV, the precursor polypeptides are proteolytically processed at a conserved asparagine residue by vacuolar processing enzyme, VPE (Hara-Nishimura *et al.*, 1993; reviewed by Müntz, 1996). Each 60 kDa monomer is cleaved into two chains with molecular masses of 40 (α-chain) and 20 (β-chain) kDa. These two chains remain linked via an intramolecular disulfide bridge co-translationally formed in the ER by protein disulfide isomerase. The disulfide bridge is not important for sorting but processing and hexamer formation is impaired by deletion of these disulfide bridges (Jung *et al.*, 1997). After processing, the two trimers form a 11S hexamer with a molecular mass of 360 kDa. In contrast, the 7S proteins are not usually post-translationally processed but are often co-translationally glycosylated and the glycan may be further modified in the Golgi prior to deposition in the vacuole. The 7S globulins also form trimer oligomers in the ER but remain as trimers after deposition in the PSV.

Other proteins are sequestered with the storage proteins, and these may be in sufficient concentration to constitute auxiliary storage proteins. The best characterized of these auxiliary proteins are the *classic* seed lectins, which in some legumes can account for 10% or more of the total protein (see Etzler, 1985 for review). Closely related to lectins are the α-amylase inhibitors that may function to defend mature dry seeds against insect feeding (see Chrispeels & Raikhel, 1991 for review). Other PSV-localized defense proteins include the Kunitz-type trypsin inhibitor (Horisberger & Volanthen, 1983) and P34 (Kalinski *et al.*, 1992), a distantly related member of the papain superfamily that binds an elicitor derived from *Pseudomonas* (Cheng *et al.*, 1998). Storing defense proteins in dry seeds may anticipate insects feeding on the seed when it is unable to respond with an inducible reaction. The other ancillary proteins stored in the PSVs include an array of hydrolytic enzymes capable of completely degrading macromolecules. Many of these hydrolases are accumulated simultaneously with the storage proteins and include glycosidases, phosphatases, phospholipase D and nucleases (Nishimura & Beevers, 1978; Mettler & Beevers, 1979; Chappell *et al.*, 1980; Herman & Chrispeels, 1980; Van der Wilden *et al.*, 1980). It is a paradox that the vacuole is both the cellular compartment that stores reserve proteins and the cell's lytic compartment that is significant in view of more recent observations that have shown that the PSV of maturing seeds can function as general lytic compartment.

The PSV deposition of the 7S globulins is not influenced by its glycosylation. Experimentally blocking glycosylation either by inhibitors or by mutating the N-glycosylation sites has been tested for its effect on targeting. Tunicamycin, an inhibitor that blocks core N-glycosylation does not block storage protein exit from the ER and sorting into the PSV (Bollini *et al.*, 1985). The one notable exception is phaseolin from *Phaseolus lunatus*, the lima bean, which is specifically retained in the ER after tunicamycin treatment due to misfolding and subsequent aggregation (Sparvoli *et al.*, 2000). In contrast, site-directed mutagenesis to eliminate one or both of the glycosylation sites of the seed lectin PHA did not impair the correct targeting of the mutant protein to the vacuole when expressed in transgenic tobacco seeds (Sturm *et al.*, 1988).

9.4   Golgi-mediated sorting of storage proteins into the PSVs

In the seeds of dicotyledonous legumes such as pea or bean the storage proteins pass through the Golgi apparatus, where they become segregated from other proteins of the secretory pathway, and are sorted into Golgi-derived transport vesicles. The movement of proteins from ER to Golgi apparatus appears to be specific for the tubular domain of the ER network. Harris (1979) observed developing seeds with thick section high voltage EM using Zn-I-Os-impregnated tissue that densely stained the endomembrane system. The Golgi apparatus was found closely aligned to the tubular ER, which indicates that this subdomain may have specific functions in secretion. Freeze fracture observations (Herman *et al.*, 1984) also showed that the *cis*-Golgi is aligned to the tubular ER. Both thin section and freeze fracture electron microscopy showed the Golgi apparatus adjacent to the tubular ER with 50 nm (putative) transit vesicles bridging the small space between the ER and the Golgi apparatus. Whether these vesicles are identical to COPII vesicles, the transport vesicles sorting proteins from the ER to the *cis*-Golgi conserved in all eukaryotic cells (Robinson *et al.*, 1998b) remain to be elucidated. The Golgi apparatus in cotyledon cells is comprised of five or six closely apposed flattened cisternae that exhibit *cis* to *trans* polarity, with increasing electron density in parallel with decreasing width of the cisternae. At the *cis* end of the Golgi there is a close association with the ER (see Harris *et al.*, 1989 for example), and at the *trans* end, there are electron-dense secretion vesicles as well as an extended *trans*-Golgi-network (TGN) that consists of elongated smooth membranes that interact with clathrin-coated vesicles (Pesacreta & Lucas, 1985).

The first indications that storage proteins are deposited in the vacuole *via* the Golgi apparatus were based on the interpretations of conventional electron micrographs showing proteinaceous electron-dense vesicles that were postulated to carry storage proteins exiting the *trans* face of the Golgi. This hypothesis was confirmed in the early 1980s, when several investigators published papers using immunogold techniques to show that storage proteins are present in the ER, Golgi apparatus, and secretory vesicles forming PSVs (see Craig & Goodchild, 1984; Herman & Shannon, 1984a,b; Greenwood & Chrispeels, 1985; Kim *et al.*, 1988; reviewed in Robinson & Hinz, 1999). The first biochemical evidence for this transport route was based on pulse-chase experiments followed by subsequent subcellular fractionation (Chrispeels *et al.*, 1982a,b; Chrispeels, 1983). Small dense vesicles containing post-Golgi processed proteins were isolated and they appeared to be homologous to vesicles exiting the Golgi observed under the electron microscope (Fig. 9.1). This route is substantiated by the fact that several storage proteins in legume seeds are complex glycosylated (as summarised in Robinson & Hinz, 1999). The Golgi apparatus processes the ER-derived high mannose glycan side chains of vacuolar glycoproteins (Chrispeels, 1983; Faye *et al.*, 1986; see Staehelin & Moore, 1995; Lerouge *et al.*, 1998; Chapter 7, for reviews). These glycan side chains are sequentially trimmed to remove mannosyl residues, and additional xylosyl, fucosyl and/or galactosyl residues may be added. Glycan processing is most often assayed by testing for the acquisition of endoglyosidase H resistance due to fucosyl attachment to *N*-acetylglucosamine #1 of the side chain. Subcellular fractionation of lysates from maturing seeds using isopycnic sucrose gradient centrifugation showed that the UDP-sugar transfer occurs in the Golgi apparatus fractions (Sturm *et al.*, 1988). The position and accessibility of the glycan side chain is a significant factor in the subsequent glycan processing in the Golgi apparatus (Faye *et al.*, 1986). Immunogold labeling of the plant Golgi apparatus with antibodies specific for various glycosyl residues has shown that the *cis*, medial and *trans* domains are differentiated (see Moore *et al.*, 1991) to mediate progressive steps in glycan processing, protein packaging, and targeting. The acquisition of xylosyl residues on a storage protein of soybean aleurone cells was visualized by immunogold EM, and it was shown that attachment of the terminal glycosyl residues occurs in the medial to *trans* domain of the ER (Yaklich & Herman, 1995). The Golgi apparatus also mediates the formation of cell wall glycans and O-linked glycosylation of cell wall proteins, such as extensin and arabinogalactan proteins. However, *O*-linked glycans have not been observed on major seed PSV proteins, although the sweet potato tuber vacuolar storage protein, sporamin, does possess *O*-linked glycan side chains (Matsuoka *et al.*, 1995).

Experimental manipulation of Golgi transport functions using drugs that disrupt the Golgi function like brefeldin A (Satiat-Jeunemaitre *et al.*, 1996) and monensin (Mollenhauer *et al.*, 1990) show that both inhibitors effectively block the sorting

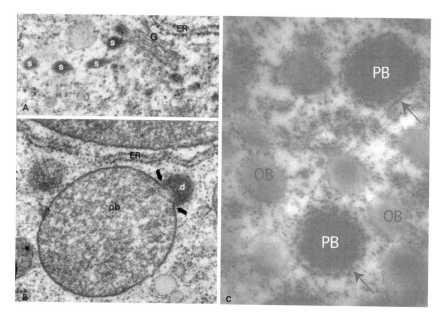

Figure 9.1 (A) The Golgi and ER of midmaturation soybean cotyledon is shown in a conventional electron micrograph. The Golgi apparatus is aligned with an adjacent segment of ER. Golgi secretion vesicles are shown associated with the *trans* cisternae and in the adjacent cytoplasm. The Golgi secretion vesicles are highly electron dense that results from the packaging of storage proteins. G – Golgi, ER – endoplasmic reticulum, S, secretion vesicles. (B) A conventional electron micrograph that shows the fusion of a Golgi-derived vesicle with the tonoplast of the protein storage vacuole (PSV) of a midmaturation soybean cotyledon cell. Note the continuity between the limiting membrane of the dense vesicle and the PSV. (C) Gene silencing of the 7S storage protein conglycinin in transgenic soybeans results in the upregulation of 11S glycinin to replace the lost conglycinin. Much of the upregulated glycinin accumulated remains as proglycinin that is accumulated in ER-derived protein bodies (PB). The PBs possess a limited membrane with bound ribosomes and a dense protein core. OB – oil body.

into the PSV (Craig & Goodchild, 1984; Gomez & Chrispeels, 1993; Matsuoka *et al.*, 1995) and disturb the ultrastructure of the Golgi apparatus (Craig & Goodchild, 1984; Robinson *et al.*, 1997, 1998a).

9.5 Vacuolar sorting signals are recognized by a Golgi-localized sorting machinery

The difference between Golgi-sorted vacuolar proteins and those bypassing the Golgi apparatus as well as proteins that are secreted is the presence of positive vacuolar sorting information in the form of specific polypeptide targeting signals that are both necessary and sufficient for correct sorting into the PSV. In mammalian cells as well as in yeast, segregation of soluble lysosomal/vacuolar enzymes from the bulk flow of secretory proteins occurs at the *trans*-Golgi network of the Golgi apparatus, and is mediated by the specific interaction of a sorting-determinant in the cargo protein with receptor proteins in the cisternal membrane. These ligand/receptor complexes are then recruited into a specific class of vacuolar transport vesicles, the clathrin-coated vesicles (CCV). This highly coordinated and regulated formation and budding of these vesicles by successive recruitment of distinct classes of coat proteins out of cytosolic pools seems to be conserved between eukaryotic cells and is thus also present in plant cells (as reviewed in Robinson *et al.*, 1998b; Rouillé *et al.*, 2000; Chapter 8).

Vacuolar hydrolases possess a sequence-specific vacuolar sorting sequence located at the N-terminus, the C-terminus, or internally (ssVSS) (Neuhaus & Rogers, 1998; Matsuoka & Neuhaus, 1999; Vitale & Raikhel, 1999; Frigerio *et al.*, 2001a; Chapter 8). This signal consists of a larger, charged amino acid (preferably N) at the first position, a non-acidic amino acid (P) at the second position, a large hydrophobic amino acid (I, L) at the third position, the fourth position is not strictly conserved, and an amino acid with a large hydrophobic side chain (L, P) at the fifth position (Matsuoka & Nakamura, 1999; Matsuoka & Neuhaus, 1999). The N-terminal NPIR motif from barley aleurain binds with high affinity (k_D of 37 nM) to a putative vacuolar sorting receptor (VSR) present in clathrin-coated vesicles (CCV) (Kirsch *et al.*, 1994; Chapter 8). Plant cells, however, unlike yeast or mammalian cells, may harbor functionally distinct vacuoles at a given time. The plant secretory system has to discriminate between these distinct vacuolar compartments and this complexity is reflected by distinct transport pathways of lumenal, soluble proteins into the different vacuoles. Among these, the transport of vacuolar hydrolases bearing ssVSS into the lytic, γ-TIP vacuole may resemble the *classical* CCV-based sorting machinery (Chapter 8). The targeting sequences and receptors that recognize the 7S and 11S storage proteins, however, have proven difficult to elucidate.

The various targeting signals identified in storage proteins do not exhibit obvious sequence homology (Matsuoka & Neuhaus, 1999). Vacuolar targeting sequences can be grouped into two classes, those that are C-terminal sorting sequences (ctVSS) with very low sequence specificity (the 7S globulin of bean seeds, phaseolin, Frigerio *et al.*, 1998), barely lectin (Dombrowski *et al.*, 1993), and those that are protein structure-dependent sorting determinants (psVSS) located in the sequence of the mature protein itself (legumin, the 11S storage globulin of pea seeds, Saalbach *et al.*, 1991, phytohemagglutinin, a storage lectin in bean seeds, van Schaewen & Chrispeels, 1993). In contrast to the ssVSS, which can be located more intrinsically inside the N- or C-terminal peptides of the protein, ctVSS are sensitive to the addition of a terminal glycin residue (Frigerio *et al.*, 1998), which means that they must be freely accessible. The sorting information present in the ctVSS is sufficient to enable the Golgi apparatus to discriminate between the storage and the lytic vacuole, *in vivo*: GFP-reporter constructs of either the ctVSS of tobacco chitinase A or the ssVSS of barley aleurain were correctly sorted into their respective target vacuole when co-expressed in the same cell (Di Sansebastiano *et al.*, 2001). A single protein may possess two distinct vacuolar sorting domains. PT20, a 20 kDa protein from potato tuber, thus possesses both a ssVSS at the N-terminus and a ctVSS at the C-terminus (Koide *et al.*, 1999). This observation may indicate that this protein may become targeted to both vacuoles.

The precursor polypeptides of the storage globulins form homotrimers soon afer their import in the lumen of the ER (Müntz, 1998), and, as has been shown by Holkeri and Vitale (2001), the vacuolar sorting determinants of such a homotrimer act cumulatively, one is not sufficient for correct sorting into the PSV. This is in contrast to the sorting of vacuolar hydrolases (Rouillé *et al.*, 2000) but is consistent with the analysis of the crystal structure of a proglycinin (the 11S storage globulin of soybean) homotrimer (Adachi *et al.*, 2001) in that all the three C-termini of this proglycinin trimer are exposed outside of the quarternary structure of the trimer and therefore accessible for interaction with the sorting-machinery.

Neither the ctVSS nor the psVSS bind with high affinity to the VSR (Kirsch *et al.*, 1994, 1996). However, a second, low affinity binding site (k_D of 100 µM) for the ctVSS has been described in the pea VSR BP-80 *in vitro* (Cao *et al.*, 2000). Therefore, because of the high concentration of precursor polypeptides of the storage globulins present in the secretory pathway of legume seeds (Hinz *et al.*, 1997), even this very low binding affinity would enable the storage globulins to compete with the binding of vacuolar hydrolases to the receptor. Because the vesicles transporting the pea storage globulins do not contain BP-80 *in situ* (Hinz *et al.*, 1999), these results may either indicate an additional segregation mechanism preventing this competition, or the presence of a receptor-retrieval system which removes the misrouted receptor proteins, or both.

9.6   Golgi apparatus transport vesicles in seeds

CCVs are a constitutive part of the secretory pathway (Robinson *et al.*, 1998b) and as such they are also detectable on the *trans*-Golgi cisternae of seed tissue, as has been shown in developing pea cotyledons (Hinz *et al.*, 1993; Hohl *et al.*, 1996). Accordingly, members of the VSR protein family (Paris & Rogers, 1996) are also expressed in seed storage tissue of dicotyledonous (Kirsch *et al.*, 1994) as well as monocotyledonous (Shy *et al.*, 2001) plants. The first member of this family to be identified has been purified out of a CCV fraction isolated from developing pea seeds (Kirsch *et al.*, 1994). However, biochemical as well as ultrastructural evidence led to the conclusion that CCV are not involved in the sorting of storage proteins.

Phosphorylated phosphoinositol-lipids play a key role in the formation of membrane buds during vesicle formation and they are also important for the fusion between the two lipid-bilayers of a docked vesicle and the acceptor membrane (Corvera, 2001). Thus, interfering with the phosphorylation-status of these lipids in turn interferes with vesicle-mediated protein sorting. Wortmannin is an inhibitor blocking the activity of phosphatidylinositol-3-kinase and thus the formation of phosphoinositol-3-phosphate. It is known to block CCV-mediated sorting in mammalian cells (Schu *et al.*, 1993). Matsuoka and co-workers (Matsuoka *et al.*, 1995) also demonstrated that, when co-expressed in the same cell, the transport of reporter constructs with either a ctVSS or the ssVSS was differentially inhibited by the inhibitor Wortmannin, which was the first experimental evidence for two distinct Golgi-based vacuolar sorting machineries. A direct proof for the participation of phosphoinositol-3-phosphate in vacuolar sorting in plants has been published quite recently (Kim *et al.*, 2001).

The presence of two distinct Golgi-based vacuolar sorting pathways has been confirmed ultrastructurally. Storage proteins and PSV lectins and enzymes were localized to characteristic Golgi-derived vesicles, so called dense vesicles (DV) (Craig & Goodchild, 1984; Herman & Shannon, 1984b; zur Nieden *et al.*, 1984; Herman & Shannon, 1985; Krishnan *et al.*, 1986; Kim *et al.*, 1988; Hohl *et al.*, 1996), which were clearly distinguishable from the CCV present on the same Golgi stack, as shown in Fig. 9.2 (Hohl *et al.*, 1996). Based on refined quantitative

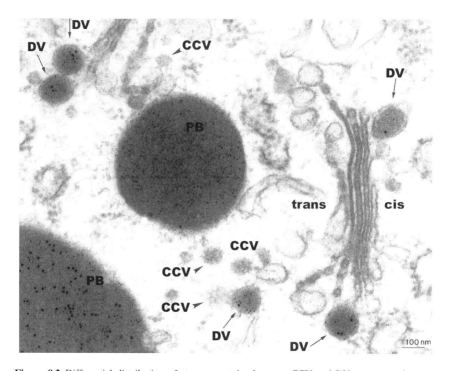

Figure 9.2 Differential distribution of storage proteins between CCV and DV present at the same Golgi stack of developing pea cotyledons. Postembedding immunogold labeling of pea cotyledons with antisera against vicilin (12 nm gold) and legumin (5 nm gold). CCV (arrowheads) and DV (arrows) are both present but only the DV are labeled with storage protein antisera. PSV are marked by PB. c and t, respectively, indicates the *cis* and *trans* side of the Golgi stack. The bar is 100 nm. (Modified from Hohl *et al.*, 1996.)

immuno electronmicroscopy in combination with biochemical methods, it was further shown that vacuolar storage proteins are not detectable in purified CCV and that, in turn, the vacuolar storage receptor BP-80 is not detectable in the DV (Hinz *et al.*, 1999). These two sorting mechanisms enable the Golgi to recognize the distinct sorting information present in the different vacuolar proteins and thus to discriminate between lytic and storage vacuoles in a given cell (Di Sansebastiano *et al.*, 2001).

9.7   Putative sorting mechanisms

These observations raise the question as to how these proteins are sorted into the DV? Are the storage proteins sorted into the budding DV by a receptor or is it sufficient that they are simply excluded from the CCV-mediated sorting? However, the latter assumption would not be the presumed mechanism because the default compartment of the secretory pathway in seeds is the cell wall and not the vacuole. Perturbing the Golgi function with the ionophore monensin (Craig & Goodchild, 1984; Robinson *et al.*, 1998a) as well as deleting the ctVSS (Neuhaus *et al.*, 1994; Frigerio *et al.*, 1998) leads to a secretion of storage proteins. Storage protein sorting appears to be an active process based on the interaction of well defined, but poorly understood signals in the protein with some kind of receptor. Several observations suggest that such a receptor might not function as the VSR in the CCV-mediated pathway (as reviewed in Robinson *et al.*, 1998b; Rouillé *et al.*, 2000).

The observation made by Holkeri and Vitale (2001) that the three sorting signals present in a homotrimer act in a cumulative manner is quite important, because it is the first direct experimental evidence for an *unusual* type of sorting mechanism. Indirect evidence also comes from electron microscopy in combination with data obtained by structural biology. Pea cotyledon DV have a diameter of about 150 nm and are filled with compact storage protein lumps which are characteristically osmiophilic and electron opaque (Hohl *et al.*, 1996). This structure is not a fixation-artifact, but instead reflects real protein aggregation as is indicated by the fact that the electron-opacity is visible in samples fixed and processed in the absence of osmium, e.g. by high-pressure freezing or cryosectioning (Hinz *et al.*, 1999; Hillmer *et al.*, 2001). The proglycinin trimer, the precursor of the 11S storage globulin of soybean, is arranged around a threefold symmetry axis with dimensions of $9.5 \times 9.5 \times 4.5$ nm (Adachi *et al.*, 2001). It seems very unlikely that such a large structure ligand can bind to a membrane receptor when located inside a large protein conglomerate. It is more likely that such proteins tend to condense (Hinz *et al.*, 1999).

Because storage protein condensation and DV formation seem to be coordinated events (Robinson *et al.*, 1997) and are spatially restricted to the rim of the cisternae (Robinson *et al.*, 1997), condensation may be necessary in the sorting process. This has been discussed earlier in comparison to the DV-mediated sorting with the sorting of regulated secreted proteins in mammalian cells (Vitale &

Chrispeels, 1992; Hinz *et al.*, 1999; Robinson & Hinz, 1999; Hillmer *et al.*, 2001). The relatively high hydrophobicity of the prolegumin-trimer as compared to the mature legumin hexamer (Duranti *et al.*, 1992; Hinz *et al.*, 1997; Adachi *et al.*, 2001) may play an important part in this condensation process. Due to differences in hydrophobicity the hydrophobic prolegumin behaves like a peripheral membrane protein, whereas the more hydrophilic mature legumin dissociates from the membrane after being processed in the PSV (Hinz *et al.*, 1997). The second storage protein to be transported via the DV in pea seeds, the 7S globulin vicilin, shows a similar membrane-binding behavior. Membrane association is relatively tight with the non-processed forms remaining bound even after three consecutive carbonate washings, whereas the processed forms are completely removed from the membrane after the first washing step (Hinz, unpublished). Therefore, it is possible that binding of precursor trimers to the membrane might trigger further aggregate formation and thus sorting (Thiele *et al.*, 1997; Tooze, 1998; Hinz *et al.*, 1999). In mammalian cells, condensation-mediated sorting into the secretory granules is also enhanced by the slightly acidic environment in the *trans*-Gogi cisternae (Thiele *et al.*, 1997; Tooze, 1998; Hinz *et al.*, 1999). Therefore, the observation that the ionophore monensin which abolishes acidification of the *trans*-Golgi also causes disruption of the DV and secretion of the storage proteins (Craig & Goodchild, 1984; Robinson *et al.*, 1998a) may further support this model.

Whether this sorting depends on some kind of receptor or whether aggregation alone is sufficient is still a matter of discussion (Thiele *et al.*, 1997; Tooze, 1998; Robinson & Hinz, 1999). That there might indeed exist a sorting machinery specific for storage proteins is indicated by two observations which demonstrate their secretion when expressed in non-storage tissue: when the synthesis of conglutin γ, a storage glycoprotein in *Lupinus albus* seeds, is induced artificially during seed germination, the protein becomes completely secreted (Duranti *et al.*, 1994). Phyto-hemagglutinin-E, a storage lectin in bean seeds, is also expressed in root tips during early germination. In this tissue, however, this protein is secreted into the cell wall and not sorted into the vacuole (Kjemtrup *et al.*, 1995). Furthermore, Tanchak and Chrispeels (1989) identified *via* a cross-linking approach two proteins of the endomembrane system with molecular masses of 9 and 67 kDa associated with both phaseolin and phytohemagglutinin in the secretory pathway. These proteins, however, have not been characterized any further.

The high concentration of storage proteins in the Golgi apparatus and the low affinity binding site of the VSR BP-80 suggest that there is an additional segregation mechanism or a retrieval system for misrouted receptor proteins. Such mechanism has indeed been demonstrated for developing pea seeds *in situ*. In this tissue the sorting of storage proteins is spatially segregated from the VSR in the same Golgi stack (Hillmer *et al.*, 2001). While the VSR is localized in the *trans*-Golgi cisternae the storage proteins are effectively sorted into DVs in the *cis*-half of the stack, as shown in Fig. 9.3. This observation indicates that BP-80 has a limited, if any, role in sorting. Therefore, DV-sorted storage proteins may not compete with other proteins destined for CCV-based sorting. The cargo transported by the CCV

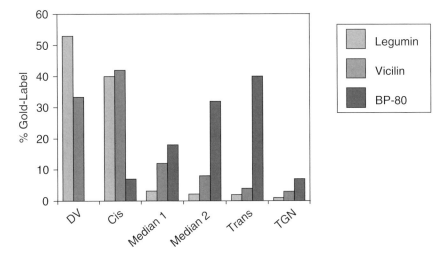

Figure 9.3 Differential distribution of storage proteins and the vacuolar sorting receptor across the Golgi stack. Quantitative analysis of the distribution of immunogold labeling with anti-vicilin, anti-legumin and anti-BP80 antisera across Golgi stacks of developing pea cotyledons. Whereas about 50% of the legumin and 30% of the vicilin label is present in the DV, no BP-80 is detectable in the DV. About 40% of both storage proteins is present in the first *cis*-cisternae of the stack but only about 5% of the BP-80. 70% of the BP-80 is detectable in the *trans* half of the stack but only less than 10% of the storage proteins. (Modified from Hillmer *et al.*, 2001.)

remains to be elucidated, but it cannot be excluded that CCV may be involved in the transport of other PSV proteins excluded from the DV or requiring the function of the BP-80 receptor. In order to test this hypothesis, transgenic plants would have to be constructed which express reporter constructs bearing the ssVSS. This experimental approach might also provide an explanation for a second morphological observation that there are abundant CCV associated with the DV at the *trans*-Golgi of pea cotyledons (Hohl *et al.*, 1996). Because the presence of a clathrin coat is an indicator of vesicle budding and not fusion (Kirchhausen, 2000), CCV might act as a kind of fail-safe mechanism to retrieve either receptor proteins or misrouted cargo. This mechanism has been described, both morphologically and biochemically, in terms of the biogenesis of the secretory granules in mammals (as reviewed in Tooze, 1998; Robinson & Hinz, 1999). But, whereas the lysosomal sorting receptor is present on the immature secretory granules in mammalian cells (Kuliawat *et al.*, 1997), a quantitative analysis of cryosectioned pea cotyledon tissue has shown that DV do not contain VSRs, although the antibody used intensely labeled *trans*-Golgi. (Hinz *et al.*, 1999; Hillmer *et al.*, 2001). However, VSRs are coded for by a multigene family (reviewed in Paris & Rogers, 1996; Hadlington & Denecke, 2000), and hence it is possible that the antibody used did not recognize all members of this family and that a second, more distantly related member of this family might be expressed at the same time, which is not recognized by the antibody. The antibody in question was raised against BP-80, the receptor originally isolated from pea cotyledon CCV. In order to test this hypothesis it might be useful to examine the expression of the different receptor proteins during seed development.

9.8   Vacuolar sorting pathways are not exclusive for a certain cell type

Seeds appear to possess a wide variety of different sorting mechanisms for storage proteins that may be active in parallel. PBs as well as PSVs are present in rice endosperm. The prolamins in rice are stored in ER-derived PBs while the glutelins of rice as well as the storage globulins are stored in PSVs (Yamagata *et al.*, 1982) and sorted via the Golgi apparatus (Krishnan *et al.*, 1986). This Golgi-mediated sorting machinery is also capable of sorting foreign globulins into the correct compartment. In one experiment phaseolin, the 7S storage globulin in garden bean, was expressed in rice seeds and imported into the second, globulin-containing PSV. Because the mature phaseolin present in the PSV was processed to possess complex glycans, it must have been transported via the Golgi apparatus (Zheng *et al.*, 1995).

In the endosperm of wheat the prolamins are sorted via a Golgi-independent route into the PSV and are taken up by autophagy. These cells are strongly reduced in Golgi activity, as determined by the relative expression-level of BP-80 as compared to the expression of ER marker proteins (Shy *et al.*, 2001). Storage globulins are sorted via the Golgi apparatus into the PSV (Kim *et al.*, 1988). When legumin is co-expressed in wheat endosperm it is also targeted to the same PSV (Stöger *et al.*, 2001). These results strengthen the assumption that both the Golgi-independent as well as the Golgi-dependant pathway are active at the same time and that both are led to the same compartment. However, since pea legumin is not glycosylated, it is not possible to decide unequivocally whether the legumin construct was sorted via a Golgi-dependent or Golgi-independent mechanism.

KDEL-bearing proteins may also reach the vacuole via a Golgi-independent pathway. Investigations on the transport of a trimeric phaseolin-KDEL construct led to the observation of an additional Golgi-independent vacuolar delivery pathway in dicotyledonous cells (Frigerio *et al.*, 2001b). A small proportion of this construct escaped the KDEL-retrieval system and became degraded in the vacuole. Because these proteins did not possess Golgi-modified glycans and their transport was inhibited by brefeldin A, but not by monensin, the sorting might indeed have occurred independently of the Golgi apparatus.

9.9   Autophagy constitutes an alternative, novel route bypassing progression through the Golgi apparatus for the vacuolar accumulation of storage proteins during seed maturation

In some cereals autophagy is used to accumulate storage proteins, bypassing the conserved mechanism of Golgi-mediated targeting and transport to the vacuole.

In wheat endosperm, the formation of the PSVs that contain proteins that other-wise would be contained in PBs proceeds by a novel process that uses autophagy (Levanony *et al.*, 1992). Storage proteins assembled in the ER are polymerized into higher order structures (Shimoni & Galili, 1996) that are directly secreted from the ER. The wheat protein PBs do not remain as separate cytosolic PBs but are instead sequestered into provacuoles (Rubin *et al.*, 1992). The origin of the provacu-oles, whether from the Golgi apparatus or directly from the ER has not been determined; however, these vacuoles do carry tonoplast marker proteins, α-TIP (tonoplast integral protein) and pyrophosphatase, which are characteristic of vacuole formation mediated by the Golgi apparatus (G. Galili, Weizmann Institute of Science, Rehovot, Israel, personal communication). The provacuoles containing sequestered PBs fuse with one another, forming one or more large central vacuoles that contain numerous storage protein aggregates. The limiting membrane of the sequestered PB appears to be digested by vacuolar enzymes, releasing the naked prolamin aggregate into the vacuolar sap. The prolamin protein accretions subse-quently aggregate, forming larger PBs.

Autophagy of prolamin-containing PBs can be observed in transgenic plants (Fig. 9.4). Coleman *et al.* (1996) showed that coexpression of α- and γ-zein in tobacco seeds resulted in the production of PBs that appear to be structurally indistin-guishable from maize PBs. The cytosolic PBs possess a rough ER-derived membrane that sequesters a matrix of γ-zein with deposits of α-zein. EM immunocytochemistry showed that the PBs also become sequestered within PSVs by apparent autophagy. Quantitative differences in γ- and α-zein content of the cytoplasmic and PSV-sequestered PBs, respectively, indicate that the α-zein, at least, is unstable once the PB is taken into the vacuole. Bagga *et al.* (1995) described similar results for β/δ-zein-containing PBs formed in transgenic tobacco seeds and leaves. Taken together, these results indicate the autophagic process that occurs in wheat endosperm, where ER-derived PBs are sequestered in vacuoles, can be duplicated by producing PBs in transgenic tobacco, even though tobacco plants do not normally produce ER-derived PBs. Autophagy does not necessarily explain all instances where hydrophobic storage proteins are sequestered in the vacuole.

A novel process for deposition of water-soluble storage proteins has been discovered in developing pumpkin cotyledons. Electron-dense transport vesicles are assembled by ER rather than by Golgi and it is these vesicles that appear to trans-port the storage protein precursor to the storage vacuole. Hara-Nishimura *et al.* (1998) have described DVs that transport the storage protein 2S albumin from the ER to the PSVs. The albumin is synthesized as a precursor that is proteolytically processed after it arrives in the PSVs. These vesicles possess a matrix of pro-2S albumin and are surrounded by an ER-derived membrane with its bound ribosomes. The pro-2S albumin is processed to the mature form in the vacuole and as the consequence of the action of vacuole-specific enzyme(s). However, there may be further complexities in the biogenesis of PAC vesicles because they appear to contain proteins with complex glycans, indicating processing by Golgi enzymes. Further evidence of Golgi interaction with the PAC vesicles is seen in the presence of VSR PV72 in the PAC vesicle membrane (Shimada *et al.*, 1997). PV72 binds with a k_D of 200 nM to the internal sorting sequence of 2S albumin (Watanabe *et al.*, 2002). But, in contrast to the binding of BP-80, the pro-aleurain sorting signal binding of PV72 to 2S-albumin is pH independent but Ca^{2+} sensitive (Kirsch *et al.*, 1994; Watanabe *et al.*, 2002). VSRs are normally localized in the Golgi where they function to sort proteins directed to the vacuole from those destined for secretion. Whether this indicates a Golgi contribution to the PAC vesicles after their assembly or whether Golgi-derived material is transported in a retrograde manner into the ER and then incorporated into the PAC vesicles still needs to be resolved. Vegetative expression under the control of the 35S promoter of the entire 2S-albumin coding sequence fused to phosphinothricin acetyltransferase (PAT) or the N-terminal half of albumin fused to PAT in *Arabidopsis* plants replicates the intrinsic PAC body formation in pumpkin seeds (Hayashi *et al.*, 1999). Aggregates of the pro-2S-albumin fusion protein surrounded by an ER-derived membrane are present in the cytoplasm. This may indicate that aggregation of the 2S protein is an intrinsic property of this protein, and it may be the cause of the *early* (prior to reaching the Golgi) formation of PAC transport vesicles.

Altering storage protein composition of seeds can also induce the formation of ER-derived compartments that resemble the PB vesicles (Kinney *et al.*, 2001) redirecting those proteins from progression through the Golgi. Silencing of the 7S conglycinin of soybean by a transgene that includes the conglycinin promoter, results in transgenic seeds where glycinin accumulation is increased to compensate for the loss of conglycinin. The compensating glycinin is accumulated in the form of proglycinin and is the product of only one of the six glycinin genes. The proglycinin is accumulated as (~0.5 µm diameter) dense PBs surrounded by an ER-derived membrane with bound ribosomes (Fig. 9.1). The membrane of the dense bodies is not labeled with antibodies to the PSV-specific tonoplast protein α-TIP. These vesicles are not present in wild type soybeans. Normally, glycinin and conglycinin are synthesized together and both are present in the ER and are deposited in the PSV by Golgi-mediated trafficking. Why the absence of β-conglycinin causes glycinin-PB vesicles to form is not clear. That only a single glycinin gene is induced to compensate for the loss of conglycinin, and that this gene product aggregates in the ER to form the PB raises a host of unanswered questions. The sequence of all the glycinin gene products is very similar and there is no readily apparent reason why one of the six simultaneously produced glycinin gene products should be segregated from the other five aggregated to form the PB is an interesting and unresolved question. The induced PB accumulate in the cytoplasm of the storage parenchyma cells and are not taken into the PSV by autophagy.

9.10   Perspectives of future research

There are considerable complexities in the movement of nascent proteins from ER and their eventual assembly into protein storage organelles. There are at least three

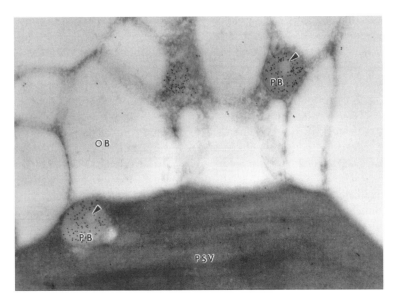

Figure 9.4 The autophagy of zein-containing protein bodies (PB) is shown. Transgenic tobacco seeds synthesizing α- and γ-zein proteins that accumulates in ER-derived PB. The α-zein forms discrete domains or locules within a matrix of γ-zein. The PBs are accumulated in the cytoplasm and are also sequestered in the protein storage vacuole (PSV) where the PBs and their contents are degraded.

primary pathways, two of which are Golgi-mediated employing either the CCV or DV pathway and the third where proteins are assembled in ER-derived protein bodies that are directly deposited into the vacuole, perhaps without interacting with the Golgi. Further complicating this is that two or more different pathways may be employed within the same cell and at the same time to move different proteins to the same vacuolar destination or to different vacuoles. Whether the utilization of these pathways remains constant during seed maturation or instead modulated during the course of development remains a question. There is considerable literature on the effects of plant nutrition and environmental conditions on the composition of seed proteins as well as membrane lipids. Whether development or growth conditions further modulates the pathways used for deposition of storage proteins remains an open question. Much research remains to be undertaken that is directed at elucidating the complexities and control of how various proteins are directed through the different pathways used to produce protein storage compartments.

As a temporally specialized unit of the secretory pathway the proteins involved in routing and mediating the assembly of protein storage vacuoles may be functionally equivalent to the gene products used in vegetative cells and yet appear to be distinct embryo specific genes. For instance, the so-called vesicular sorting receptors, the SNAREs (Jahn & Südhof, 1999) are one example. As discussed by Raikhel and Sanderfoot (Sanderfoot *et al.*, 2000; Sanderfoot & Raikhel, 2001) the small plant *Arabidopsis thaliana* has significantly more SNARE proteins than a human! These proteins are not redundant in their function (Sanderfoot *et al.*, 2001), thus strengthening the hypothesis of a more complex organisation of the plant secretory pathway as compared to mammals or yeast. In order to avoid a *mixing* of vacuolar content, perhaps the DV- and the CCV-based vacuolar sorting pathways are characterized by different sets of SNARE-pairs when both pathways are active in the same cell. It would be of interest to identify CCV- and DV-specific SNAREs, and to study their expression in different tissues and during cell development and their protein–protein interactions to determine how this specificity may be explained. There are opportunities derived from progress in whole genome sequencing of *Arabidopsis* and rice as well as the genome programs on other plants to determine which genes are specifically used during seed development. With the added investigative tools of insertion knockouts, expression of transgenes, and gene silencing it may be feasible to alter the balance of the pathways from the ER to the vacuole and determine the underlying reasons for the apparent redundancy of deposition mechanisms.

References

Adachi, M., Takenaka, Y., Gidamis, A.B., Mikami, B. & Utsumi, S. (2001) Crystal structure of soybean proglycinin A1aB1b homotrimer. *J. Mol. Biol.*, **305**, 291–305.

Adler, K. & Müntz, K. (1983) Origin and development of protein bodies in cotyledons of *Vicia faba*. Proposal of a uniform mechanism. *Planta*, **157**, 401–410.

Bagga, S., Adams, H., Kemp, J.D. & Sengupta-Gopalan, C. (1995) Accumulation of 15-kilodalton zein in novel protein bodies in transgenic tobacco. *Plant Physiol.*, **107**, 13–23.

Bassham, D.C., Gal, S., Conceicao, A.D. & Raikhel, N.V. (1995) An *Arabidopsis* syntaxin homologue isolated by functional complementation of a yeast *pep12* mutant. *Proc. Natl. Acad. Sci. USA*, **92**, 7262–7266.

Bollini, R., Ceriotti, A., Daminati, M.G. & Vitale, A. (1985) Glycosylation is not needed for intracellular transport of phytohemagglutinin in developing *Phaseolus vulgaris* cotyledons and for maintenance of its biological activity. *Physiol. Plant.*, **65**, 15–22.

Brown, M.J., Hor, Y.L. & Greenwood, J.S. (2001) Reserve accumulation and protein storage vacuole formation during development of recalcitrant seeds of *Durio zibethinus* L. *Seed Sci. Res.*, **11**, 293–303.

Cao, X., Rogers, S.W., Butler, J., Beevers, L. & Rogers, J.C. (2000) Structural requirements for ligand binding by a plant vacuolar sorting receptor. *Plant Cell*, **12**, 493–506.

Casey, R. (1979) Genetic variability in the structure of the α-subunits of legumin from *Pisum* – a two dimensional electrophoresis study. *Heredity*, **43**, 265–272.

Casey, R. & Domoney, C. (1999) Pea globulins, in *Seed Proteins* (eds P.R. Shewry & R. Casey), Kluwer Academic Publishers, Dordrecht, The Netherlands, pp. 171–208.

Chappell, J., Van der Wilden, W. & Chrispeels, M.J. (1980) The biosynthesis of ribonuclease and its accumulation in protein bodies in the cotyledons of mung beans. *Dev. Biol.*, **76**, 115–125.

Cheng, J.I., Boyd, C., Slaymaker, D., Okinaka, Y., Herman, E.M. & Keen, N.T. (1998) Purification and characterization of a 34 kDa syringolide binding protein from soybean. *Proc. Natl. Acad. Sci. USA*, **95**, 3306–3311.

Chrispeels, M.J. (1983) The Golgi apparatus mediates the transport of phytohemagglutinin to the protein bodies in bean cotyledons. *Planta*, **158**, 140–151.

Chrispeels, M.J. & Raikhel, N.V. (1991) Sorting of proteins in the secretory system. *Annu. Rev. Plant Physiol. Plant Mol. Biol.*, **42**, 21–53.

Chrispeels, M.J., Higgins, T.J.V., Craig, S. & Spencer, D. (1982a) Role of the endoplasmic reticulum in the synthesis of reserve proteins and their kinetics of their transport to protein bodies in developing pea cotyledons. *J. Cell Biol.*, **93**, 5–14.

Chrispeels, M.J., Higgins, T.J.V. & Spencer, D. (1982b) Assembly of storage protein oligomers in the endoplasmic reticulum and processing of the polypeptides in the protein bodies of developing pea cotyledons. *J. Cell Biol.*, **93**, 306–313.

Coleman, C.E., Herman, E.M., Takasaki, K. & Larkins, B.A. (1996) γ-Zein sequesters α-zein and stabilizes its accumulation in transgenic tobacco endosperm. *Plant Cell*, **8**, 2335–2345.

Corvera, S. (2001) Phosphoinositol 3-kinase and the control of endosome dynamics: new players defined by structural motifs. *Traffic*, **2**, 859–866.

Craig, S. & Goodchild, D.J. (1984) Golgi-mediated vicilin accumulation in pea cotyledon cells is redirected by monensin and nigericin. *Protoplasma*, **122**, 91–97.

Craig, S., Goodchild, D.J. & Hardham, A.R. (1979) Structural aspects of protein accumulation in developing pea cotyledons I: qualitative and quantitative changes in parenchyma cell vacuoles. *Aust. J. Plant Physiol.*, **6**, 81–98.

Craig, S., Goodchild, D.J. & Millerd, A. (1980) Structural aspects of protein accumulation in developing pea cotyledons II: three-dimensional reconstitutions of vacuoles and protein bodies from serial sections. *Aust. J. Plant Physiol.*, **7**, 329–337.

Di Sansebastiano, G.-P., Paris, N., Marc-Martin, S. & Neuhaus, J.-M. (1998) Specific accumulation of GFP in a non-acidic vacuolar compartment via a C-terminal propeptide-mediated sorting pathway. *Plant J.*, **15**, 449–457.

Di Sansebastiano, G.P., Paris, N., Marc-Martin, S. & Neuhaus, J.-M. (2001) Regeneration of a lytic vacuole and of neutral peripheral vacuoles can be visualized by green fluorescent proteins targeted to either type of vacuole. *Plant Physiol.*, **126**, 78–86.

Dombrowski, J.E., Schroeder, M.R., Bednarek, S.Y. & Raikhel, N.V. (1993) Determination of the functional elements within the vacuolar targeting signal of barley lectin. *Plant Cell*, **5**, 587–596.

Duranti, M., Guerrieri, N., Cerletti, P. & Vecchio, G. (1992) The legumin precursor from lupin white seeds: identity of the subunites, assembly and proteolysis. *Eur. J. Biochem.*, **206**, 941–947.

Duranti, M., Scarafoni, A., Gius, C., Negri, A. & Faoro, F. (1994) Heat-induced synthesis and tunicamycine-sensitive secretion of the putative storage glycoprotein conglutin gamma from mature lupin seeds. *Eur. J. Biochem.*, **222**, 387–393.

Etzler, M.E. (1985) Plant lectins: molecular and biological aspects. *Annu. Rev. Plant Physiol. Plant Mol. Biol.*, **36**, 209–234.

Faye, L., Sturm, A., Bollini, R., Vitale, A. & Chrispeels, M.J. (1986) The position of the oligosaccharide side-chains on phytohemagglutinin and their accesibility to glycosidases determines their subsequent processing in the Golgi. *Eur. J. Biochem.*, **158**, 741–745.

Frigerio, L., de Virgilio, M., Prada, A., Faoro, F. & Vitale, A. (1998) Sorting of phaseolin to the vacuole is saturable and requires a short C-terminal peptide. *Plant Cell*, **10**, 1031–1042.

Frigerio, L., Jolliffe, N.A., Di Cola, A., Felipe, D.H., Paris, N., Neuhaus, J.-M., Lord, J.M., Ceriotti, A. & Roberts, L.M. (2001a) The internal propeptide of the ricin precursor carries a sequence-specific determinant for vacuolar sorting. *Plant Physiol.*, **126**, 167–175.

Frigerio, L., Pastres, A., Prada, A. & Vitale, A. (2001b) Influence of KDEL on the fate of trimeric or assembly-defective phaseolin: selective use of an alternative route to vacuoles. *Plant Cell*, **13**, 1109–1126.

Gomez, L. & Chrispeels, M.J. (1993) Tonoplast and soluble vacuolar proteins are targeted by different mechanisms. *Plant Cell*, **5**, 1113–1124.

Greenwood, J.S. & Chrispeels, M.J. (1985) Correct targeting of the bean storage protein phaseolin in the seeds of transformed tobacco. *Plant Physiol.*, **79**, 65–71.

Hadlington, J.L. & Denecke, J. (2000) Sorting of soluble proteins in the secretory pathway of plants. *Curr. Opin. Plant Biol.*, **3**, 461–468.

Häger, K.P., Braun, H., Czihal, A., Müller, B. & Bäumlein, H. (1995) Evolution of seed storage protein genes: legumin genes of *Ginkgo biloba. J. Mol. Evol.*, **41**, 455–466.

Hara-Nishimura, I., Nishimura, M. & Akazawa, T. (1985) Biosynthesis and intracellular transport of 11S globulin in developing pumpkin cotyledons. *Plant Physiol.*, **77**, 747–752.

Hara-Nishimura, I., Takeuchi, Y. & Nishimura, M. (1993) Molecular characterization of a vacuolar processing enzyme related to a putative cysteine proteinase of *Schistosoma mansoni. Plant Cell*, **5**, 1651–1659.

Hara-Nishimura, I., Shimada, T., Hatano, K., Takeuchi, Y. & Nishimura, M. (1998) Transport of storage proteins to protein storage vacuoles is mediated by large precursor-accumulating vesicles. *Plant Cell*, **10**, 825–836.

Hayashi, M., Toriyama, K., Kondo, M., Hara-Nishimura, I. & Nishimura, M. (1999) Accumulation of a fusion protein containing 2S albumin induces novel vesicles in vegetative cells of *Arabidopsis. Plant Cell Physiol.*, **40**, 263–272.

Harris, N. (1979) Endoplasmic reticulum in developing seeds of *Vicia faba*. A high voltage electron microscope study. *Planta*, **146**, 63–69.

Harris, N., Grindley, H., Mulchrone, J. & Croy, J.D. (1989) Correlated *in situ* hybridization and immuno-chemical studies of legumin storage protein deposition in pea (*Pisum sativum* L.). *Cell Biol. Int. Rep.*, **13**, 23–35.

Herman, E.M. & Chrispeels, M.J. (1980) Phospholipase D and phosphatidic acid phosphatase: acid hydrolases involved in phospholipid catabolism in mung bean cotyledons. *Plant Physiol.*, **66**, 1001–1007.

Herman, E.M. & Shannon, L.M. (1984a) Immunocytochemical localization of concanavalin A in developing jack bean cotyledons. *Planta*, **161**, 97–104.

Herman, E.M. & Shannon, L.M. (1984b) Immunocytochemical evidence for the involvement of Golgi-apparatus in the deposition of seed lectin of *Bauhinia purpurea* (Leguminoseae). *Protoplasma*, **121**, 163–170.

Herman, E.M. & Shannon, L.M. (1985) Accumulation and subcellular localization of α-galactosidase in developing soybean cotyledons. *Plant Physiol.*, **77**, 886–890.

Herman, E.M., Platt-Aloeia, K.A., Thomson, W.W. & Shannon, L.M. (1984) Freeze fracture and filipin cytochemical observations of developing soybean cotyledon protein bodies and Golgi apparatus. *Eur. J. Cell Biol.*, **35**, 1–7.

Higuchi, W. & Fukazawa, C. (1987) A rice glutelin and a soybean glycinin have evolved from a common ancestral gene. *Gene*, **55**, 245–253.

Hillmer, S., Movafeghi, A., Robinson, D.G. & Hinz, G. (2001) Vacuolar storage proteins are sorted at the cis-cisternae of the pea cotyledon Golgi-apparatus. *J. Cell Biol.*, **152**, 41–50.

Hinz, G., Hoh, B. & Robinson, D.G. (1993) Strategies in the recognition and isolation of storage protein receptors. *J. Exp. Bot.*, **44** (Suppl.), 351–357.

Hinz, G., Menze, A., Hohl, I. & Vaux, D. (1997) Isolation of prolegumin from developing pea seeds: its binding to endomembranes and assembly into prolegumin hexamers in the protein storage vacuole. *J. Exp. Bot.*, **48**, 139–149.

Hinz, G., Hillmer, S., Bäumer, M. & Hohl, I. (1999) Vacuolar storage proteins and the putative vacuolar sorting receptor BP-80 exit the Golgi apparatus of developing pea cotyledons in different transport vesicles. *Plant Cell*, **11**, 1509–1524.

Horisberger, M. & Volanthen, M.T. (1983) Ultrastructural localization of Kunitz trypsin inhibitor on thin sections of *Glycine max* (soybean) cv Maple Arrow by the gold method. *Histochemistry*, **77**, 313–321.

Hoh, B., Hinz, G., Jeong, B.-K. & Robinson, D.G. (1995) Protein storage vacuoles form *de novo* during pea cotyledon development. *J. Cell Sci.*, **108**, 299–310.

Hohl, I., Robinson, D.G., Chrispeels, M.J. & Hinz, G. (1996) Transport of storage proteins to the vacuole is mediated by vesicles without a clathrin coat. *J. Cell Sci.*, **109**, 2539–2550.

Holkeri, H. & Vitale, A. (2001) Vacuolar sorting determinants within a plant storage protein trimer act cumulatively. *Traffic*, **2**, 737–741.

Jahn R. & Südhof, T.C. (1999) Membrane fusion and exocytosis. *Annu. Rev. Biochem.*, **68**, 863–911.

Jiang, L.W., Phillips, T.E., Hamm, C.A., Drozdowicz, Y.M., Rae, P.A., Maeshima, M., Rogers, S.W. & Rogers, J.C. (2001) The protein storage vacuole: a unique compound organelle. *J. Cell Biol.*, **155**, 991–1002.

Jauh, G.-Y., Phillips, T.E. & Rogers, J.C. (1999) Tonoplast intrinsic proteins isoforms as markers for vacuolar functions. *Plant Cell*, **11**, 1867–1882.

Jung, R., Nam, Y.W., Saalbach, I., Müntz, K. & Nielsen, N.C. (1997) Role of the sulfhydryl redox state and disulfide bonds in processing and assembly of 11S seed globulins. *Plant Cell*, **9**, 2037–2050.

Kalinski, A.J., Melroy, D.L., Dwivedi, R.S. & Herman, E.M. (1992) A soybean vacuolar protein (P34) related to thiol proteases which is synthesized as a glycoprotein precursor during seed maturation. *J. Biol. Chem.* **267**, 12068–12076.

Kim, W.T., Franceschi, V.R., Krishnan, H.B. & Okita, T.W. (1988) Formation of wheat protein bodies: involvement of the Golgi apparatus in gliadin transport. *Planta*, **176**, 173–182.

Kim, D.H., Eu, Y.-J., Yoo, C.M., Kim, Y.-W., Pih, K.T., Jin, J.B., Kim, S.J., Stenmark, H. & Hwang, I. (2001) Trafficking of phsophatidylinositol 3-phosphate from the *trans*-Golgi network to the lumen of the central vacuole in plant cells. *Plant Cell*, **13**, 287–301.

Kinney, A.J., Jung, R. & Herman, E.M. (2001) Cosuppression of the α subunits of β-conglycinin in transgenic seeds induces the formation of endoplasmatic reticulum-derived protein bodies. *Plant Cell*, **13**, 1165–1178.

Kirchhausen, T. (2000) Clathrin. *Annu. Rev. Biochem.*, **69**, 699–727.

Kirsch, T., Paris, N., Butler, J.M., Beevers, L. & Rogers, J.C. (1994) Purification and initial characterization of a potential plant vacuolar targeting receptor. *Proc. Natl. Acad. Sci. USA*, **91**, 3403–3407.

Kirsch, T., Saalbach, G., Raikhel, N.V. & Beevers, L. (1996) Interaction of a potential vacuolar targeting receptor with amino- and carboxyl-terminal targeting determinants. *Plant Physiol.*, **111**, 469–474.

Kjemtrup, S., Borksenhious, O. & Raikhel, N.V. (1995) Targeting and release of phytohemagglutinin from the roots of bean seedlings. *Plant Physiol.*, **109**, 603–610.

Koide, Y., Matsuoka, K., Ohto, M. & Nakamura, K. (1999) The N-terminal propeptide and the C-terminus of the precursor to the 20-kilo-dalton potato tuber protein can function as different types of vacuolar sorting signals. *Plant Cell Physiol.*, **40**, 1152–1159.

Krishnan, H.B., Franceschi, V.R. & Okita, T.W. (1986) Immunochemical studies on the role of the Golgi complex in protein body formation in rice cells. *Planta*, **169**, 471–480.

Kuliawat, R., Klumperman, J., Ludwig, P. & Arvan, P. (1997) Differential sorting of lysosomal enzymes out of the regulated secretory pathway in pancreatic β-cells. *J. Cell Biol.*, **37**, 595–608.

Lending, C.R. & Larkins, B.A. (1989) Changes in the zein composition of protein bodies during maize endosperm development. *Plant Cell*, **1**, 1011–1023.

Lending, C.R., Kriz, A.L., Larkins, B.A. & Bracker, C.E. (1988) Structure of maize protein bodies and immunocytochemical localization of zeins. *Protoplasma*, **143**, 51–62.

Lerouge, P., Cananes-Macheteau, M., Rayon, C., Fischette-Lainé, A.C., Gomord, V. & Faye, L. (1998) N-glycoprotein biosynthesis in plants: recent developments and future trends. *Plant Mol. Biol.*, **38**, 31–48.

Levanony, H., Rubin, R., Altschuler, Y. & Galili, G. (1992) Evidence for a novel route of wheat storage proteins to the vacuoles. *J. Cell Biol.*, **119**, 1117–1128.

Marty, F. (1999) Plant vacuoles. *Plant Cell*, **11**, 587–599.

Matsuoka, K. & Nakamura, K. (1999) Large alkyl side-chains of isoleucine and leucine in the NPIR region constitute the core of the vacuolar sorting determinant of sporamin precursor. *Plant Mol. Biol.*, **41**, 825–835.

Matsuoka, K. & Neuhaus, J.-M. (1999) *Cis*-elements of protein transport to the plant vacuoles. *J. Exp. Bot.*, **50**, 165–174.

Matsuoka, K., Bassham, D.C., Raikhel, N.V. & Nakamura, K. (1995) Different sensitivity to wortmannin of two vacuolar sorting signals indicates the presence of distinct sorting machineries in tobacco cells. *J. Cell Biol.*, **130**, 1307–1318.

Mettler, I.J. & Beevers, H. (1979) Isolation and characterization of the protein body membrane of castor beans. *Plant Physiol.*, **64**, 506–511.

Mollenhauer, H.H., Morré, D.J. & Rowe, L.D. (1990) Alteration of intracellular traffic by monensin: mechanism, specificity and relationship to toxicity. *Biochim. Biophys. Acta*, **1031**, 225–246.

Moore, P.J., Swords, K.M.M., Lynch, M.A. & Staehelin, L.A. (1991) Spatial organization of the assembly pathways of glycoproteins and complex polysaccharides in the Golgi apparatus of plants. *J. Cell Biol.*, **112**, 589–602.

Müntz, K. (1996) Proteases and proteolytic cleavage of stored proteins in developing and germinating dicotyledonous seeds. *J. Exp. Bot.*, **47**, 605–622.

Müntz, K. (1998) Deposition of storage proteins. *Plant Mol. Biol.*, **38**, 77–99.

Neuhaus, J.M. & Rogers, J.C. (1998) Sorting of proteins to vacuoles in plants. *Plant Mol. Biol.*, **38**, 127–144.

Neuhaus, J.-M., Pjetrzak, M. & Boller, T. (1994) Mutation analysis of the C-terminal vacuolar targeting peptide of tobacco chitinase: low specificity of the sorting system, and gradual transition between intracellular retention and secretion into the extracellular space. *Plant J.*, **5**, 45–54.

Nishimura, M. & Beevers, H. (1978) Hydrolases in vacuoles from castor beans. *Plant Physiol.*, **62**, 44–48.

Paris, N. & Rogers, J.C. (1996) The role of receptors in targeting soluble proteins from the secretory pathway to the vacuole. *Plant Physiol. Biochem.*, **34**, 223–227.

Paris, N., Stanley, M.C., Jones, R.L. & Rogers, J.C. (1996) Plant cells contain two functionally distinct vacuolar compartments. *Cell*, **85**, 563–572.

Pelham, H.R.B. (1988) Evidence that luminal ER proteins are sorted from secreted proteins in a post-ER compartment. *EMBO J.*, **7**, 913–918.

Pesacreta, T.C. & Lucas, W.J. (1985) Presence of a partially-coated reticulum and a plasma membrane coat in angiosperms. *Protoplasma*, **125**, 173–184.

Robinson, D.G. & Hinz, G. (1997) Vacuole biogenesis and protein transport to the plant vacuole: a comparison with the yeast vacuole and the mammalian lysosome. *Protoplasma*, **197**, 1–25.

Robinson, D.G. & Hinz, G. (1999) Golgi-mediated transport of seed storage proteins. *Seed Sci. Res.*, **9**, 263–283.

Robinson, D.G., Hoh, B., Hinz, G. & Hohl, I. (1995) One vacuole or two vacuoles: do protein storage vacuoles arise *de novo* during pea cotyledon development? *J. Plant Physiol.*, **45**, 654–664.

Robinson, D.G., Bäumer, A., Hinz, G. & Hohl, I. (1997) Ultrastructure of the pea cotyledon Golgi apparatus: origin of dense vesicles and action of brefeldine A. *Protoplasma*, **200**, 198–209.

Robinson, D.G., Bäumer, M., Hinz, G. & Hohl, I. (1998a) Vesicle transfer of storage proteins to the vacuole. The role of the Golgi apparatus and multivesicular bodies. *J. Plant Physiol.*, **152**, 659–667.

Robinson, D.G., Hinz, G. & Holstein, S.E.H. (1998b) The molecular biology of transport vesicles. *Plant Mol. Biol.*, **38**, 49–76.

Robinson, D.G., Rogers, J.C. & Hinz, G. (2000) Post-Golgi, prevacuolar compartments, in *Annual Plant Reviews, Vol. 5: Vacuolar Compartments*, (eds. D.G. Robinson & J.C. Rogers), Sheffield Academic Press, Sheffield England, pp. 270–298.

Rubin, R., Levanony, H. & Galili, G. (1992) Evidence for the presence of two different types of protein bodies in wheat endosperm. *Plant Physiol.*, **99**, 718–724.

Rouillé, Y., Rohn, W. & Hoflack, B. (2000) Targeting of lysosomal proteins. *Cell Dev. Biol.*, **11**, 165–171.

Saalbach, G., Jung, R., Kunze, G., Saalbach, I., Adler, K. & Müntz, K. (1991) Different legumin protein domains act as vacuolar targeting signals. *Plant Cell*, **3**, 695–708.

Sanderfoot, A.A. & Raikhel, N.V. (2001) *Arabidopsis* could shed light on human genome. *Nature*, **410**, 299.

Sanderfoot, A.A., Ahmed, S.U., Marty-Mazars, F., Rapoport, I., Kirchhausen, T., Marty, F. & Raikhel, N.V. (1998) A putative vacuolar cargo receptor partially colocalizes with AtPEP12p on a prevacuolar compartment in *Arabidopsis* roots. *Proc. Natl. Acad. Sci. USA*, **95**, 9920–9925.

Sanderfoot, A.A., Assaad, F.F. & Raikhel, N.V. (2000) The *Arabidopsis* genome: an abundance of soluble *N*-ethylmaleimide-sensitive factor adaptor protein receptors. *Plant Physiol.*, **124**, 1558–1569.

Sanderfoot, A.A., Pilgrim, M., Adam, L. & Raikhel, N.V. (2001) Disruption of individual members of *Arabidopsis* syntaxin gene families indicates each has essential functions. *Plant Cell*, **13**, 659–666.

Satiat-Jeunemaitre, B., Steele, C. & Hawes, C. (1996) Golgi-membrane dynamics are cytoskeleton dependent. A study on Golgi stack movement induced by brefeldin A. *Protoplasma*, **191**, 21–33.

von Schaewen, A. & Chrispeels, M.J. (1993) Identification of vacuolar sorting information in phytohemagglutinin, an unprocessed vacuolar protein. *J. Exp. Bot.*, **44** (Suppl.), 339–342.

Schu, P.V., Takegawa, K., Fry, M.J., Stack, J.H., Waterfield, M.D. & Emr, S.D. (1993) Phosphatidyl-inositol 3-kinase encoded by yeast VPS34 gene essential for protein sorting. *Science*, **260**, 88–91.

Shewry, P.R. (1995) Plant storage proteins. *Biol. Rev.*, **70**, 375–426.

Shewry, P.R., Napier, J.A. & Tatham, A.S. (1995) Seed storage proteins: structures and biosynthesis. *Plant Cell*, **7**, 945–956.

Shimada, T., Kuroyanagi, M., Nishimura, M. & Hara-Nishimura, I. (1997) A pumpkin 72-kDa membrane protein of precursor-accumulating vesicles has characteristics of a vacuolar sorting receptor. *Plant Cell Physiol.*, **38**, 1414–1420.

Shimoni, Y. & Galili, G. (1996) Intramolecular disulfide bonds between conserved cysteines in wheat gliadins control their deposition into protein bodies. *J. Biol. Chem.*, **271**, 18869–18874.

Shutov, A.D., Kakhovskaya, I.D., Braun, H., Bäumlein, H. & Mütz, K. (1995) Legumin-like and vicilin-like seed storage proteins: evidence for a common single-domain ancestral gene. *J. Mol. Evol.*, **41**, 1057–1069.

Shy, G., Ehler, L., Herman, E.M. & Galili, G. (2001) Expression patterns of genes encoding endomembrane proteins support a reduced function of the Golgi in wheat endosperm during the onset of storage protein deposition. *J. Exp. Bot.*, **52**, 2387–2388.

Sparvoli, F., Faoro, F., Gloria Damiati, M., Ceriotti, A. & Bollini, R. (2000) Misfolding and aggregation of vacuolar glycoproteins in plant cells. *Plant J.*, **24**, 825–836.

Staehelin, L.A. & Moore, I. (1995) The plant Golgi apparatus: structure, functional organization and trafficking mechanisms. *Annu. Rev. Plant Physiol. Plant Mol. Biol.*, **46**, 261–288.

Stöger, E., Parker, M., Christou, P. & Casey, R. (2001) Pea legumin overexpressed in wheat endosperm assembles into an ordered paracrystalline matrix. *Plant Physiol.*, **125**, 1732–1742.

Sturm, A., Voelker, T.A., Herman, E.H. & Chrispeels, M.J. (1988) Correct glycosylation, Golgi-processing, and targeting to protein bodies of the vacuolar protein phytohemagglutinin in transgenic tobacco. *Planta*, **175**, 170–183.

Swanson, S.J., Bethke, P.C. & Jones, R.L. (1998) Barley aleuron cells conatin two types of vacuoles: characterization of lytic organelles by use of fluorescent probes. *Plant Cell*, **10**, 685–698.

Tanchak, M.A. & Chrispeels, M.J. (1989) Crosslinking of microsomal proteins identifies P-9000, a protein that is co-transported with phaseolin and phytohemagglutinin in bean cotyledons. *Planta*, **179**, 495–505.

Thiele, C., Gerdes, H.-H. & Huttner, W.B. (1997) Protein secretion: puzzling receptors. *Curr. Biol.*, **7**, R496–R500.

Tooze, S.A. (1998) Biogenesis of secretory granules in the *trans*-Golgi network of neuroendocrine and endocrine cells. *Biochim. Biophys. Acta*, **1404**, 231–244.

Torres, E., Gonzalez-Melendi, P., Stoger, E., Shaw, P., Twyman, R.M., Nicholson, L., Vaquero, C., Fischer, R., Christou, P. & Perrin, Y. (2001) Native and artificial reticuloplasmins co-accumulate in distinct domains of the endoplasmic reticulum and in post-endoplasmic reticulum compartments. *Plant Physiol.*, **127**, 1212–1223.

Van der Wilden, W., Herman, E.M. & Chrispeels, M.J. (1980) Protein bodies of mung bean cotyledons as autophagic organelles. *Proc. Natl. Acad. Sci. USA*, **77**, 428–432.

van Schaewen, A. & Chrispeels, M.J. (1993) Identification of vacuolar sorting information in phytohemagglutinin: an unprocessed vacuolar protein. *J. Exp. Bot.*, **44** (Suppl.), 339–342.

Vitale, A. & Chrispeels, M.J. (1992) Sorting of proteins to the vacuoles of plant cells. *Bioessays*, **14**, 151–160.

Vitale, A. & Raikhel, N.V. (1999) What do proteins need to reach different vacuoles? *Trends Plant Sci.*, **4**, 149–155.

Watanabe, E., Shimada, T., Kuroyanagi, M., Nishimura, M. & Hara-Nishimura, I. (2002) Calcium-mediated association of a putative vacuolar sorting receptor PV72 with the propeptide of 2S albumin. *J. Biol. Chem.*, **277**, 8708–8715.

Wink, M. (1993) The plant vacuole: a multifunctional compartment. *J. Exp. Bot.*, **44** (Suppl.), 231–246.

Yaklich, B. & Herman, E.M. (1995) Protein storage vacuoles of soybean aleurone cells accumulate a unique glycoprotein as well as proteins thought to be embryo specific. *Plant Sci.*, **107**, 57–67.

Yamagata, H., Sugimoto, T., Tanaka, K. & Kasai, Z. (1982) Biosynthesis of storage proteins in developing rice seeds. *Plant Physiol.*, **70**, 1094–1100.

Zheng, Z., Sumi, K., Tanaka, K. & Murai, N. (1995) The bean storage protein β-phaseolin is synthesized, processed, and accumulated in the vacuolar type-II protein bodies of transgenic rice endosperm. *Plant Physiol.*, **109**, 777–786.

zur Nieden, U., Manteuffel, R., Weber, E. & Neumann, D. (1984) Dictyosomes participate in the intracellular pathway of storage proteins in developing *Vicia faba* cotyledons. *Eur. J. Cell Biol.*, **34**, 9–17.

10 Rab proteins and the Golgi apparatus

Stephen Rutherford and Ian Moore

10.1 Introduction

Golgi membranes must have mechanisms to ensure that they retain appropriate proteins and lipids while cargo molecules are packaged into budding transport vesicles that are ultimately delivered to the correct destinations. These destinations can be other Golgi membranes, the ER, the PM or one or more pre-vacuolar compartments. This problem can be divided broadly into two stages; sorting of cargo, and targeting of vesicles. Correct sorting requires that a transport vesicle forming on an endomembrane organelle incorporates cargo molecules destined for export while excluding other molecules that either reside in the organelle or are destined to be packaged into a different vesicle with a different destination. Once vesicles are produced, they must be correctly targeted to ensure that they deliver their contents to the appropriate organelle after membrane fusion. Two independent estimates of the rate of secretory vesicle production by the Golgi in rapidly growing coleoptile epidermis suggested that the quantity of vesicle membrane that is delivered to (and probably recycled from) the PM is sufficient to cause the entire surface area of the PM to be turned over once every 3–4 h (Phillips *et al.*, 1988; Thiel *et al.*, 1998). In hypersecretory maize root cap cells, the turn over time may be as little as 10 min (Shannon & Steer, 1984). That the Golgi and the PM retain their distinct molecular identities in the face of such rapid membrane flux is a testament to the efficiency of the sorting and targeting functions of the endomembrane compartments and their associated vesicles. But, how is this achieved? This chapter deals with the Rab family of GTP-binding proteins that make an important contribution to vesicle targeting.

10.2 The molecular basis of vesicle targeting at the plant Golgi

Although functional studies of the vesicle trafficking apparatus in plants are few, the *Arabidopsis* genome appears to confirm that the core machinery for formation and targeting of vesicles in plants is similar to that described for yeast (Bischoff *et al.*, 1999; Sanderfoot *et al.*, 2000). However, these studies have also revealed that in plants, as in mammals, the gene families such as Rab GTPases that encode key determinants of vesicle targeting specificity are considerably more diverse than in yeast (Lazar *et al.*, 1997; Pereira-Leal & Seabra, 2000, 2001; Sanderfoot *et al.*, 2000; Zerial & McBride, 2001). In mammals, this greater diversity appears to reflect the greater complexity of the membrane trafficking pathways of these organisms, particularly in specialised cells such as polarised epithelial cells, neurons and endocrine cells (Chen & Scheller, 2001; Zerial & McBride, 2001). In plants, it is unclear to what extent the increased sequence diversity reflects the evolution of new trafficking functions.

Rab GTPases are members of the *ras* superfamily of regulatory GTPases and are emerging as key regulators of targeting specificity in eukaryotic membrane traffic (Lazar *et al.*, 1997; Segev, 2001; Zerial & McBride, 2001). In yeast and mammalian cells, they are known to regulate the activity of the tethering factors and possibly also the *SNARE* complexes that combine to promote the initial docking and subsequent fusion of specific vesicle and organellar membranes. Some Rab GTPases also promote interactions between transport vesicles and the cytoskeleton (Segev, 2001; Zerial & McBride, 2001). Individual members of the Rab GTPase family each appear to be responsible for distinct vesicle targeting events though in some cases a single Rab GTPase can act in two consecutive transport steps (Jedd *et al.*, 1995). To perform these regulatory roles, they interact with a large array of regulatory and effector molecules that couple the cycle of GTP-binding and GTP-hydrolysis to the process of vesicle formation, docking and fusion (Segev, 2001; Zerial & McBride, 2001). Guanine-nucleotide exchange factors promote dissociation of bound GDP thereby facilitating the binding of GTP. Nucleotide exchange follows the recruitment of the Rab GTPase to a particular membrane, by virtue of targeting information in its carboxy-terminal region and a geranyl-geranyl isoprenyl post-translational modification at the carboxy-terminus (Chavrier *et al.*, 1991; Zerial & McBride, 2001). GTPase activating proteins in turn stimulate the low endogenous GTPase activity of the Rab protein, converting the nucleotide to the GDP form. Nucleotide exchange and hydrolysis is regulated in space and time such that the GTP-bound form of the Rab protein is able to recruit and activate effector molecules that play various roles in vesicle targeting and possibly also vesicle formation (Zerial & McBride, 2001). In cases of heterotypic membrane fusion, the Rab protein is likely to reside on the target membrane after fusion, which makes it necessary to recycle the Rab protein if it is to participate in further rounds of membrane traffic. This is achieved after GTP hydrolysis by the action of Rab GDI (Rab GDP Dissociation Inhibitor) which extracts the Rab from the membrane and maintains it in a soluble complex capable of being delivered to the donor membrane (Lazar *et al.*, 1997; Segev, 2001; Zerial & McBride, 2001).

Rab GTPases can be assigned to distinct subclasses based on sequence comparisons and, to a first approximation, each subclass appears to be responsible for a particular vesicle-targeting event in the cell. There appears to be a core set of six Rab subclasses that represent a minimal complement of eukaryotic Rab functions. Indeed the available functional evidence suggests that one member of this set can account for the principal transport events between the major organelles of the biosynthetic and endocytic pathways. The Rab complements of *Saccharomyces cerevisiae* and *Schizosaccharomyces pombe* appear to be restricted essentially to this basic eukaryotic set. In marked contrast, humans have at least 60 Rab protein sequences, which have been assigned to 41 subclasses by sequence analysis (numbered HsRab1 to HsRab41), and hence it is possible that mammals have about 40 different Rab functions (Pereira-Leal & Seabra, 2000; Zerial & McBride, 2001).

In the *Arabidopsis* genome sequence (The Arabidopsis Genome Initiative, 2000), we have identified 57 loci that can encode Rab GTPases. 57 Rab loci were also identified in an independent analysis of the *Arabidopsis* genome sequence using search criteria that were used to identify invertebrate, fungal and human Rab families (Pereira-Leal & Seabra, 2001). Forty-eight of the 57 *Arabidopsis* Rab sequences are known as cDNAs or ESTs. In spite of their large number, all plant Rab sequences can be assigned to just eight phylogenetic groups, six of which are clearly related to the six subclasses common to yeasts and animals; the other two are similar to mammalian Rab2 and Rab18 that are not present in yeasts (Lazar *et al.*, 1997). The *Arabidopsis* genome sequence has not revealed any sequence that cannot be easily assigned to one of these eight groups. Of these eight groups, five are related to mammalian proteins that have been implicated in vesicle traffic to or from Golgi membranes.

These phylogenetic analyses show that 33 of the 41 mammalian subclasses have no clear orthologue in *Arabidopsis*, indicating that the diversification of mammalian Rab GTPase functions has not occurred in the plant lineage. In some cases such as Rab9 which promotes recycling from endosomal compartments to the Golgi, the missing Rab subclasses perform functions that one might reasonably expect to be common to other eukaryotes. In such cases it may be that similar trafficking events do not occur in plant cells or it may be that analogous functions are performed by other molecules, for example, members of the large plant RabA family (see below).

It is possible that the eight phylogenetic groups in the plant Rab GTPase family represent the principle functional subclasses. However, these eight plant Rab groups have an unusually large number of individual members (between 3 and 26) which often exhibit greater sequence dissimilarity than is usual for Rab isoforms in other organisms (usually less than 30%). Using criteria of over-all sequence similarity and of conservation in the sequence motifs that appear to define Rab identity (Moore *et al.*, 1995; Pereira-Leal & Seabra, 2001), we as well as others have proposed that *Arabidopsis* has perhaps 18 Rab GTPase sub-classes. It may be that each of these 18 subclasses represents a distinct Rab GTPase function as do, apparently, most mammalian subclasses. If this view is borne out by functional studies, it will indicate that the basic eukaryotic Rab GTPase set has undergone a quite distinct adaptive radiation in the angiosperm lineage. However, in yeast, members of two subclasses defined by the same criteria have overlapping functions (Ypt51 and Ypt52) and two are apparently redundant under laboratory conditions (Ypt31 and Ypt32) (Lazar *et al.*, 1997; Pereira-Leal & Seabra, 2001). Consequently, substantially more functional data will be required before we can predict with any confidence how many Rab functions

Arabidopsis encodes. At present the best estimate is *at least 8 but possibly 18 or more*.

Even if the 18 subtypes we recognise do represent distinct subclasses with distinct trafficking functions, many plant Rab subclasses are uncommonly large. Nine of the 18 *Arabidopsis* subclasses contain three or more members compared to only 5 of 111 subclasses in other organisms. Only 15 of the 57 sequences can be simply accounted for as passive participants in genome evolution. It thus remains possible that most of the 57 *Arabidopsis* Rab sequences have been maintained by natural selection. An explanation for the maintenance of this diversity is that specialisation has occurred within each subclass to generate isoforms that act in the same membrane trafficking step but are adapted to function in specialised cell types or under particular environmental or physiological conditions. Alternatively, maintenance of Rab isoforms may simply reflect the diversification of gene regulatory sequences so that a single trafficking function is provided by a group of genes, each of which is expressed in a different pattern or under different physiological conditions. There is some evidence to suggest that individual Rab genes may function in specific cell types or in response to particular physiological or environmental stimuli (Bolte *et al.*, 2000; Moore *et al.*, 1997; O'Mahoney and Oliver, 1999; Terryn *et al.*, 1993). However, there is nothing to suggest that these specific expression patterns reflect significant functional specialisation; they may serve only to provide a common cellular function to all cells of the organism through the combined activities of genes with different but essentially arbitrary expression patterns.

Here we compare the complement of *Arabidopsis* Rab GTPases with those Rab proteins that have been implicated in Golgi function in other eukaryotes. We also summarise some recent limited progress in understanding the function of GTPases that may be involved in plant Golgi function. In this chapter, we have adopted a classification system that was proposed for *Arabidopsis* in a recent comprehensive summary of Rab GTPase families in organisms with complete genome sequences (Pereira-Leal & Seabra, 2001). This hierarchical classification recognises the phylogenetic structure of the plant Rab GTPase family and can be applied successfully to other dicot species. In this way, it helps to clarify the relationships between plant Rab GTPases which have been named according to a variety of *ad hoc* systems.

10.3   The RabD group (related to Rab1)

This subclass is closely related to the mammalian Rab1 subclass which is involved in ER to Golgi transport and the initial stages of intra-Golgi transport. The homologue in *S. cerevisiae*, Ypt1p, was the first Rab GTPase to be discovered though its role in membrane traffic was identified only after the second member of the family, the yeast Sec4p protein, was characterised. Mutations at the yeast *Ypt1* locus are lethal and result in extensive accumulation of ER membrane. *In vitro* assays have identified a role for Ypt1p in recruitment of the tethering factor Uso1p to membranes to facilitate docking of ER-derived vesicles with Golgi membranes (Cao *et al.*, 1998). It has also emerged that the multimeric Golgi-bound TRAPP complex acts as a guanine nucleotide exchange factor for Ypt1p (Wang *et al.*, 2000) and that Ypt1p can fulfil its vesicle targeting role when bound exclusively to Golgi membranes (Cao & Barlowe, 2000). Consequently, it may be that Ypt1p can be recruited to vesicles and activated by nucleotide exchange relatively late in the vesicle formation and targeting process. In mammalian cells, dominant inhibitory mutations in the conserved GTP-binding motifs have proven effective in elucidating Rab1 function in intact cells. The most commonly used mutations (equivalent to the S17N and N116I mutants of Ras) are believed to act by titrating the Rab1 nucleotide exchange activity thereby inhibiting the conversion of wild-type Rab1 protein to the active GTP-bound form (Segev, 2001; Zerial & McBride, 2001). Such mutations in either of the two mammalian Rab1 isoforms (Rab1a or Rab1b) inhibit ER-to-Golgi transport and result in the disruption of Golgi organisation. Biochemical studies have shown that mammalian Rab1 acts similarly to Ypt1 in that it recruits the tethering factor p115, a homologue of Uso1 in yeast. However, it appears that in mammals this process must take place on the ER and that the inhibition of p115 recruitment to nascent COPII vesicles prevents vesicle targeting and/or fusion with the Golgi (Allan *et al.*, 2000). Thus, it may be that despite the similar roles of Rab1 and Ypt1 in directing membrane traffic, the spatial regulation and interactions of these proteins may differ in yeast and mammalian cells. Recently, a role for Ypt1p in cargo sorting at the ER has been identified in yeast (Morsomme & Riezman, 2002), indicating that both yeast and mammalian proteins appear to have a function on the ER.

In *Arabidopsis* the RabD subclass contains four members divided into two major branches, termed RabD1 and RabD2 by Pereira-Leal and Seabra (2001). This division of the Rab1 group into two distinct branches is unique to higher plants and is conserved across the angiosperms. The minor branch RabD1 has a single representative in all the monocot and dicot species we have investigated, while the RabD2 group is known to contain several sequences in most of these species. The maize RabD1 cDNA Zm-RabD1 (Yptm1) complements the growth defect and the membrane trafficking phenotype of *Ypt1* null strains but does so less than Zm-RabD2 (Palme *et al.*, 1993 and our unpublished data). At present, it is unclear whether RabD1 and RabD2 proteins perform similar or distinct functions nor is it known whether they are expressed in similar cells.

Using an antisense approach it was shown that the inhibition of RabD function in soya bean nodules prevented nodule development, and hence RabD function appears to be essential in these cells (Cheon *et al.*, 1993). The affected nodules exhibited various aberrant membrane structures but owing to the severe growth defect they exhibit, it is not clear whether they represent a primary consequence of RabD deficiency or secondary consequence of poor cell viability. Using a dominant-inhibitory mutant form of *At*-RabD2a (also named ARA5 and AtRab1b) and a GFP-based membrane trafficking assay, it has been shown that the RabD subclass is required for normal ER-to-Golgi transport in tobacco (Batoko *et al.*, 2000). This assay is based on the observation that a secreted form of GFP exhibits poor stability and fluorescence in the cell wall, but fluoresces brightly if its transport to the PM is inhibited. Using an efficient *Agrobacterium*-mediated transient expression system to co-express the secreted GFP marker with a dominant-inhibitory *At*-RabD2a in tobacco leaves, it was shown that transport of the secreted GFP was inhibited resulting in intracellular accumulation of GFP. Confocal imaging of the accumulated GFP revealed that it was present in the ER but not in the Golgi, a conclusion supported by the observation that an N-glycosylated form of GFP accumulated an endoglycosidase-H resistant species in the presence of the Rab mutant. Importantly, it was shown that the effect of the Rab mutant could be rescued by co-expression with wild-type *At*-RabD2a but not by an unrelated *Arabidopsis* Rab GTPase. These observations each indicate that the transport defect induced by the dominant-inhibitory *At*-RabD2a mutant resulted from specific inhibition of RabD2 function.

Although the results above indicate a role for RabD function in transport between ER and Golgi compartments, it is more difficult to ascribe a directionality to the transport event in question. Transport between the ER and Golgi is almost certainly a bi-directional affair, though this has not been demonstrated directly in plants. As GFP accumulation in the assay of Batoko *et al.* (2000) takes place over several hours, whereas rounds of vesicular transport in these cells probably take place in a few minutes (Brandizzi *et al.*, 2002b), it is not clear whether the primary effect of the *At*-RabD1a mutant is in forward or retrograde vesicle traffic: inhibition of retrograde traffic could result secondarily in perturbation of anterograde traffic, owing to a failure to recycle one or more essential components to the ER. Evidence in favour of a primary role in anterograde traffic from ER to Golgi comes with the observation that the *At*-RabD2a dominant-inhibitory mutant does not inhibit the redistribution of Golgi to ER upon application of brefeldin A, but does inhibit recovery of Golgi stacks following withdrawal of this drug (Saint-Jore *et al.*, 2002).

Intriguingly, the normal movement of the Golgi apparatus over the ER was inhibited in samples expressing the *At*-RabD2a mutant (Batoko *et al.*, 2000). Since the discovery that plant Golgi stacks move rapidly over an actin network that is co-extensive with the ER, there has been discussion about the location and mechanism of vesicle exchange between these two organelles (see Chapters 4 and 5). In one model, it was proposed that ER-derived vesicles remained associated with the actin network and were captured by passing Golgi stacks, while another model, based on the observation of Golgi behaviour, proposed that Golgi moved to localised sites of vesicle production, pausing to effect membrane exchange. However, there was no direct experimental evidence to suggest that Golgi motility played any role in the process of membrane exchange with the ER. The observation that a dominant-inhibitory *At*-RabD2a mutant resulted in the inhibition of both Golgi motility and membrane traffic provided the first evidence that there may indeed be a mechanistic link of some sort between these two processes. However, recent work (H. Batoko and I. Moore, unpublished) casts doubt on this idea. Further investigation of the *At*-RabD2a mutant has failed to identify a clear correlation between the accumulation of a secretory marker and the degree to which Golgi motility is affected in individual cells, while initial results suggest that the mitochondrial and Golgi motility are affected similarly. These observations suggest a more widespread inhibition of actin-based motility rather than a specific and strict coupling between Golgi motility and ER-to-Golgi traffic. These observations are consistent with other lines of evidence suggesting that normal Golgi motility is not a prerequisite for membrane exchange with the ER. Firstly, redistribution and reformation of the Golgi apparatus during brefeldin-A treatment and washout are unaffected when long-range Golgi motility is abolished with the inhibitors of actin polymerisation (Saint-Jore *et al.*, 2002). Secondly, FRAP (fluorescence recovery after photobleaching) studies of plants treated with the same actin polymerisation inhibitors indicate that transport to Golgi can occur rapidly in the absence of Golgi movement (Brandizzi *et al.*, 2002b), even though it cannot be excluded that the process would be more faster in the presence of active Golgi motility. The first cellular phenotype described for yeast *Ypt1* strains was disruption of the microtubule network (Schmitt *et al.*, 1986) but this is not now considered to be the primary effect of this mutation.

10.4   The RabB group (related to Rab2)

The RabB subclass is related to mammalian Rab2. There is no homologue in either *S. cerevisiae* or *S. pombe* though it is present in invertebrates and algae. In mammals, Rab2 is implicated in the maturation of VTCs (vesicular tubular clusters) which are formed by homotypic fusion of ER-derived COPII transport vesicles and travel along microtubules to the Golgi apparatus at the microtubule-organising centre near the nucleus (Tisdale & Balch, 1996). As they move towards the Golgi, COPI vesicles assemble on the VTC membrane and selectively package molecules that must be returned to the ER. In plants, there is no equivalent of the VTC and the Golgi is organised entirely differently (dozens of discrete Golgi stacks tracking over the ER network on actin microfilaments), so the role of Rab2 in plants is of interest. Transcripts of *At*RabB1c (originally published as AtRab2a (Palme *et al.*, 1993; Moore *et al.*, 1997)) are found in all organs, but its promoter is activated in mature pollen tubes and responds to rapid growth in the organs of young seedlings (Moore *et al.*, 1997). Dominant-inhibitory mutants of a tobacco RabB homologue, that is also highly expressed in pollen, have recently been shown to inhibit the pollen tube growth, the transport of GFP markers to the plasma membrane, and to cause redistribution of a predominantly Golgi localised ERD2-GFP marker to the ER (Cheung *et al.*, 2002). These observations suggest that RabB function is required to sustain normal membrane traffic from the ER to the pollen tip, though the precise site of action of the dominant-inhibitory mutant is unclear. Again, it will be more difficult to establish whether the primary site of action is in anterograde or retrograde traffic. GFP fusions to the tobacco RabB homologue labelled the Golgi stacks (Cheung *et al.*, 2002), consistent with a role in anterograde or retrograde transport between the ER and Golgi. It was found that when the GFP-RabB fusion was bombarded into tobacco epidermal cells, it failed to localise to the Golgi resulting instead in a cytoplasmic localisation. It was suggested that expanded epidermal cells, which presumably exhibit low rates of transport between ER and Golgi relative to pollen tubes, lacked the ability to localise the RabB fusion to the Golgi (Cheung *et al.*, 2002). However, using *Agrobacterium* mediated transient expression in the same cell type, we have observed good co-localisation of GFP-RabB fusions to the Golgi apparatus, though over-expression of this or any other Rab fusion we have tested resulted in cytoplasmic accumulation (U. Neumann, J. Johansen, H. Batoko, C. Hawes, and I. Moore unpublished). Therefore, while it is reasonable to suggest that epidermal cells exhibit lower rates of ER-Golgi exchange than pollen tubes and a lower capacity to recruit RabB to Golgi membranes, we suspect that the failure to localise RabB fusions to the Golgi in epidermal cells by particle bombardment reflects the technical limitations of the method more than the biochemical capabilities of the cells under study.

10.5   The RabE group (related to Rab8)

This branch of the plant Rab family is related to proteins that are known or suspected to be involved in post-Golgi transport to the PM. Technically, the plant RabE subclass is most similar to the mammalian Rab8 subclass but it is misleading to single out this subclass as the true orthologue of the plant group. Rab8 is a member of a large and complex group of mammalian Rab subclasses that have acquired specific functions in post-Golgi transport in the mammalian lineage and have no clear orthologues in either yeast or plants (Pereira-Leal & Seabra, 2001; Segev, 2001; Zerial & McBride, 2001). Indeed, the closest cross-kingdom similarities in this group exist between the plant RabE subclass and the single Ypt2 locus of *S. pombe* which is required for the transport from Golgi to PM in this organism. The fact that all plant RabE sequences appear to belong to a single subclass suggests that post-Golgi transport to the PM has not undergone the same diversification and specialisation in angiosperms as it has in mammals, or that other Rab subclasses have acquired analogous functions in angiosperms.

10.6   The RabH group (related to mammalian Rab6)

The RabH group is related to mammalian Rab6 and yeast Ypt6. In mammals, Rab6a is implicated in retrograde transport through the Golgi stack and is also required for a slow, COPI-independent, retrograde transport pathway from Golgi to ER (Storrie *et al.*, 2000). This pathway may allow Golgi residents to be recycled through the ER for scrutiny by the ER quality control systems. The over-expression of wt Rab6a or a GTPase deficient mutant resulted in the redistribution of Golgi glycosyltransferases to the ER (Martinez *et al.*, 1997). Rab6a interacts with a kinesin-like motor Rabkinesin-6 to effect these transport events (Segev, 2001; Zerial & McBride, 2001). Given that the plant Golgi is associated with the actin cytoskeleton rather than with microtubules as in mammals, it is not clear whether similar interactions can be expected for RabH proteins. In yeast, the RabH/Rab6 homologue Ypt6 is required for retrograde transport from a pre-vacuolar/endosomal compartment to the Golgi and possibly beyond (Siniossoglou & Pelham, 2001) suggesting a difference in the trafficking function of the yeast and mammalian homologues (Lazar *et al.*, 1997; Segev, 2001). This apparent discrepancy may be resolved by the recent demonstration that a splice variant of mammalian Rab6a, differing by only three amino acids in a relatively non-conserved region of the protein, is required for retrograde transport from the early recycling endosomes to the *trans*-Golgi network (Mallard *et al.*, 2002). Despite the minimal difference, the over-expression of wild-type and mutant forms of the two splice variants resulted in quite different effects on membrane traffic and the Golgi organisation indicating that Rab *isoforms* with very similar sequence may have distinct cellular roles. This underlines the need for functional studies of Rab function to test predictions based on the sequence.

A member of the *Arabidopsis* RabH subclass has been isolated as a cDNA and was shown to complement the temperature sensitive growth defect associated with deletion of the *S. cerevisiae* homologue *Ypt6* (Bednarek *et al.*, 1994). We have found only one EST derived from any of the other four RabH loci, although 75% of all *Arabidopsis* Rabs are represented. In a previous survey of 105 plant Rab GTPase sequences, only two RabH sequences were identified (Borg *et al.*, 1997). It would be remarkable if three of the five *Arabidopsis* RabH sequences were pseudogenes. So it may be worth using promoter-GUS fusions to test the hypothesis that they exhibit developmentally or physiologically restricted expression patterns. If this family performs functions similar to Rab6 and Ypt6, given the diversity of plant vacuolar compartments, it may be that certain RabH genes are expressed in cells with high demand for particular pre-vacuolar sorting events.

10.7 The RabA group (related to mammalian Rab11)

The available cDNA and EST data from *Arabidopsis* and other angiosperms have been hinting for several years that an abundance of RabA sequences is a common property of the angiosperm Rab families. The *Arabidopsis* genome sequence confirms that, for this species at least, the abundance of sequenced RabA cDNAs reflects a true multiplicity of genes rather than simply their ease of cloning or the abundance of their mRNAs. This branch of the *Arabidopsis* family has 26 genes accounting for almost half the total Rab complement in this species. Given their numbers, the sequence diversity between these 26 proteins, we consider it unlikely that all 26 perform the same role in membrane traffic. We have divided the group into six provisional subclasses numbered RabA1 to RabA6 which have between one and nine members each. AtRabA6b is known to be expressed and a RabA6 cDNA is known in *Lotus japonicus*, and hence it seems likely that at least one of the two *Arabidopsis* RabA6 loci is functional. The largest group, RabA1, contains nine members that segregate into two major branches that are conserved in other dicot RabA1 families. This may reflect further functional specialisation in this branch. Similarly, AtRabA5a forms a distinct branch within the RabA5 family and orthologues of both branches exist in other species. One sequence, AtRabA4e, is rather unusual in that it has a lysine in the position of the highly conserved glutamic acid of the WDTAGQE motif and has the sequence CVAAQ at its carboxy-terminus rather than the conventional CCXX(X) or CXC motifs that promote geranylgeranylation and membrane anchoring.

The closest mammalian homologues, Rab11a, Rab11b, and Rab25 have all been localised to apical recycling endosomes in polarised epithelial cells, where they have been studied most extensively although they are present in other cell types (Cassanova *et al.*, 1999; Segev, 2001). Despite localising to the same organelle, over-expression of wild-type and dominant mutant forms of each subclass can result in different phenotypes prompting the suggestion that they may control distinct transport routes between the recycling endosome and the Golgi or PM (Cassanova *et al.*, 1999; Rodman and Wandinger-Ness, 2000). The closest RabA homologues in *S. cerevisiae*, Ypt31 and Ypt32, have been implicated in the transport between a late Golgi compartment and a pre-vacuolar compartment (Lazar *et al.*, 1997; Segev, 2001). Although the different organisation of endosomal compartments in *S. cerevisiae* and mammalian epithelial cells precludes a direct comparison, it seems that *S. cerevisiae* and mammalian sequences in this branch of the Rab family are all involved in transport to or from endosomal/pre-vacuolar compartments. On sequence criteria alone, AtRabA2 would appear to be the prime suspect for an analogous role in plants. However, the phylogenetic relationships between the plant RabA subgroups and their mammalian and yeast homologues are not clear. Furthermore, as the diversification of plant RabA subclasses has occurred only in the plant lineage, it is not clear that any of the AtRabA subclasses is truly orthologous to the yeast and/or mammalian homologues.

Nevertheless, it seems likely that some of the AtRabA subclasses function in transport between trans-Golgi, endosomal/pre-vacuolar, and PM membranes, but localisation and transport data *in planta* is now clearly essential. Plant endosomal compartments have been identified in the Electron Microscope by following the uptake of cationised ferritin by protoplasts but these compartments are poorly described other than by their morphology, and molecular markers have been lacking (Hawes *et al.*, 1995). In recent years however, vacuolar protein sorting receptors and SNARE proteins have been localised to pre-vacuolar compartments that may also be intermediates in the endosomal pathway(s) (Brandizzi *et al.*, 2002a; Humair *et al.*, 2001; Sanderfoot *et al.*, 2000). Furthermore, AtPIN1, a component of the auxin efflux carrier, appears to recycle through an intracellular compartment (Steinmann *et al.*, 1999). It will therefore be possible to establish whether *At*RabA proteins co-localise with these markers (Inaba *et al.*, 2002). In mammalian cells, antibodies raised against the variable carboxy-termini of Rab GTPases and GFP-Rab fusions have been used successfully to localise proteins in fixed and live cells respectively (Chavrier *et al.*, 1991; Segev, 2001; Zerial & McBride, 2001). In the only published immuno-EM study using antibodies to a plant Rab GTPase, monoclonal antibodies raised against *At*RabA5c (ARA4) labelled the Golgi as well as structures that were identified as trans-Golgi network and Golgi-derived vesicles in pollen of wild-type plants, and plants over-expressing the protein (Ueda *et al.*, 1996). This is consistent with a role in post-Golgi/endosomal transport but co-localisation studies with the newer markers of Golgi and prevacuolar compartments would be valuable. The GFP-fusion approach also seems promising (Cheung *et al.*, 2002; Inaba *et al.*, 2002; Ueda *et al.*, 2001) and the availability of spectrally distinguishable GFP variants should facilitate rapid co-localisation studies in live cells.

Studies of plant RabA function are minimal. Over-expression and antisense expression of members of several RabA subclasses in tobacco resulted in a suite of abnormal morphological and developmental phenotypes but revealed little about the cellular functions of any of these proteins (Aspuria *et al.*, 1995; Lu *et al.*, 2001; Sano *et al.*, 1994). Although complementation of yeast mutants by plant sequences can indicate aspects of protein function that are conserved across kingdoms, one study showed, curiously, that the cold-sensitive phenotype of the yeast *Ypt1-1* mutant, which is defective in the yeast RabD homologue, could be suppressed by a *Brassica* RabA5 clone (*Bra*) prompting the conclusion that the plant sequence was a functional homologue of yeast Ypt1 (Park *et al.*, 1994). Considering the taxonomic positions of the *Brassica* and yeast sequences, this finding is surprising. However, an alternative explanation is provided by the observation that negative regulators of yeast Ypt1 (the Gyp3 and Gyp7 GAP activities) also interact with Ypt31/Ypt32 (Zerial & McBride, 2001), while a positive regulator of Ypt1 (the TRAPP complex which has Ypt1 GEF activity) appears to be substantially more active on Ypt1 than Ypt31/Ypt32 (Wang *et al.*, 2000). Consequently, over-expression of the *Brassica* Ypt31/Ypt32 homologue may have preferentially titrated a negative regulator resulting in alleviation of the low-temperature phenotype of *Ypt1-1 ts* mutant. This study makes two important general points. Firstly, complementation studies should be performed with non-conditional deletion mutants rather than partial loss-of-function mutants to ensure a more stringent test for complementation (or suppression) of the yeast phenotype. Secondly, yeast complementation studies should be used cautiously to infer the function of a plant protein *in planta*. Two innovative studies of *At*RabA5c (ARA4) and various *At*RabA5c mutants defective in aspects of GTP binding or hydrolysis exploited the toxicity of this protein in yeast *Ypt* mutants to screen for interactions with regulatory factors (Ueda *et al.*, 2000). This resulted in the identification of an *Arabidopsis* homologue of the Rab regulatory protein RabGDI and a novel potential interacting protein, SAY1, from *Arabidopsis*. In this study, genetic interactions with the functionally dissimilar *YPT1* locus were also investigated but, here, they were used to infer non-specific interaction with the yeast membrane trafficking apparatus. This approach may be applicable to members of other subclasses.

The RabA3 subclass is unusual in that it is represented by a single gene in *Arabidopsis* and has a CXC prenylation motif rather than CCXX(X). A study of Pra2, a RabA3 protein from pea (*Ps*-RabA3) concluded that the function of this subclass is to integrate light and brassinosteroid signalling pathways in the etiolation response. It is proposed to do so by stimulating the activity of a cytochrome P450 that can catalyse a step in brassinolide biosynthesis (Kang *et al.*, 2001). *Ps*-RabA3 is normally expressed in a light repressible manner in the elongation zone of etiolated pea epicotyls (Inaba *et al.*, 2000), suggesting a role in etiolated growth. Dark-grown transgenic tobacco seedlings expressing *Ps*-RabA3, in sense or antisense fashion, exhibited short hypocotyls that could be rescued by exogenous brassinosteroid or its immediate biosynthetic precursor. The effect of these heterologous transgenes was attributed to an observed reduction in the steady-state transcript level of a member of the tobacco RabA family. The phenotype was specific to *Ps*-RabA3 as sense and antisense constructs derived from a member of the pea RabA4 subclass, *Ps*-RabA4 (Pra3) failed to alter hypocotyl elongation.

A role for a Rab GTPase in a process such as the brassinosteroid biosynthetic pathway would be unique and hence it will be important to address some curiosities in the observations. Chief amongst these is the fact that the proposed target of the *Ps*RabA3 antisense constructs in tobacco is not in fact a member of the RabA3 subclass but is clearly a member of the RabA4 subclass (*Nt*Rab11d) along with the pea protein *Ps*-RabA4 that acted as the negative control in the two-hybrid, biochemical, and transgenic analysis of *Ps*-RabA3 (Pra2). Secondly, a recent careful analysis of *Ps*-RabA3 (Pra2) localisation in tobacco cells has found that it is not localised to the ER along with DDWF1 as originally reported, but that *Ps*-RabA3 (Pra2) and *Ps*-RabA4 (Pra3) are each found on distinct populations of punctate structures that are likely to be pre-vacuolar or endosomal compartments (Inaba *et al.*, 2002). These observations call into question the original conclusion, based on low resolution imaging of bombarded onion cells, that both DDWF1 and *Ps*RabA3 localise to the ER where they may interact. They are, however, consistent with the suggestion that members of the RabA3 and RabA4 subclasses perform distinct cellular functions and that the RabA subclasses are involved in aspects of pre-vacuolar/endosomal traffic. Finally, the stimulation of DDWF1 activity by *Ps*RabA3 is modest (up to 1.3 fold), as are the changes in the endogenous concentrations of brassinosteroid metabolites in the transgenic lines. The statistical and biological significance of these findings needs to be investigated further. We also suggest an alternative interpretation of the observations. If the RabA3 and RabA4 subclasses are involved in transport to the vacuoles, we propose that their apparent role in brassinosteroid signalling and de-etiolation may result indirectly from a defect in vacuole function as it has been shown that *Arabidopsis* mutants that exhibit the short hypocotyl phenotype in the dark can arise through mutation of a subunit of the vacuolar ATPase (*det3*) as well as defects in brassinosteroid metabolism

(Schumacher *et al.*, 1999). It may be that when exogenous brassinosteroid was applied to the Pra2 transgenic lines, it could alleviate the effect of impaired vacuole function.

10.8 Conclusions

The last two years have seen some initial progress in understanding the role of Rab proteins in membrane traffic to and from Golgi membranes. This progress is likely to be sustained through the use of inhibitory mutants, Rab-GFP fusions, and the development of transport assays to investigate membrane trafficking events *in vivo*. Assays for transport within the Golgi and for retrograde traffic would be of considerable value. Genome sequences show that higher plants have a distinct complement of Rab GTPases though the functional divisions in the family are yet to be defined. *Arabidopsis* genetics will surely be important in determining function and resolving redundancy issues. A recent analysis of the plant syntaxin family, an important component of the SNARE complex required for vesicle fusion, suggested the existence of ten subclasses each with isoforms. It will be of interest to establish how the ten putative syntaxin families and 18 putative Rab GTPase families interact to regulate membrane traffic in plant cell types. The existence of more Rab families than syntaxin families may reflect the ability of syntaxins and other SNAREs to act in more than one vesicle transport pathway and to generate combinatorial specificity (Parlati *et al.*, 2002) – for example Sed5 resides on the early Golgi membranes and is involved in the fusion of both ER-derived anterograde transport vesicles and Golgi-derived retrograde vesicles that recycle material from later Golgi compartments. With the genetic and molecular tools available, we can look forward to rapid progress in understanding the significance of the unique complement of Rabs and syntaxins that have been revealed by genome sequencing.

References

Allan, B., Moyer, B.D. & Balch, W.E. (2000) Rab1 recruitment of p115 into a cis-SNARE complex: programming budding COPII vesicles for fusion. *Science*, **289**, 444–448.

Aspuria, E.T., Anai, T., Fujii, N., Ueda, T., Miyoshi, M., Matsui, M. *et al.* (1995) Phenotypic instability of transgenic tobacco plants expressing *Arabidopsis thaliana* small GTP-binding protein genes. *Mol. Gen. Genet.*, **246**, 509–513.

Batoko, H., Zheng, H., Hawes, C. & Moore, I. (2000) A Rab1 GTPase is required for transport between the endoplasmic reticulum and Golgi apparatus and for normal Golgi movement in plants. *Plant Cell*, **12**, 2201–2217.

Bednarek, S.Y., Reynolds, T.L., Schroeder, M., Grabowski, R., Hengst, L., Gallwitz, D. & Raikhel, N.V. (1994) A small GTP-binding protein from *Arabidopsis thaliana* functionally complements the yeast *YPT6i* null mutant. *Plant Physiol.*, **104**, 591–596.

Bischoff, F., Molendijk, A., Rajendrakumar, C.S.V. & Plame, K. (1999) GTP-binding proteins in plants. *Cell Mol. Life Sci.*, **55**, 233–256.

Bolte, S., Scheine, K. & Deitz, K. (2000) Characterisation of a small GTP-binding protein of the Rab5 family in *Mesembryanthemum crystalinum* with increased level of expression during early salt stress. *Plant Mol. Biol.*, **42**, 923–936.

Borg, S., Brandstrup, B., Jensen, T.J. & Poulsen, C. (1997) Identification of new protein species among 33 different small GTP-binding proteins encoded by cDNAs from *Lotus japonicus*, and expression of corresponding mRNAs in developing root nodules. *Plant J.*, **11**, 237–250.

Brandizzi, F., Frangne, N., Marc-Martin, S., Hawes, C., Neuhaus, J.M. & Paris, N. (2002a) The destination for single-pass membrane proteins is influenced markedly by the length of the hydrophobic domain. *Plant Cell*, **14**, 1077–1092.

Brandizzi, F., Snapp, E.L., Roberts, A.L., Lippincott-Schwartz, J. & Chris Hawes (2002b) Membrane protein transport between the endoplasmic reticulum and the Golgi in tobacco leaves is energy dependent but cytoskeleton independent: evidence from selective photobleaching. *Plant Cell*, **14**, 1293–1309.

Cao, X., Ballew, N. & Barlowe, C. (1998) Initial docking of ER-derived vesicles requires Uso1p and Ypt1p but is independent of SNARE proteins. *EMBO J.*, **17**, 2156–2165.

Cao, X. & Barlowe, C. (2000) Asymmetric requirements for a Rab GTPase and SNARE proteins in fusion of COPII vesicles with acceptor membranes. *J. Cell Biol.*, **149**, 55–66.

Cassanova, J.E., Wang, X., Kumar, R., Bhartur, S., Navarre, J., Woodrum, J., Altschuler, Y., Ray, G.S. & Goldenring, J.R. (1999) Association of Rab25 and Rab11a with the apical recycling system of polarised Madin-Darby Canine Kindeny cells. *Mol. Biol. Cell*, **10**, 47–61.

Chavrier, P., Gorvel, J.-P., Stelzer, E., Simons, K., Gruenberg, J. & Zerial, M. (1991) Hypervariable C-terminal domain of a Rab protein acts as a targeting signal. *Nature*, **353**, 769–772.

Chen, Y.A. & Scheller, R.H. (2001) SNARE-mediated membrane fusion. *Nat. Rev. Mol. Cell Biol.*, **2**, 98–106.

Cheon C.-I., Lee, N.-G., Siddique, A.-B.M., Bal, A.K. & Verma, D.P.S. (1993) Roles of plant homologues of Rab1p and Rab7p in the biogenesis of the peribacteroid membrane in soybean root nodules. *EMBO J.*, **12**, 4125–4135.

Cheung, A.Y., Chen C.Y.-H., Glaven, R.H., de Graaff, B.H.J., Vidali, L., Hepler, P.K. & Wu, H.-M. (2002) Rab2 GTPase regulates trafficking between the endoplasmic reticulum and the Golgi bodies and is important to pollen tube growth. *Plant Cell*, **14**, 945–962.

Hawes, C., Crooks, K., Coleman, J. & Satiat-Jeunemaitre, B. (1995) Endocytosis in plants: fact or artefact? *Plant Cell Environ.*, **18**, 1245–1252.

Humair, D., Hernandez Felipe, D., Neuhaus, J.M. & Paris, N. (2001) Demonstration in yeast of the function of BP-80, a putative plant vacuolar sorting receptor. *Plant Cell*, **13**, 781–792.

Inaba, T. Nagano, Y., Sakakibara, T. & Sasaki, Y. (2000) DE1, a 12-base-pair *cis* regulatory element sufficient to confer dark-inducible and light down-regulated expression to a minimal promoter in pea. *J. Biol. Chem.*, **275**, 19723–19727.

Inaba, T., Nagano, Y., Nagasaki, T. & Sasaki, Y. (2002) Distinct localization of two closely related Ypt3/Rab11 proteins on the trafficking pathway in higher plants. *J. Biol. Chem.*, **277**, 9183–9188.

Jedd, G., Richardson, C., Litt, R. & Segev, N. (1995) The Ypt1 GTPase is essential for the first two steps of the yeast secretory pathway. *J. Cell Biol.*, **131**, 583–590.

Kang, J.-G., Yun, J., Kim, D.H., Chung, K.S., Fujioka, S., Kim, J.I., Dae, H.W., Yoshida, S., Takatsuto, S., Song, P.S. & Park, C.M. (2001) Light and brassinosteroid signals are integrated via a dark-induced small G protein in etiolated seedling growth. *Cell*, **105**, 625–636.

Lazar, T., Goette, M. & Gallwitz, D. (1997) Vesicular transport: how many Ypt/Rab-GTPases make a eukaryotic cell? *Trends Biol. Sci.*, **22**, 468–472.

Lu, C., Zainal, Z., Tucker, G. & Lycett, G. (2001) Developmental abnormalities and reduced fruit softening in tomato plants expressing an antisense Rab11 GTPase gene. *Plant Cell*, **13**, 1819–1833.

Mallard, F., Luen Tang, B., Galli, T., Tenza, D., Saint-Pol, A., Yue, X., Antony, C., Hong, W., Goud, B. & Johannes, L. (2002) Early/recycling endosomes-to-TGN transport involves two SNARE complexes and a Rab6 isoform. *J. Cell Biol.*, **156**, 653–664.

Martinez, O., Antony, C., Pehu-Arnaudet, G., Berger, E.G., Salamero, J. & Goud, B. (1997) GTP-bound forms of rab6 induce the redistribution of Golgi proteins into the endoplasmic reticulum. *Proc. Natl. Acad. Sci. USA*, **94**, 1828–1833.

Moore. I., Schell, J. & Palme, K. (1995) Subclass-specific sequence motifs identified in Rab GTPases. *Trends Biochem. Sci.*, **20**, 10–12.

Moore, I., Diefenthal, T., Zarsky, V., Schell, J. & Palme, K. (1997) A homologue of mammalian Rab2 is present in *Arabidopsis* and is expressed predominantly in pollen grains and seedlings. *Proc. Natl. Acad. Sci. USA*, **94**, 762–767.

Morsomme, P. & Riezman, H. (2002) The Rab GTPase Ytp1p and tethering factors couple protein sorting at the ER to vesicle targeting to the Golgi apparatus. *Dev. Cell*, **2**, 307–317.

O'Mahoney, P. & Oliver, M. (1999) Characterisation of a desiccation-responsive small GTP-binding protein (Rab2) from the desiccation tolerant grass *Sporobolus stapfianus*. *Plant Mol. Biol.*, **39**, 809–821.

Palme, K., Diefenthal T. & Moore, I. (1993) The *ypt* gene family from maize and *Arabidopsis*: structural and functional analysis. *J. Exp. Bot.*, **44** (Suppl.), 183–195.

Park, Y.S., Song, O.-K., Kwak, J., Hong, S.W., Lee, H.H. & Nam, H.G. (1994) Functional complementation of a yeast vesicular transport mutation *ypt1-1* by a *Brassica napus* cDNA clone encoding a small GTP-binding protein.

Parlati, F., Varlamov, O., Paz, K., McNew, J., Hurtado, D., Soellner, T. & Rothman, J.E. (2002) Distinct SNARE complexes mediating membrane fusion in Golgi transport based on combinatorial specificity. *Proc. Natl. Acad. Sci. USA*, **99**, 5424–5429.

Pereira-Leal, J.B. & Seabra, M.C. (2000) The mammalian Rab family of small GTPases: definition of family and subfamily sequence motifs suggests a mechanism for functional specificity in the Ras superfamily. *J. Mol. Biol.*, **301**, 1077–1087.

Pereira-Leal, J.B. & Seabra, M.C. (2001) Evolution of the Rab family of small GTP-binding proteins *J. Mol. Biol.*, **313**, 889–901.

Phillips, D.G., Preshaw, C. & Steer, M.W. (1988) Dicytosome vesicle production and plasmamembrane turnover in auxin-stimulated outer epidermal cells of coleoptile segments from *Avena sativa* (L.). *Protoplasma*, **145**, 59–65.

Rodman, J.S. & Wandinger-Ness, A. (2000) Rab GTPases coordinate endocytosis. *J. Cell Sci.*, **113**, 183–192.

Saint-Jore, C., Evins, J., Batoko, H., Brandizzi, F., Moore, I. & Hawes, C. (2002) Redistribution of membrane proteins between the Golgi apparatus and endoplasmic reticulum in plants is reversible and not dependent on cytoskeletal networks. *Plant J.*, **29**, 661–678.

Sanderfoot, A.A., Assad, F. & Raikhel, N.V. (2000) The *Arabidopsis* genome. An abundance of soluble *N*-ethylmaleimide-sensitive factor adapter protein receptors. *Plant Phys.*, **124**, 1558–1569.

Sano, H., Seo, S., Orudgev, E., Youssefian, S., Ishizuka, K. & Ohashi, Y. (1994) Expression of a gene for a small GTP-binding protein in transgenic tobacco elevates endogenous cytokinin levels, abnormally induces salicylic acid in response to wounding, and increases resistance to tobacco mosaic virus infection. *Proc. Natl. Acad. Sci. USA*, **91**, 10556–10560.

Schmitt, H.-D., Wagner, P., Pfaff, E. & Gallwitz, D. (1986) The *ras*-related *YPT1* gene product in yeast: a GTP-binding protein that might be involved in microtubule organization. *Cell*, **47**, 401–412.

Schumacher, K., Vafeados, D., McCarthy, M., Sze, H., Wilkins, T. & Chory, J. (1999) The *Arabidopsis* det3 mutant reveals a central role for the vacuolar H$^+$-ATPase in plant growth and development. *Genes Dev.*, **13**, 3259–3270.

Segev, N. (2001) Ypt and Rab GTPases: insigt into functions through novel interactions. *Curr. Opis. Cell Biol.*, **13**, 500–511.

Shannon, T.M. & Steer, M.W. (1984) The root cap as a test system for the evaluation of Golgi inhibitors. I. Structure and dynamics of the secretory system and response to solvents. *J. Exp. Bot.*, **35**, 1697–1707.

Siniossoglou, S. & Pelham, H.R. (2001) An effector of Ypt6p binds the SNARE Tlg1p and mediates selective fusion of vesicles with late Golgi membranes. *EMBO J.*, **20**, 5991–5998.

Steinmann, T., Geldner, N., Grebe, M., Mangold, S., Jackson, C.L., Paris, S., Gaelweiler, L., Palme, K. & Juergens, G. (1999) Coordinated polar localisation of auxin efflux carrier PIN1 by GNOM ARF GEF. *Science*, **286**, 316–318.

Storrie, B., Pepperkok, R. & Nilsson, T. (2000) Breaking the COPI monopoly on Golgi recycling. *Trends Cell Biol.*, **10**, 385–391.

Terryn, N., Arias, M.B., Engler, G., Tire, C., Villaroel, R., Van Montague, M. & Inze, D. (1993) *rha1*, a gene encoding a small GTP-binding protein from *Arabidopsis*, is expressed primarily in developing guard cells. *Plant Cell*, **5**, 1761–1769.

The Arabidopsis Genome Initiative (2000) Analysis of the genome sequence of the flowering plant *Arabidopsis thaliana*. *Nature*, **408**, 796–826.

Thiel, G., Kreft, M. & Zorec, R. (1998) Unitary exocytotic and endocytotic events in *Zea mays* coleoptile protoplasts. *Plant J.*, **13**, 101–104.

Tisdale, E. & Balch, W. (1996) Rab2 is essential for the maturation of pre-Golgi intermediates. *J. Biol. Chem.*, **271**, 29372–29379.

Ueda, T., Anai, T., Tsukaya, H., Hirata, A. & Uchimaya, H. (1996) Characterisation and subcellular localisation of a small GTP-binding protein (*Ara-4*) from *Arabidopsis*: conditional expression under control of the promoter of the gene for heat-shock protein HSP81–1. *Mol. Gen. Genet.*, **250**, 533–539.

Ueda, T., Matsuda, N., Uchimiya, H. & Nakano, A. (2000) Modes of interaction between the *Arabidopsis* Rab protein, Ara4, and its putative regulator molecules revealed by a yeast expression system. *Plant J.*, **21**, 341–349.

Ueda, T., Yamaguchi, M., Uchimaya, H. & Nakano, A. (2001) Ara6, a plant-unique novel type Rab GTPase, functions in the endocytic pathway of *Arabidopsis thaliana*. *EMBO J.*, **20**, 4730–4741.

Wang, W., Sacher, M. & Ferro-Novick, S. (2000) TRAPP stimulates guanine nucleotide exchange on Ypt1p. *J. Cell Biol.*, **151**, 289–295.

Zerial, M. & McBride, H. (2001) Rab proteins as membrane organisers. *Nat. Rev. Mol. Cell Biol.*, **2**, 107–117.

11 Glycosyltransferases in the plant Golgi

Herta Steinkellner and Richard Strasser

11.1 Introduction

As in animal cells, the Golgi apparatus is a major biosynthetic organelle of the plant cell (see Chapter 2). A common feature of mammalian and plant Golgi is their central role in N- and/or O-glycosylation of glycoproteins, as well as in the synthesis of glycolipids. The basic features of the N-glycosylation pathway are highly conserved in plant and animal cells. Both oligomannosidic and complex-type *N*-linked glycans are contained within animal and plant glycoproteins (reviewed in Kornfeld & Kornfeld, 1985; Lerouge *et al.*, 1998). However, the structures of mature complex-type *N*-glycans differ between plants and mammals because of differences in the final steps of the biosynthetic pathway. Plant complex-type *N*-glycans are generally of smaller size, do not bear sialic acid, but instead contain β1,2-xylose and/or core α1,3-fucose.

Glycosidases and glycosyltransferases (GT) that catalyse respectively the stepwise trimming and addition of sugar residues are generally considered to act in a co-ordinated and highly ordered fashion to form mature *N*-glycans. Localisation studies on a number of enzymes of the N-glycosylation pathway in mammals revealed that these are arranged in a sequential, but overlapping, manner within the Golgi stacks, indeed, in the same order in which they act to modify oligosaccharide substrates (for review see Rabouille *et al.*, 1995). Considerable effort has been directed at understanding the basis of the subcellular localisation of Golgi GTs in mammals (for reviews see Colley, 1997; Gleeson, 1998). However, the lack of sequence similarities between different GTs suggests that multiple signals may be involved in their specific subcellular localisation.

All known Golgi GTs are N_{in}/C_{out} (type II) membrane proteins and display a common domain structure, namely, a short amino-terminal cytoplasmic (C) tail, a single hydrophobic transmembrane (T) domain, a luminal stem (S) region and a large carboxy-terminal catalytic domain (Fig. 11.1). By analysis of different glycosyltransferase chimeras it has been shown that the T domain plays a central role in Golgi retention in mammalian cells (Munro, 1991; Nilsson *et al.*, 1991; Aoki *et al.*, 1992; Tang *et al.*, 1992; Teasdale *et al.*, 1992; Wong *et al.*, 1992). In addition, the sequences flanking the T domain and the neighbouring C and S regions were demonstrated to be auxiliary mediators (Munro, 1991; Colley *et al.*, 1992; Burke *et al.*, 1994).

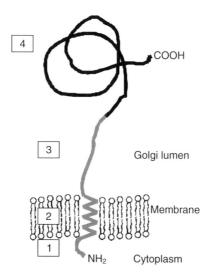

Figure 11.1 Schematic presentation of the domain structure of a typical Golgi resident glycosyl-transferase: 1 = cytoplasmic tail (C); 2 = transmembrane region (T); 3 = stem region (S); 4 = catalytic domain.

As a result of these localisation studies in mammals, two different models have been proposed as to how the differential distribution of Golgi-bound enzymes may be explained: the *bilayer thickness* model (Bretscher & Munro, 1993; Munro, 1995) postulates that the length of the transmembrane domain mediates Golgi retention with the importance of luminal sequences, whereas the *kin recognition* model (Nilsson *et al.*, 1993) assumes that different enzymes of a particular Golgi compartment interact with each other forming large hetero-oligomeric structures. Recent findings of different oligomerisation states of medial- and late-acting Golgi GTs indicated again that more than one mechanism might determine Golgi retention of GTs. Further, Opat *et al.* (2000) describe that medial Golgi enzymes exist in large molecular weight complexes, in contrast to the late-acting enzymes that are present as monomers and dimers. This difference in behaviour between medial- and late-acting Golgi enzymes may contribute to their differential localisation and their ability to glycosylate efficiently in the correct Golgi sub-compartment.

In recent years considerable evidence has been provided in support of a conservation of the basic features of N-glycosylation machinery in plant and mammalian cells. Indirect evidence for a differential Golgi sub-compartmentation of GT activities in plant cells was provided by Fitchette-Lainé *et al.* (1994) by showing the asymmetric distribution of reaction products of β1,2-xylosyltransferase and core α1,3-fucosyltransferase across the Golgi stack. Interestingly, heterologously expressed rat sialyltransferase, which has no equivalent counterpart in plant cells, was detected in the *trans* cisternae of the Golgi in transgenic plants (Wee *et al.*,

1998). This suggested that Golgi retention signals are conserved from plant to mammals. Here we describe the current knowledge of plant GTs and their localisation in the plant Golgi. Since little is known about the location of GTs involved in the O-glycosylation (see Chapter 7) only GTs involved in the N-glycosylation were considered.

11.2   β1,2-N-acetylglucosaminyltransferase I

In both plant and animal cells, a highly conserved multistep biosynthetic pathway covalently links carbohydrate to asparagine residues of newly synthesised proteins. This N-linked protein glycosylation is initiated in the endoplasmatic reticulum (ER) by the transfer of $Glc_3Man_9GlcNAc_2$ glycans to asparagine residues in the sequence Asn-X-Ser/Thr of the nascent polypeptide. Glycosidases and GTs in the ER and in the Golgi compartment subsequently convert typical $Glc_3Man_9GlcNAc_2$ first to oligo-mannose N-glycans ($Man_5GlcNAc_2$) and further to complex and hybrid N-glycans (Fig. 11.2). The first step that initiates the formation of hybrid or complex N-glycans in both plants and animals is catalysed by the enzyme β1,2-N-acetyl-glucosaminyltransferase I (GnTI) (for review see Kornfeld & Kornfeld, 1985). GnTI transfers a GlcNAc residue from the donor substrate UDP-GlcNAc to $Man_5GlcNAc_2$-Asn.

Several GnTI genes from different animal and invertebrate species have been cloned and characterised (for review see Schachter, 2000; Mucha *et al.*, 2001) and localised in the medial and *trans* cisternae of the mammalian Golgi (Burke *et al.*, 1994; Rabouille *et al.*, 1995). Recently GnTI genes from three plant species, *Nicotiana tabacum, Arabidopsis thaliana* and *Solanum tuberosum* (potato), respectively, have been identified and characterised (Bakker *et al.*, 1999; Strasser *et al.*, 1999a; Wenderoth & von Schaewen, 2000). Like its mammalian counterpart, plant GnTI is a type II transmembrane protein with a globular C-terminal part containing the catalytic domain. Furthermore, a typical short amino-terminal cytoplasmic (C) tail, a hydrophobic transmembrane (T) domain and a luminal stem (S) region are indicated (Strasser *et al.*, 1999a). These reports were the first that described the cDNA sequence of a plant GT and thus provided the necessary tool for its localisation and characterisation at the molecular level. Since human and *A. thaliana* derived GnTI can complement each other in GnTI deficient mutants (Gomez & Chrispeels, 1994; Bakker *et al.*, 1999), there was some indirect evidence that the plant GnTI is indeed localised in the Golgi and that targeting signals from this enzyme are conserved from plants to mammals. However, the first direct evidence of the existence of plant-derived Golgi targeting and retention sequences was given by Essl *et al.* (1999). Seventy-seven amino acids from the N-terminus of tobacco GnTI containing the cytoplasmic-transmembrane-stem (CTS-) region of the enzyme were fused to the reporter molecule green fluorescent protein (GFP) and transiently expressed in the tobacco-related model plant species *Nicotiana benthamiana*. Confocal laser scanning microscopy showed small fluorescent

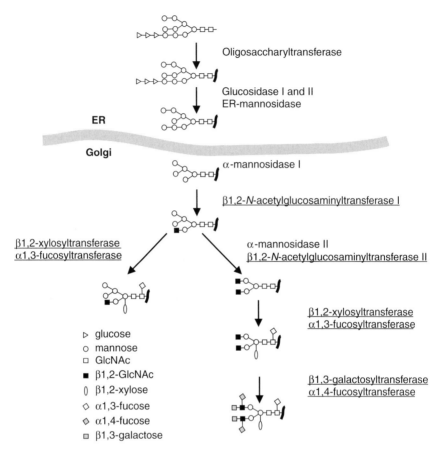

Figure 11.2 Schematic presentation of the N-glycosylation pathway in plants. Enzymes that are considered in this chapter are underlined.

vesicular bodies, typical for Golgi localisation, in CTS-GFP expressing cells (Fig. 11.3B), while GFP alone expressed in control plants and was uniformly distributed throughout the cytoplasm (Fig. 11.3A). The CTS-GFP fusion protein colocalised in immunolabelling with an antibody specific for the Golgi located plant Lewis a epitope (Essl *et al.*, 1999). Furthermore, treatment with brefeldin A, resulted in the formation of large fluorescent vesiculated areas (Fig. 11.3C). These results provided evidence for a Golgi location of CTS-GFP and, as a consequence, revealed that the N-terminal 77 amino acids of tobacco GnTI are sufficient to target and retain a reporter protein in the plant Golgi. These results are in good agreement with results obtained in localisation studies of mammalian GnTI (Burke *et al.*, 1994; Rabouille *et al.*, 1995). However, subcompartmentation studies of the plant GnTI have not yet been reported.

GFP   **PC**

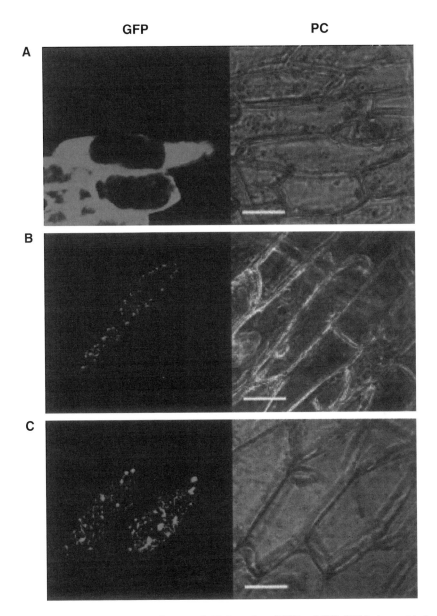

Figure 11.3 Localisation of *N. benthamiana* GnTI: Detection of GFP and CTS-GFP fusion protein in epidermal *N. benthamiana* cells by confocal laser scanning microscopy. (A) homogenous labelling throughout the cytoplasm in GFP expressing cells. (B) punctuate labelling in a CTS-GFP expressing epidermal cell. (C) Disruption and clustering of CTS-GFP labelled structures in epidermal cells after brefeldin A treatment. GFP: Recording of GFP associated fluorescence. PC: Corresponding phase contrast view. Bars in A, B and C = 50 μm (from Essl *et al.*, 1999).

11.3   β1,2-*N*-acetylglucosaminyltransferase II

After the action of GnTI, Golgi α-mannosidase II removes two mannose residues to form the substrate for β1,2-*N*-acetylglucosaminyltransferase II (GnTII) that adds a GlcNAc residue from UDP-GlcNAc to $GlcNAcMan_3GlcNAc_2$-Asn (Fig. 11.2). GnTII was cloned from several animal species (for review see Schachter, 2000) and like GnTI it has been localised in the medial and *trans*-Golgi compartments of mammalian cells (for review see Colley, 1997).

Although activity of the corresponding plant enzyme was originally detected in extracts from developing bean cotyledon (Johnson & Chrispeels, 1987), it was first purified from mung bean seedlings (Szumilo *et al.*, 1987). Strasser *et al.* (1999b) reported the identification and characterisation of a GnTII cDNA from *A. thaliana*. However, no localisation studies have been performed so far. Indirect evidence for the Golgi location of the enzyme was provided by Tezuka *et al.* (1992) through the detection of activity in Golgi fractions from suspension cultured sycamore cells.

11.4   β1,2-xylosyltransferase

An enzyme that has no counterpart in mammals and acts further downstream of the N-glycosylation pathway is β1,2-xylosyltransferase (XylT). XylT catalyses the transfer of D-xylose from UDP-xylose to the core β-linked mannose of *N*-linked oligosaccharides (Fig. 11.2). Activity of XylT has been identified in different plants (Johnson & Chrispeels, 1987; Rodgers & Bolwell, 1992; Tezuka *et al.*, 1992) and the protein has been purified from soya bean (Zheng *et al.*, 1997). By employing antibodies raised against β1,2-xylose containing plant *N*-glycans, it was shown that the reaction product of XylT was distributed asymmetric-ally across the Golgi compartments (Laine *et al.*, 1991). According to these results the transfer of xylose starts in the *cis* cisternae and increasing amounts of xylose-containing glycoproteins can be detected in the medial and also *trans* cisternae.

Meanwhile, the first and so far only plant cDNA coding for XylT was cloned from *A. thaliana* (Strasser *et al.*, 2000). Sequence analysis revealed a type II membrane protein topology typical for all previously cloned Golgi-located GTs. Since XylT is not present in animals it is of particular interest to investigate whether the Golgi retention and targeting mechanism are similar to that of mammalian GTs. Dirnberger *et al.* (2002) have investigated the contribution of different sequences in Golgi tar-geting and retention. Several deletion constructs of the putative cytoplasmic (C), transmembrane (T) and stem (S) regions of the enzyme were transiently expressed in *N. benthamiana*. Subcellular localisation of fusion proteins between CTS, CT, T, or C domains and the reporter molecule GFP using fluorescence microscopy and density-gradient centrifugation has revealed that the CT region alone is sufficient to sustain Golgi retention of XylT without the contribution of any luminal sequences. The finding of an incomplete retention by the T region alone suggests an important auxiliary role of the C domain in Golgi retention of the protein. However, the C segment did not confer any Golgi retention by itself, as the respective fusion protein was found exclusively in the cytoplasm. These results provide the first direct evidence that plant and mammalian cells rely on similar mechanisms to deliver GTs to the Golgi apparatus.

How do these data relate to the findings on mammalian GTs? The attempts to elucidate the distinct contributions of C, T and luminal sequences in Golgi localisation were focused on three enzymes, namely β1,4-galactosyltransferase, GnTI and α2,6-sialyltransferase. In the case of β1,4-galactosyltransferase, a *trans*-Golgi cisternal enzyme, the primary requirement for Golgi retention was shown to be the T domain (Nilsson *et al.*, 1991; Aoki *et al.*, 1992; Teasdale *et al.*, 1992; Masibay *et al.*, 1993; Teasdale *et al.*, 1994; Yamaguchi & Fukuda, 1995). Similar to XylT, the deletion of the C sequence led to incomplete Golgi retention, resulting in cell surface expression of the respective reporter construct (Nilsson *et al.*, 1991). However, contrary to the findings on XylT, evidence has been provided for an auxiliary role of luminal β1,4-galactosyltransferase sequences (Teasdale *et al.*, 1994). Interestingly, and in contrast to other investigated GTs, correct subcellular localisation of medial to *trans*-Golgi localised GnTI depends on the presence of the T, S and C domains (Burke *et al.*, 1992; Tang *et al.*, 1992; Burke *et al.*, 1994; Nilsson *et al.*, 1996). For α2,6-sialyltransferase, it appears that different regions independently lead to Golgi retention as it has been demonstrated that the T and flanking sequences are individually sufficient for retention (Munro, 1991; Colley *et al.*, 1992; Dahdal & Colley, 1993; Tang *et al.*, 1995). In comparison with the results obtained from mammalian GTs, the properties of XylT resemble most closely those of β1,4-galactosyltransferase with the transmembrane domain playing a central role in defining its subcellular distribution.

11.5 Core α1,3-fucosyltransferase

Core α1,3-fucosyltransferase (FucT) is involved in the synthesis of *N*-linked glycans, which are known to occur only in plants and invertebrates (Staudacher *et al.*, 1999). The enzyme catalyses the transfer of a fucose residue from GDP-Fuc in α1,3-linkage to the asparagine-linked GlcNAc of glycoproteins (Fig. 11.2). Enzyme activity has been detected in partially purified fractions from bean cotyledons (Johnson & Chrispeels, 1987), suspension cultured sycamore cells (Tezuka *et al.*, 1992) and in a 3000-fold purification from mung bean seedlings (Staudacher *et al.*, 1995). The first cDNA sequences from mung bean and *A. thaliana* FucT were reported by Leiter *et al.* (1999), Wilson *et al.* (2001a) and Bakker *et al.* (2001). The domain structure of FucT resembles the type II transmembrane structure. The location of the FucT in the plant Golgi has so far only been indirectly monitored by detection of its product: core α1,3-linked fucose. The asymmetric distribution of the reaction products across the Golgi stack was shown in sycamore cells using antibodies specific for *N*-glycans harbouring core α1,3-fucose (Fitchette-Lainé *et al.*, 1994). Core α1,3-fucose residues were mainly found in the *trans* cisternae and the TGN. Thus, based on this indirect immunolocalisation data, the fucosylation of the core GlcNAc residue occurs either shortly after, or in parallel with xylosylation of *N*-glycans (see 11.4).

11.6   Lewis a type GTs: α1,4-fucosyltransferase and β1,3-galactosyltransferase

After the transfer of β1,2-xylose and core α1,3-fucose, complex *N*-glycans can be further modified by the attachment of terminal fucose and galactose. The resulting Galβ1-3(Fucα1-4)GlcNAc structure is known as Lewis a antigen (Le[a]) and results from the subsequent action of β1,3-galactosyltransferase and α1,4-fucosyltransferase. Le[a] antigens are usually found on cell-surface glycoconjugates in mammals and are involved in cell–cell recognition and cell adhesion processes (Feizi, 1993). The Le[a] structure is very widespread among plants (Fitchette-Lainé *et al.*, 1997; Melo *et al.*, 1997; Fitchette *et al.*, 1999; Wilson *et al.*, 2001b), but has not been detected in *A. thaliana* and in other members of the Brassicaceae family so far (Fitchette-Lainé *et al.*, 1997; Fitchette *et al.*, 1999; Rayon *et al.*, 1999; Wilson *et al.*, 2001b) using structural analysis. However, Le[a] epitopes have been detected on *A. thaliana* glycoproteins using Le[a] specific antibodies (Leonard *et al.*, 2002).

Several reports have demonstrated the activity of α1,4-fucosyltransferase (Crawley *et al.*, 1989; Standacher *et al.*, 1995) and recently the corresponding cDNA sequences from several plant species have been reported (Bakker *et al.*, 2001; Wilson, 2001; Wilson *et al.*, 2001a). Although β1,3-linked galactose has been identified in a great number of plants (Wilson *et al.*, 2001b), no reports are available which demonstrate the corresponding enzyme activity.

The distribution of glycoproteins carrying Le[a] structures within the cell has been monitored by immunofluorescence experiments with anti-plant Le[a] antibodies in different plant cells (Fitchette *et al.*, 1999). Labelling was found on the plasma membrane and in the Golgi, but not in the ER, the tonoplast and in the vacuole. Furthermore, it has been shown that the monoclonal antibody JIM 84, which serves frequently as a Golgi marker (Satiat-Jeunemaitre & Hawes, 1992; Horsley *et al.*, 1993), had the same specificity as anti-plant Le[a] antibodies (Fitchette *et al.*, 1999). Labelling of the plant Golgi apparatus with anti-Le[a] was observed over the *trans* face of the dictyosome. This suggests that β1,3-galactosyltransferase and α1,4-fucosyltransferase are localised within the *trans*-Golgi, where they catalyse the transfer of galactose and fucose residues to terminal glucosamine residues of mature plant *N*-glycans.

11.7   Conclusion and future perspective

In recent years, indirect evidence was provided for a conservation of the basic features of the N-glycosylation machinery in plant and mammalian cells. In particular, reaction products of late-acting plant GTs have been monitored as being asymmetrically distributed over the Golgi stacks by employing anti-xylose (Laine *et al.*, 1991) and anti-plant Le[a] antibodies (Fitchette *et al.*, 1999). Indirect evidence for a conservation of targeting and retention mechanisms of animal and plant GTs has also been provided by the fact that human and plant GnTIs can complement each other in GnTI deficient mutants (Gomez & Chrispeels, 1994; Bakker *et al.*, 1999). Only two specific molecular studies on the localisation of plant GTs have been described (GnTI and XylT) and these have demonstrated the importance of the N-terminal sequences of the enzymes (Essl *et al.*, 1999; Dirnberger *et al.*, 2002).

In animals, as well as in plants, the basis for the distinct yet overlapping distribution of GTs in the Golgi is clearly complex and not fully understood. As discussed above, analysis of GT-chimeras from transfected mammalian and plant cells has demonstrated that the transmembrane domain plays a central role in their Golgi localisation, in addition to neighbouring sequences. However, further information is required concerning the organisation of Golgi-resident GTs within Golgi membranes and the mechanisms that may lead to the asymmetric distribution of GTs across the Golgi stack. Since the cloning of several plant GTs brings a number of useful reagents (e.g. antibodies or chimeric GTs) within reach, characterisation of the precise localisation of individual GTs in the plant Golgi is now realisable.

Acknowledgements

We would like to thank Prof. Iain B.H. Wilson for reading the manuscript.

References

Aoki, D., Lee, N., Yamaguchi, N., Dubois, C. & Fukuda, M.N. (1992) Golgi retention of a *trans*-Golgi membrane protein, galactosyltransferase, requires cysteine and histidine residues within the membrane-anchoring domain. *Proc. Natl. Acad. Sci. USA*, **89**, 4319–4323.

Bakker, H., Lommen, A., Jordi, W., Stiekema, W. & Bosch, D. (1999) An *Arabidopsis thaliana* cDNA complements the *N*-acetylglucosaminyltransferase I deficiency of CHO Lec1 cells. *Biochem. Biophys. Res. Commun.*, **261**, 829–832.

Bakker, H., Schijlen, E., de Vries, T., Schiphorst, W.E., Jordi, W., Lommen, A., Bosch, D. & van Die, I. (2001) Plant members of the α1 → 3/4-fucosyltransferase gene family encode an α1 → 4-fucosyltransferase, potentially involved in Lewis(a) biosynthesis, and two core α1 → 3-fucosyltransferases. *FEBS Lett.*, **507**, 307–312.

Bretscher, M.S. & Munro, S. (1993) Cholesterol and the Golgi apparatus. *Science*, **261**, 1280–1281.

Burke, J., Pettitt, J.M., Humphris, D. & Gleeson, P.A. (1994) Medial-Golgi retention of *N*-acetylglucosaminyltransferase I. Contribution from all domains of the enzyme. *J. Biol. Chem.*, **269**, 12049–12059.

Burke, J., Pettitt, J.M., Schachter, H., Sarkar, M. & Gleeson, P.A. (1992) The transmembrane and flanking sequences of β-1,2-*N*-acetylglucosaminyltransferase I specify *medial*-Golgi localization. *J. Biol. Chem.*, **267**, 24433–24440.

Colley, K.J. (1997) Golgi localization of glycosyltransferases: more questions than answers. *Glycobiology*, **7**, 1–13.

Colley, K.J., Lee, E.U. & Paulson, J.C. (1992) The signal anchor and stem regions of the beta-galactoside α-2,6-sialyltransferase may each act to localize the enzyme to the Golgi apparatus. *J. Biol. Chem.*, **267**, 7784–7793.

Crawley, S.C., Hindsgaul, O., Ratcliffe, R.M., Lamontagne, L.R. & Palcic, M.M. (1989) A plant fucosyltransferase with human Lewis blood-group specificity. *Carbohydr. Res.*, **193**, 249–256.

Dahdal, R.Y. & Colley, K.J. (1993) Specific sequences in the signal anchor of the beta-galactoside α-2,6-sialyltransferase are not essential for Golgi localization. Membrane flanking sequences may specify Golgi retention. *J. Biol. Chem.*, **268**, 26310–26319.

Dirnberger, D., Bencur, P., Mach, L. & Steinkellner, H. (2002) The Golgi localisation of *Arabidopsis thaliana* β1,2-xylosyltransferase in plant cells is dependent on its cytoplasmic and transmembrane sequences. *Plant Mol. Biol.*, **50**, 273–281.

Essl, D., Dirnberger, D., Gomord, V., Strasser, R., Faye, L., Glössl, J. & Steinkellner, H. (1999) The N-terminal 77 amino acids from tobacco *N*-acetylglucosaminyltransferase I are sufficient to retain a reporter protein in the Golgi apparatus of *Nicotiana benthamiana* cells. *FEBS Lett.*, **453**, 169–173.

Feizi, T. (1993) Oligosaccharides that mediate mammalian cell-cell adhesion. *Curr. Opin. Struct. Biol.*, **3**, 701–710.

Fitchette-Lainé, A.C., Gomord, V., Chekkafi, A. & Faye, L. (1994) Distribution of xylosylation and fucosylation in the plant Golgi apparatus. *Plant J.*, **5**, 673–682.

Fitchette-Lainé, A.C., Cabanes-Macheteau, M., Marvin, L., Martin, B., Satiat-Jeunemaitre, B., Gomord, V., Crooks, K., Lerouge, P., Faye, L. & Hawes, C. (1999) Biosynthesis and immuno-localization of Lewis a-containing N-glycans in the plant cell. *Plant Physiol.*, **121**, 333–344.

Fitchette-Lainé, A.C., Gomord, V., Cabanes, M., Michalski, J.C., Saint Macary, M., Foucher, B., Cavelier, B., Hawes, C., Lerouge, P. & Faye, L. (1997) N-glycans harbouring the Lewis a epitope are expressed at the surface of plant cells. *Plant J.*, **12**, 1411–1417.

Gleeson, P.A. (1998) Targeting of proteins to the Golgi apparatus. *Histochem. Cell Biol.*, **109**, 517–532.

Gomez, L. & Chrispeels, M.J. (1994) Complementation of an *Arabidopsis thaliana* mutant that lacks complex asparagine-linked glycans with the human cDNA encoding *N*-acetylglucosaminyltrans-ferase I. *Proc. Natl. Acad. Sci. USA*, **91**, 1829–1833.

Horsley, D., Coleman, J., Evans, D., Crooks, K., Peart, J., Satiat-Jeunemaitre, B. & Hawes, C. (1993) A monoclonal antibody, JIM 84, recognizes the Golgi apparatus and plasma membrane in plant cells. *J. Exp. Bot.*, **44**, 223–229.

Johnson, A. & Chrispeels, M.J. (1987) Substrate specificities of *N*-acetylglucosaminyl-, fucosyl-, and xylosyltransferases that modify glycoproteins in the Golgi apparatus of bean cotyledons. *Plant Physiol.*, **84**, 1301–1308.

Kornfeld, R. & Kornfeld, S. (1985) Assembly of asparagine-linked oligosaccharides. *Annu. Rev. Biochem.*, **54**, 631–664.

Laine, A.C., Gomord, V. & Faye, L. (1991) Xylose-specific antibodies as markers of subcompartmentation of terminal glycosylation in the Golgi apparatus of sycamore cells. *FEBS Lett.*, **295**, 179–184.

Leiter, H., Mucha, J., Staudacher, E., Grimm, R., Glössl, J. & Altmann, F. (1999) Purification, cDNA cloning, and expression of GDP-L-Fuc: Asn-linked GlcNAc α1,3-fucosyltransferase from mung beans. *J. Biol. Chem.*, **274**, 21830–21839.

Léonard, R., Costa, G., Darrambide, E., Lhernould, S., Fleurat-Lessard, P., Carlvé, M., Gomord, V., Faye, L. & Maftah, A. (2002) The presence of Lewis a epitopes in *Arabidopsis thaliana* glyco-conjugates depends on an active 4-fucosyl-transferase gene. *Glycobiology*, **12**, 299–306.

Lerouge, P., Cabanes-Macheteau, M., Rayon, C., Fitchette-Laine, A.C., Gomord, V. and Faye, L. (1998) *N*-glycoprotein biosynthesis in plants: recent developments and future trends. *Plant Mol. Biol.*, **38**, 31–48.

Masibay, A.S., Balaji, P.V., Boeggeman, E.E. & Qasba, P.K. (1993) Mutational analysis of the Golgi retention signal of bovine β-1,4-galactosyltransferase. *J. Biol. Chem.*, **268**, 9908–9916.

Melo, N.S., Nimtz, M., Conradt, H.S., Fevereiro, P.S. & Costa, J. (1997) Identification of the human Lewis(a) carbohydrate motif in a secretory peroxidase from a plant cell suspension culture (*Vaccinium myrtillus* L.). *FEBS Lett.*, **415**, 186–191.

Mucha, J., Svoboda, B., Fröhwein, U., Strasser, R., Mischinger, M., Schwihla, H., Altmann, F., Hane, W., Schachter, H., Glössl, J. & Mach, L. (2001) Tissues of the clawed frog *Xenopus laevis* contain two closely related forms of UDP-GlcNAc: α3-D-mannoside β-1,2-N-acetylglucosaminyltransferase I. *Glycobiology*, **11**, 769–778.

Munro, S. (1991) Sequences within and adjacent to the transmembrane segment of α-2,6-sialyltransferase specify Golgi retention. *EMBO J.*, **10**, 3577–3588.

Munro, S. (1995) An investigation of the role of transmembrane domains in Golgi protein retention. *EMBO J.*, **14**, 4695–4704.

Nilsson, T., Lucocq, J.M., Mackay, D. & Warren, G. (1991) The membrane spanning domain of β-1,4-galactosyltransferase specifies *trans* Golgi localization. *EMBO J.*, **10**, 3567–3575.

Nilsson, T., Rabouille, C., Hui, N., Watson, R. & Warren, G. (1996) The role of the membrane-spanning domain and stalk region of *N*-acetylglucosaminyltransferase I in retention, kin recognition and structural maintenance of the Golgi apparatus in HeLa cells. *J. Cell Sci.*, **109**, 1975–1989.

Nilsson, T., Slusarewicz, P., Hoe, M.H. & Warren, G. (1993) Kin recognition. A model for the retention of Golgi enzymes. *FEBS Lett.*, **330**, 1–4.

Opat, A.S., Houghton, F. & Gleeson, P.A. (2000) *Medial* Golgi but not late Golgi glycosyltransferases exist as high molecular weight complexes. Role of luminal domain in complex formation and localization. *J. Biol. Chem.*, **275**, 11836–11845.

Opat, A.S., van Vliet, C. & Gleeson, P.A. (2001) Trafficking and localisation of resident Golgi glycosylation enzymes. *Biochimie*, **83**, 763–773.

Rabouille, C., Hui, N., Hunte, F., Kieckbusch, R., Berger, E.G., Warren, G. & Nilsson, T. (1995) Mapping the distribution of Golgi enzymes involved in the construction of complex oligosaccharides. *J. Cell Sci.*, **108**, 1617–1627.

Rayon, C., Cabanes-Macheteau, M., Loutelier-Bourhis, C., Salliot-Maire, I., Lemoine, J., Reiter, W.D., Lerouge, P. & Faye, L. (1999) Characterization of N-glycans from *Arabidopsis*. Application to a fucose-deficient mutant. *Plant Physiol.*, **119**, 725–734.

Rodgers, M.W. & Bolwell, G.P. (1992) Partial purification of Golgi-bound arabinosyltransferase and two isoforms of xylosyltransferase from French bean (*Phaseolus vulgaris* L.). *Biochem. J.*, **288**, 817–822.

Satiat-Jeunemaitre, B. & Hawes, C. (1992) Redistribution of a Golgi glycoprotein in plant cell treated with brefeldin A. *J. Cell Sci.*, **103**, 1153–1166.

Schachter, H. (2000) The joys of HexNAc. The synthesis and function of N- and O-glycan branches. *Glycoconj. J.*, **17**, 465–483.

Staudacher, E., Altmann, F., Wilson, I.B.H. & März, L. (1999) Fucose in N-glycans: from plant to man. *Biochim. Biophys. Acta*, **1473**, 216–236.

Staudacher, E., Dalik, T., Wawra, P., Altmann, F. & März, L. (1995) Functional purification and characterization of a GDP-fucose: β-N-acetylglucosamine (Fuc to Asn linked GlcNAc) α1, 3-fucosyltransferase from mung beans. *Glycoconj. J.*, **12**, 780–786.

Strasser, R., Mucha, J., Mach, L., Altmann, F., Wilson, I.B.H., Glössl, J. & Steinkellner, H. (2000) Molecular cloning and functional expression of β1,2-xylosyltransferase cDNA from *Arabidopsis thaliana*. *FEBS Lett.*, **472**, 105–108.

Strasser, R., Mucha, J., Schwihla, H., Altmann, F., Glössl, J. & Steinkellner, H. (1999a) Molecular cloning and characterization of cDNA coding for β1,2-N-acetylglucosaminyltransferase I (GlcNAc-TI) from *Nicotiana tabacum*. *Glycobiology*, **9**, 779–785.

Strasser, R., Steinkellner, H., Boren, M., Altmann, F., Mach, L., Glössl, J. & Mucha, J. (1999b) Molecular cloning of cDNA encoding *N*-acetylglucosaminyltransferase II from *Arabidopsis thaliana*. *Glycoconj. J.*, **16**, 787–791.

Szumilo, T., Kaushal, G.P. & Elbein, A.D. (1987) Purification and properties of the glycoprotein processing *N*-acetylglucosaminyltransferase II from plants. *Biochemistry*, **26**, 5498–5505.

Tang, B.L., Low, S.H., Wong, S.H. & Hong, W. (1995) Cell type differences in Golgi retention signals for transmembrane proteins. *Eur. J. Cell Biol.*, **66**, 365–374.

Tang, B.L., Wong, S.H., Low, S.H. & Hong, W. (1992) The transmembrane domain of *N*-glucosaminyltransferase I contains a Golgi retention signal. *J. Biol. Chem.*, **267**, 10122–10126.

Teasdale, R.D., D'Agostaro, G. & Gleeson, P.A. (1992) The signal for Golgi retention of bovine β1,4-galactosyltransferase is in the transmembrane domain. *J. Biol. Chem.*, **267**, 4084–4096.

Teasdale, R.D., Matheson, F. & Gleeson, P.A. (1994) Post-translational modifications distinguish cell surface from Golgi-retained β1,4-galactosyltransferase molecules. Golgi localization involves active retention. *Glycobiology*, **4**, 917–928.

Tezuka, K., Hayashi, M., Ishihara, H., Akazawa, T. & Takahashi, N. (1992) Studies on synthetic pathway of xylose-containing N-linked oligosaccharides deduced from substrate specificities of the processing enzymes in sycamore cells (*Acer pseudoplatanus* L.). *Eur. J. Biochem.*, **203**, 401–413.

Wee, E.G., Sherrier, D.J., Prime, T.A. & Dupree, P. (1998) Targeting of active sialyltransferase to the plant Golgi apparatus. *Plant Cell*, **10**, 1759–1768.

Wenderoth, I. & von Schaewen, A. (2000) Isolation and characterization of plant *N*-acetyl glucosaminyltransferase I (GntI) cDNA sequences. Functional analyses in the *Arabidopsis cgl* mutant and in antisense plants. *Plant Physiol.*, **123**, 1097–1108.

Wilson, I.B.H. (2001) Identification of a cDNA encoding a plant Lewis-type α1,4-fucosyltransferase. *Glycoconj. J.*, **18**(6), 439–447.

Wilson, I.B.H., Rendić, D., Freilinger, A., Dumić, J., Altmann, F., Mucha, J., Müller, S. & Hauser, M.T. (2001a) Cloning and expression of cDNAs encoding α1,3-fucosyltransferase homologues from *Arabidopsis thaliana*. *Biochim. Biophys. Acta*, **1527**, 88–96.

Wilson, I.B.H., Zeleny, R., Kolarich, D., Staudacher, E., Stroop, C.J., Kamerling, J.P. & Altmann, F. (2001b) Analysis of Asn-linked glycans from vegetable foodstuffs: widespread occurrence of Lewis a, core α1,3-linked fucose and xylose substitutions. *Glycobiology*, **11**, 261–274.

Wong, S.H., Low, S.H. & Hong, W. (1992) The 17-residue transmembrane domain of beta-galactoside α 2,6-sialyltransferase is sufficient for Golgi retention. *J. Cell Biol.*, **117**, 245–258.

Yamaguchi, N. & Fukuda, M.N. (1995) Golgi retention mechanism of β-1,4-galactosyltransferase. Membrane-spanning domain-dependent homodimerization and association with α- and β-tubulins. *J. Biol. Chem.*, **270**, 12170–12176.

Zeng, Y., Bannon, G., Thomas, V.H., Rice, K., Drake, R. & Elbein, A. (1997) Purification and specificity of β1,2-xylosyltransferase, an enzyme that contributes to the allergenicity of some plant proteins. *J. Biol. Chem.*, **272**, 31340–31347.

12 Perturbation of ER-Golgi vesicle trafficking

David G. Robinson and Christophe Ritzenthaler

12.1 Introduction

It has long been established that the organelles of the endomembrane system are not rigid, but highly dynamic structures that undergo continuous membrane and protein trafficking and recycling. This is especially true for the Golgi apparatus, which receives newly synthesized proteins and lipids from the ER, modifies them, and then distributes them to compartments downstream in the secretory pathway. To compensate this flow of cargo molecules, anterograde ER-to-Golgi traffic is constantly balanced by a retrograde traffic of membranes and proteins back to the ER. In a way, therefore, the existence of the Golgi apparatus is dependent on the efficient operation of the vesiculating machinery that is responsible for the bi-directional traffic. These, as we know are the COP-coated vesicles (see Chapter 3). Indeed, as recent research on mammalian cells suggests (Ward *et al.*, 2001), the Golgi apparatus disappears if the formation of either type of COP-vesicle is selectively inhibited.

As we learn more and more about the individual proteins involved in these trafficking processes, it is becoming more and more fashionable to apply genetical methods e.g. overexpression, or the production of dominant negative mutants (see Chapters 6 and 10) in order to modify the endomembrane system. However, one can also learn much about organelles or compartments when their functioning is in some way physiologically perturbed. This can be achieved through the application of agents which are relatively unspecific in their mode of action e.g. temperature, or those influencing ATP levels or pH gradients, or through the use of drugs directed against a specific target molecule e.g. brefeldin A or the cytochalasins. As we shall report, viruses are also useful in this regard. Here we present a short overview of the plant literature pertaining to these methodological approaches.

12.2 Metabolic inhibitors

The classic work of George Palade in the 1960/70s established the energy-dependency of secretion (Jamieson & Palade, 1968). It is now known that vesiculation, irrespective of the type of vesicle involved, requires energy in the form of ATP (Rothman & Wieland, 1996). Therefore, the application of agents which lead to the lowering or depletion of cytosolic ATP-levels will obviously have an adverse effect on vesicle-mediated transport events. In pancreatic acinar cells held at 37°C, given

15 min anoxia or exposed to 2,4-dinitrophenol (DNP) significant changes in the ultrastructure of the ER/Golgi system are caused: there is a dramatic reduction in the number of *transition* (COPII ?) vesicles between ER and Golgi apparatus, there is a marked increase in the number of clathrin-coated vesicles and empty clathrin cages at the *trans*-pole of the Golgi apparatus, and there is an increase in the number and size of *trans*-like cisternae (Merisko *et al.*, 1986).

There are numerous reports in the plant literature concerning the effects of KCN, DNP ± 2-deoxyglucose, or an O_2-free atmosphere (e.g. Schnepf, 1963; Morré & Mollenhauer, 1964), as well as antipyrine (Deysson & Benbadis, 1972; Glas & Robinson, 1982; Glas-Albrecht & Robinson, 1987). These all lead to a serious reduction in cellular ATP-levels, and produce a common phenotype: so-called *cup-shaped dictyosomes*. The effects – including cyanide poisoning (remember plants possess an alternate oxidase pathway) – are reversible (Robinson & Ray, 1977). Depending upon the plane of section, cup- or beaker-shaped Golgi stacks may also have the appearance of concentric circles of cross-sectioned Golgi-cisternae.

Cup-shaped Golgi stacks have several features in common: their cisternae are identical in terms of lumenal width, inter-cisternal separation and stainability, but most importantly the cisternae are considerably longer than normal. They are therefore without polarity (see Robinson & Kristen, 1982 for parameters of polarity). It is probably correct to assume that what leads to the characteristic change in Golgi morphology is the prevention of vesicle transport from the ER, and the inability to form vesicles. But, how do the cisternae grow in length, and why do they curve (it is not really clear from the published micrographs whether this occurs in the former *cis*- or *trans*-directions, although it is usually stated that curvature is towards the *trans*-pole)? It is not unreasonable to assume that the vesiculation events are not uniform in their response to ATP-depletion, and that the production of vesicles by the Golgi apparatus is more susceptible than that of ER, i.e. the Golgi gains more membrane than it loses. One might also speculate that vesiculation in the *cis*-most cisternae is greater than at other positions in the stack, so that they will grow more than other cisternae in the stack when vesiculation is inhibited. However, until a careful morphological analysis of the time-course of these events has been carried out, we will not be able to provide an adequate explanation for the genesis of cup-shaped Golgi stacks.

Before leaving this section, the effects of barbital should be mentioned. According to Benbadis and Deysson (1975), this substance, presented to onion roots at the relatively high concentration of 5–7 mg/ml, causes the doubling of the number of cisternae per Golgi stack. We have not been able to reproduce this observation on BY-2 cells (Robinson, unpublished data).

12.3   Low temperature effects

Protein transport in the endomembrane system of mammalian cells is sensitive towards the reduction of temperature. However, different temperatures exert blockages at different sites, and as emphasized by Saraste *et al.* (1986), these effects are distinct from those induced by ATP-shortage. Pancreatic acinar cells held at 15°C for 1 h are characterized by the accumulation of transition (COPII ?) vesicles between ER and Golgi (Saraste *et al.*, 1986). A similar phenotype is observed in *Semliki forest virus*-infected cells: viral glycoproteins accumulate in vesicles proximal to the Golgi apparatus at 15°C (Saraste & Kuismanen, 1984). By contrast, movement through or release from the Golgi apparatus is more temperature sensitive: lowering the temperature to 20–22°C is sufficient to block the transfer of secretory proteins from the Golgi apparatus to condensing vacuoles in pancreas (Tartakoff, 1986), and arrests the transport of viral glycoproteins in the *trans*-Golgi network (TGN, Matlin & Simons, 1983).

Since plants are not homeopoietic organisms, low temperature effects are bound to be different. Indeed, nobody has ever recorded specific effects obtained through exposure to 20°C or 15°C. On the other hand, very low temperatures (1–4°C), do induce remarkable morphological changes in the endomembrane system of some plants (Mollenhauer *et al.*, 1975). In the mantle cells of the maize root cap which have been held at 4°C for 10 h, the Golgi stacks no longer produce typical slime-containing vesicles, and the cisternae elongate, often producing cup-shaped dictyosomes (Fig. 12.1). These extended cisternae either fuse with cisternae coming from neighboring Golgi stacks, or are seen to be contiguous with the ER. Such continuities were not restricted to particular cisternae in the stack. Amazingly, the relative amount of ER in the cells increased several fold after 24 h cold treatment (Robinson, 1981), and is often visualized as whorls in the cytoplasm or as concentric layers around the nucleus (see Fig. 12.1). ER not only responds quantitatively to cold treatment, its morphology changes as well. Quader *et al.* (1989) have shown that the cisternal-type ER of onion epidermal cells is converted to a tubular form, and long ER tubules are fragmented into smaller ones. A similar change is brought about by exposing the cells to weak organic acids (Quader & Fast, 1990), encouraging these authors to suggest that the changes resulting from cold treatment may occur via acidification of the cytosol.

12.4   Cytoskeleton inhibitors

Unlike the case with mammalian cells, microtubule inhibitors such as colchicine appear to be without noticeable effect on the organelles of the plant secretory pathway. In contrast, microfilament inhibitors such as the cytochalasins produce significant changes. As is now well known (see Chapter 4), Golgi stacks move along actin microfilaments, which run parallel to tubular ER in the cortex of higher plant cells. Application of cytochalasin stops their tumbling motion (Boevink *et al.*, 1998; Nebenführ *et al.*, 1999). Cytochalasins elicit another, very clear effect: the accumulation of secretory vesicles in the immediate vicinity of the Golgi stacks (Mollenhauer & Morré, 1976; Robinson *et al.*, 1976; Pope *et al.*, 1979; Kristen & Lockhausen, 1983) (see Fig. 12.2). Normally, the Golgi stacks themselves are not

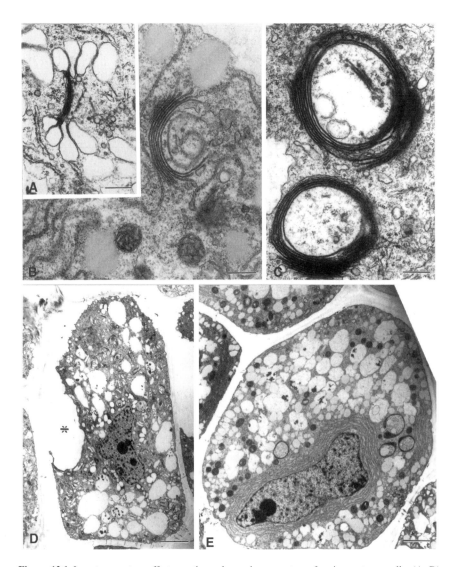

Figure 12.1 Low temperature effects on the endomembrane system of maize root cap cells. (A, D) Control cells. (B, C, E) Cells from roots held at 4°C for 16 h. (A) Dictyosome in control cell. Note the extreme dilated slime-filled vesicles. (B, C) Longitudinal and horizontal sections through cup-shaped Golgi complexes formed by neighboring dictyosomes growing and fusing. Slime-containing vesicles are no longer present. (D) Overview of root cap cell. Intense secretory activity is indicated by the withdrawal of the plasma membrane (asterisk). (E) Cold treatment leads to a significant increase in ER membrane, especially around the nucleus. Bars = 200 nm (A, B, C); 5 μm (D, E) (unpublished micrographs of the authors).

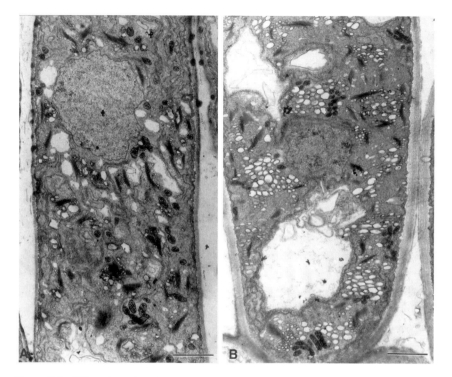

Figure 12.2   Effect of cytochalasin B (100 µg/ml; 1 h) on the Golgi apparatus of the ovary gland cells of *Aptenia cordifolia*. (A) Portion of a control cell. (B) After cytochalasin B treatment. Although Golgi stacks seem to remain intact, the massive accumulation of secretory vesicles is apparent. Bars = 1 µm (unpublished micrographs of U. Kristen, Univ. Hamburg, Germany).

affected (see however Volkmann & Czaja, 1981). The effects of cytochalasins are reversible. The usual interpretation of the cytochalasin-induced accumulation of secretory vesicles is that actin microfilaments are required for the transport of the vesicles to the plasma membrane, but an indirect effect on their capability to fuse with the plasma membrane cannot be ruled out.

12.5   pH-modulators

The ionophore monensin which exchanges Na^+/K^+ for protons has been used extensively by animal and plant cell biologists to investigate the location and function of acidic compartments. It has, therefore, been used in studies on secretion and endocytosis (Mollenhauer *et al.*, 1990). In animal cells, the application of monensin leads to a swelling of *trans*-Golgi cisternae, the inhibition of late processing events, and an arrest of Golgi export (Farquhar, 1985). In plants, monensin has been shown to block protein secretion (Heupke & Robinson, 1985), and to interfere with the targeting and processing of vacuolar storage proteins (Craig & Goodchild, 1984; Stinissen *et al.*, 1985). Ultrastructural investigations of plant cells treated with monensin have also revealed the dilation of the *trans*-most cisternae of the Golgi apparatus (Shannon & Steer, 1984; Heupke & Robinson, 1985), and an accumulation of swollen vesicles in the vicinity of the Golgi. In a later paper, Zhang *et al.* (1993) have shown that, despite the swelling of *trans*-cisternae vesicles, they at least temporarily can still bud off the Golgi stacks.

Although the pH in individual cisternae of the plant Golgi apparatus has never been determined, by analogy to mammalian cells (see Mollenhauer *et al.*, 1990 for literature) it is assumed that the *trans*-cisternae are more acidic than the rest of the Golgi. This is supported by the direct demonstration (immuno-labeling) of the presence of V-ATPase (Zhang *et al.*, 1996) and PPase (Ratajczak *et al.*, 1999) at the *trans*-pole of the plant Golgi apparatus. Swelling of the *trans*-cisternae is, therefore, usually interpreted as being the result of exchanging protons for osmotically active cations with the consequence that water is taken up from the cytoplasm (Boss *et al.*, 1984; Griffing & Ray, 1985).

12.6   Brefeldin A (BFA)

For a substance which has been so widely used, it is surprising in the plant community that so much confusion exists as to how it works. On the one hand, this situation is a result of the various morphological responses reported in the plant literature, and on the other hand due to the inability of many plant scientists to distinguish between short and long term effects of BFA. Through the availability of antisera against COPI- and COPII-vesicle coat proteins, it has only recently been possible to identify the primary target of BFA in plants, and to provide a unifying explanation for the different structural phenotypes (Nebenführ *et al.*, 2002; Ritzenthaler *et al.*, 2002a).

The first measurable response to BFA application in all eukaryotes is the release of the membrane-associated COPI-coat proteins Arf1p and coatomer into the cytosol (Donaldson *et al.*, 1990; Ritzenthaler *et al.*, 2002a). This feature is an immediate consequence of the binding of BFA to a guanine nucleotide exchange factor (GEF), which is responsible for converting Arf1p in its GDP-(unbound) form to the GTP-form which can bind to membranes. Arf1p is principally localized to the Golgi apparatus in all eukaryotes (Stearns *et al.*, 1990; Pimpl *et al.*, 2000), and, at least in mammalian and yeast cells, so too is the associated BFA-sensitive GEF (Mansour *et al.*, 1999; Spang *et al.*, 2001).

The loss of the ability to form COPI-vesicles means that ER-targeting SNAREs, which are normally concentrated in these vesicles, accumulate in an uncovered form in the Golgi cisternae (Elazar *et al.*, 1994). This enables direct fusion between ER and Golgi membranes to occur, a process which is prevented in untreated cells by the presence of recruited COPI-coat proteins. Due to the close proximity of ER and Golgi membranes in plants, this fusion event is rapid

(Ritzenthaler *et al.*, 2002a), whereas in mammalian cells it is preceded by an extensive tubulation of the Golgi apparatus (e.g. Hess *et al.*, 2000). Inevitably, however, a hybrid ER-Golgi compartment arises as a result of treating cells with BFA (see Fig. 12.3). When this happens, vesicle-mediated protein transport to the cell surface or to the vacuole is totally blocked.

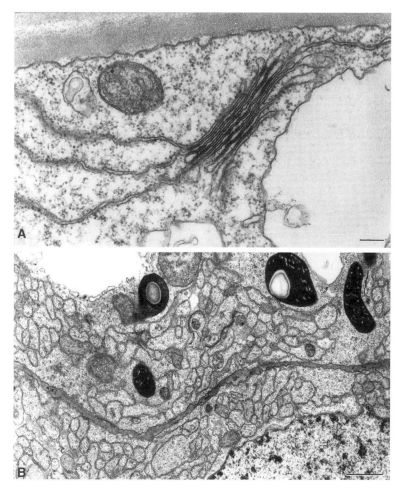

Figure 12.3 Development of an ER-Golgi hybrid membrane system in tobacco BY-2 cells as a result of treatment with 10 μg/ml brefeldin A (BFA). (A) 30 min BFA. Note the clear continuity between stacked Golgi cisternae (recognizable through patches of intercisternal filaments at various positions) and ER cisternae. These connections are not restricted to a particular pole of this hybrid structure. (B) 3 h BFA. Although other organelles do not seem to be affected, a recognizable Golgi apparatus has been lost and a highly ramifying ER-like system has developed. Bars = 200 nm (A); 1 μm (B) (unpublished micrographs of the authors).

In addition to the creation of an ER-Golgi hybrid organelle, some plants also form a post-Golgi structure known as the BFA compartment (Satiat-Jeunemaitre *et al.*, 1996). The mechanism underlying the formation of this structure, which in essence is a large aggregate of endosomal membranes plus vesicles together with individual Golgi cisternae, is unclear. However, it has been speculated that it reflects the continued sloughing of *trans*-cisternae and associated vesicles, under conditions where SNARE–SNARE interactions at the *cis*-pole of the Golgi stack are limited. This may be due to either low SNARE expression levels, or through their masking by residual COPI-coats which have not been efficiently dissociated at suboptimal BFA concentrations (Ritzenthaler *et al.*, 2002).

Interestingly, recent research on mammalian cells has shown that when COPII-mediated ER export is blocked, the Golgi apparatus disassembles and becomes incorporated into the ER in a BFA-like manner, indicating that efficient ER-export is necessary for the maintenance of Golgi structure (Miles *et al.*, 2001; Ward *et al.*, 2001). Thus, the idea emerging from these studies is that the Golgi apparatus is a highly dynamic organelle whose existence is a consequence of the dynamic interplay between two vesiculating machineries: that of COPI and that of COPII. When either is affected, the Golgi apparatus cannot be sustained and what remains of it collapses into the ER.

12.7   Perturbing ER-Golgi structure and function through viral infection

Viruses rely on basic cellular mechanisms of eukaryotic cells in order to multiply. Viral infections result in the production of a relatively small number of viral proteins, but in very large amounts. These proteins are usually highly antigenic and can be easily mutated, thus making them ideal objects for immunological and genetic studies. For these reasons, viral proteins have become very valuable as tracer molecules in cell biological investigations. In terms of membrane trafficking, probably the best example is that of the type I transmembrane glycoprotein G of *vesicular stomatitis virus* (VSV-G), which has been successfully used to investigate the mechanism of ER-export (Sevier *et al.*, 2000), and to study intra-Golgi transport *in vitro* (Orci *et al.*, 1997). Tagged with GFP and expressed as a temperature sensitive mutant, VSV-G has become an extremely powerful tool for *in vivo* studies on exocytic vesicle-mediated transport (e.g. Hirschberg *et al.*, 2000). Another example of how useful a virus protein can be in defining critical steps in the secretory process is NSP4, a non-structural membrane-anchored glycoprotein encoded by rotavirus, a non-enveloped icosahedral animal virus. NSP4 behaves as a microtubule-associated protein, and its expression selectively disrupts microtubule-mediated membrane transport (Xu *et al.*, 2000).

Unfortunately, viral proteins have so far found little use in the characterization of the secretory pathway in plants. However, the simple facts that all known RNA viruses use membranes to support their replication (David *et al.*, 1992; De Graaff & Jaspars, 1994) and that more than 90% of plant viruses are RNA-viruses, open fantastic new opportunities for plant cell biologists working on endomembrane-related problems.

The plant viral literature has virtually been ignored by those working on the secretory pathway, but contains potentially useful information. The replication of positive-strand plant RNA viruses, for example, always involves specific membranes. It is assumed that this requirement for membranes is needed to ensure the protection of the viral RNA from endogenous RNases during the replication process. These membranes differ according to the virus under consideration. For example, *turnip yellow mosaic virus* induces vesiculation of chloroplast membranes (Prod'homme *et al.*, 2001), whereas *flock house virus* RNA replication occurs on outer mitochondrial membranes (Miller *et al.*, 2001) and infection with *tombusvirus* results in the formation of multivesicular bodies, which develop from peroxisomes or mito-chondria (Rubino & Russo, 1998; Rubino *et al.*, 2001). Obviously, this entails specific interactions between viral-encoded proteins and the *host* membrane, but in most cases the viral proteins responsible for these interactions remain to be identified.

In spite of the few examples given above, the majority of plant RNA viruses seem to prefer membranes derived from organelles of the endomembrane system for their replication. Replication of some members of the *bromoviridae* family, such as *alfalfa mosaic virus*, occurs at the tonoplast (Van Der Heijden *et al.*, 2001) but most viruses appear to specifically modify membranes of the early secretory pathway. Examples can be found in different genera: *tobamovirus* (Heinlein, 2002), *bromovirus* (Restrepo-Hartwig & Ahlquist, 1996), *potyvirus* (Schaad *et al.*, 1997), *comovirus* (Carette *et al.*, 2000), and *pecluvirus* (Dunoyer *et al.*, 2002). *Tobacco mosaic virus* induces the formation of large ER aggregates (Kahn *et al.*, 1998; Reichel & Beachy, 1998; Reichel *et al.*, 1999), whereas *peanut clump virus* (PCV) infected cells reveal perturbation of perinuclear ER morphology (Dunoyer *et al.*, 2002). In addition to ER proliferation and fragmentation, malformation of the Golgi apparatus can also be observed during PCV infection (see Fig. 12.4). With *tobacco etch virus* (TEV), a *potyvirus*, a hydrophobic 6 kDa protein has been shown to become integrated into ER membranes, where it subsequently guides polyproteins to the viral replication site (Schaad *et al.*, 1997). With *brome mosaic virus* (BMV), a multicomponent positive-strand virus, two key enzymes encoded by separate genomic RNAs are brought together at the ER: the helicase 1a that unwinds the dsRNA template, and the polymerase 2a which is responsible for the synthesis of the viral RNA. 1a spontaneously associates with the ER and then recruits 2a from the cytosol. When expressed alone, P2 remains cytosolic (Restrepo-Hartwig & Ahlquist, 1999; Chen & Ahlquist, 2000; Den Boon *et al.*, 2001). In this case, replication occurs on ER-derived 50- to 100-nm cytoplasmic vesicles that accumulate in a perinuclear region (Restrepo-Hartwig & Ahlquist, 1996). Interestingly, unsaturated fatty acids production was recently shown to be important for BMV replication as revealed by the severe inhibition observed in the yeast OLE1 mutant (Lee *et al.*, 2001). This essential gene encodes a delta9 fatty acid desaturase, an integral ER membrane protein, and the first enzyme in unsaturated fatty acid synthesis. As for BMV, perinuclear ER is also needed for *cowpea mosaic*

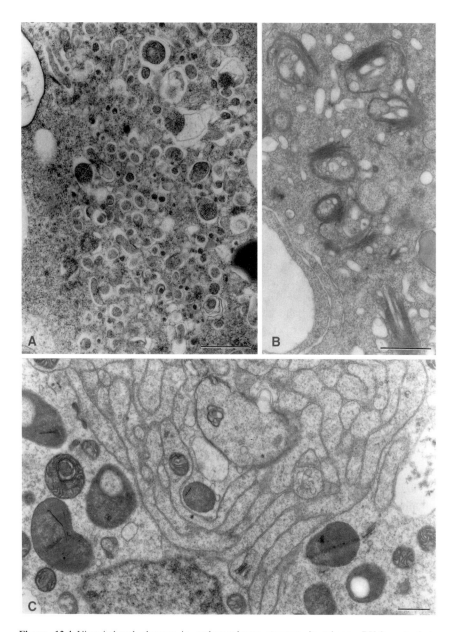

Figure 12.4 Virus-induced changes in endomembrane structure in tobacco BY-2 protoplasts. (A) Grapevine fanleaf virus (GFLV). Note the membrane proliferation and the accumulation of small vesicles. (B, C) Peanut clump virus (PCV). Malformation of the Golgi apparatus (B), proliferation of the ER (C). Bars = 1 μm (unpublished micrographs of O. Rohfritsch, Strasbourg, France; and the authors).

virus (CPMV) replication (De Zoeten *et al.*, 1974, Carette *et al.*, 2000). Here the RNA-1 encoded 60 kDa VPg precursor protein has been shown by immunological methods to be involved in the formation of the ER-derived vesicles (van Bokhoven *et al.*, 1992; Carette *et al.*, 2000). In contrast, the viral replication complex for TMV is formed in the cell cortex (Reichel & Beachy, 1998; Reichel *et al.*, 1999), but again the viral replicase and movement protein localize to the cytoplasmic surface of the ER.

An interesting case is that of the *nepovirus, grapevine fanleaf virus* (GFLV), because features of the infection process are very similar to that of the closely related poliovirus (Goldbach, 1987; Goldbach & Wellink, 1988). It has been known for many years that the replication of poliovirus involves the production of vesicles (Caliguri & Tamm, 1970; Bienz *et al.*, 1983), and that the proteins coded by the P2 genomic region are bound to these vesicles which often assemble into *rosette-like* clusters (Bienz *et al.*, 1990). Efficient viral replication is dependent upon continued membrane lipid synthesis (Guinea & Carrasco, 1990; Barco & Carrasco, 1995), although secretion is blocked in cells infected with poliovirus. Nevertheless, an active vesiculating machinery is still required since BFA (see above) inhibits poliovirus replication (Doedens & Kirkegaard, 1995; Cuconati *et al.*, 1998). The recently published data of Rust *et al.* (2001) strongly suggest that poliovirus exploits COPII-vesiculation processes for viral replication.

GFLV infection has been followed in BY-2 cells expressing Golgi- and ER-targeted GFP construct (Gaire *et al.*, 1999; Ritzenthaler, unpublished results). Contrary to infection with *artichoke yellow ringspot nepovirus* which results in severe modification of Golgi membranes (Russo *et al.*, 1978), the major detectable symptom of GFLV infection was a depletion of cortical ER accompanied by the progressive build-up of ER-derived perinuclear vesicle aggregates (see Fig. 12.4). *In situ* immuno-confocal microscopy and sucrose density gradient analysis indicate that VPg precursors are most likely present on the vesicles that organized into rosette-like clusters (Ritzenthaler *et al.*, 2002b). As with poliovirus, GFLV replication is sensitive towards BFA. It is therefore attractive to consider that GFLV uses a similar mechanism to that of poliovirus for the production of its viral replication complex.

The literature contains only few examples for the direct involvement of the Golgi apparatus in plant viral replication. *Tomato spotted wilt virus* (TSWV), the type member of the *bunyaviridae* family, is one such case (Kikkert *et al.*, 1999). Here, the plant host is induced to form *paired parallel* membranes from the Golgi apparatus. Double-enveloped virus particles are then formed by the paired-parallel membranes wrapping around cytosolic nucleocapsids. Single-enveloped particles subsequently form by fusion of the double-enveloped particles with the ER. Although this mechanism is not seen in animal cells infected with *bunyavirus* (in animal cells, the virus particles are seen inside the Golgi-derived vesicles), the processing and transport of plant TSWV glycoproteins in animal cells are identical to that of animal-infecting *bunyavirus* glycoproteins (Kikkert *et al.*, 2001).

References

Barco, A. & Carrasco, L. (1995) A human virus protein, poliovirus protein 2BC, induces membrane proliferation and blocks the exocytic pathway in the yeast *Saccharomyces cerevisiae*. *EMBO J.*, **14**, 3349–3364.

Benbadis, M. C. & Deysson, G. (1975) Morphologie des dictyosomes dans les cellules des méristemes radiculaires d'*Allium sativum* L.: éffets du barbital. *Planta*, **123**, 283–290.

Bienz, K., Egger, D., Rasser, Y. & Bossart, W. (1983) Intracellular distribution of poliovirus proteins and the induction of virus specific cytoplasmic structures. *Virology*, **131**, 39–48.

Bienz, K., Egger, D., Troxler, M. & Pasamontes, L. (1990) Structural organization of poliovirus RNA replication is mediated by viral proteins of the P2 genomic region. *J. Virol.*, **64**, 1156–1163.

Boevink, P., Oparka, K., Santa Cruz, S., Martin, B., Betteridge, A. & Hawes, C. (1998) Stacks on tracks: the plant Golgi apparatus traffics on an actin/ER network. *Plant J.*, **15**, 441–447.

Boss, W.F., Morré, D.J. & Mollenhauer, H.H. (1984) Monensin-induced swelling of Golgi apparatus cisternae mediated by a proton gradient. *Eur. J. Cell Biol.*, **34**, 1–8.

Caliguri, L.A. & Tamm, I. (1970) The role of cytoplasmic membranes in poliovirus biosynthesis. *Virology*, **42**, 100–111.

Carette, J.E., Stuiver, M., van Lent, J., Wellink, J. & van Kammen, A. (2000) Cowpea mosaic virus infection induces a massive proliferation of endoplasmic reticulum but not Golgi membranes and is dependent on *de novo* membrane synthesis. *J. Virol.*, **74**, 6556–6563.

Chen, J. & Ahlquist, P. (2000) Brome mosaic virus polymerase-like protein 2a is directed to the endoplasmic reticulum by helicase-like viral protein 1a. *J. Virol.*, **74**, 4310–4318.

Craig, S. & Goodchild, D.J. (1984) Golgi-mediated vicilin accumulation in pea cotyledon cells is re-directed by monensin and nigericin. *Protoplasma*, **122**, 91–97.

Cuconati, A., Molla, A. & Wimmer, E. (1998) Brefeldin A inhibits cell-free, *de novo* synthesis of poliovirus. *J. Virol.*, **72**, 6456–6464.

David, C., Gargouri-Bouzid, R. & Haenni, A.L. (1992) RNA replication of plant viruses containing an RNA genome. *Prog. Nucleic Acid Res. Mol. Biol.*, **42**, 157–227.

De Graaff, M. & Jaspars, E.M.J. (1994) Plant viral RNA synthesis in cell-free systems. *Ann. Rev. Phytopath.*, **32**, 311–335.

Den Boon, J.A., Chen, J. & Ahlquist, P. (2001) Identification of sequences in brome mosaic virus replicase protein 1a that mediate association with endoplasmic reticulum membranes. *J Virol.*, **75**, 12370–12381.

Deysson, G. & Benbadis, M.C. (1972) Déstruction et réconstitution des dictyosomes provoquées par l'antipyrine dans les cellules végétales. *J. de Microscopie*, **13**, 207–216.

De Zoeten, G.A., Assink, A.M. & van Kammen, A. (1974) Association of cowpea mosaic virus-induced double-stranded RNA with a cytopathological structure in infected cells. *Virology*, **59**, 351–355.

Doedens, J.R. & Kirkegaard, K. (1995) Inhibition of cellular protein secretion by poliovirus proteins 2B and 3A. *EMBO J.*, **14**, 894–907.

Donaldson, J., Lippincott-Schwartz, J., Bloom, G., Kreis, T. & Klausner, R. (1990) Dissociation of a 110-kD peripheral membrane-protein from the Golgi apparatus is an early event in brefeldin A action. *J. Cell Biol.*, **111**, 2295–2306.

Dunoyer, P., Ritzenthaler, C., Hemmer, O., Michler, P. & Fritsch, C. (2002) Intracellular localization of the peanut clump virus replication complex in tobacco BY-2 protoplasts containing green fluorescent protein-labeled endoplasmic reticulum or Golgi apparatus. *J. Virol.*, **76**, 865–874.

Elazar, Z., Orci, L., Ostermann, J., Amherdt, M., Tanigawa, G. & Rothman, J.E. (1994) ADP-ribosylation factor and coatomer couple fusion to vesicle budding. *J. Cell Biol.*, **124**, 415–424.

Farquhar, M.G. (1985) Progress in unravelling pathways of Golgi traffic. *Annu. Rev. Cell Biol.*, **1**, 447–488.

Gaire, F., Schmitt, C., Stussi-Garaud, C., Pinck, L. & Ritzenthaler, C. (1999) Protein 2A of grapevine fanleaf nepovirus is implicated in RNA2 replication and colocalizes to the replication site. *Virology*, **264**, 25–36.

Glas-Albrecht, R. & Robinson, D.G. (1987) Morphological and biochemical changes in membranes of sugar cane cells after dinitrophenol and antipyrine treatments. *Eur. J. Cell Biol.*, **45**, 116–125.

Glas, R. & Robinson, D.G. (1982) On the induction of cup-shaped dictyosomes in suspension-cultured *Acer pseudoplatanus*. *Protoplasma*, **113**, 97–102.

Goldbach, R. & Wellink, J. (1988) Evolution of plus-strand RNA viruses. *Intervirology*, **29**, 260–267.

Goldbach, R. (1987) Genome similarities between plant and animal RNA viruses. *Microbiol. Sci.*, **4**, 197–202.

Griffing, L.R. & Ray, P.M. (1985) Involvement of monovalent cations in Golgi secretion by plant cells. *Eur. J. Cell Biol.*, **39**, 24–31.

Guinea, R. & Carrasco, L. (1990) Phospholipid biosynthesis and poliovirus genome replication, two coupled phenomena. *EMBO J.*, **9**, 2011–2016.

Heinlein, M. (2002) The spread of tobacco mosaic virus infection: insights into the cellular mechanism of RNA transport. *Cell Mol. Life Sci.*, **59**, 58–82.

Hess, M., Muller, M., Debbage, P., Vetterlein, M. & Pavelka, M. (2000) Cryopreparation provides new insight into the effects of brefeldin A on the structure of the HepG2 Golgi apparatus. *J. Struct. Biol.*, **130**, 63–72.

Heupke, H.J. & Robinson, D.G. (1985) Intracellular transport of α-amylase in barley aleurone cells: evidence for the participation of the Golgi apparatus. *Eur. J. Cell Biol.*, **39**, 265–272.

Hirschberg, K., Phair, R.D. & Lippincott-Schwartz, J. (2000) Kinetic analysis of intracellular trafficking in single living cells with vesicular stomatitis virus protein G-green fluorescent protein hybrids. *Methods Enzymol.*, **327**, 69–89.

Jamieson, J.D. & Palade, G.E. (1968) Intracellular transport of secretory proteins in the pancreatic exocrine cell. IV. Metabolic requirements. *J. Cell Biol.*, **39**, 589–603.

Kahn, T.W., Lapidot, M., Heinlein, M., Reichel, C., Cooper, B., Gafny, R. & Beachy, R.N. (1998) Domains of the TMV movement protein involved in subcellular localization. *Plant J.*, **15**, 15–25.

Kikkert, M., Verschoor, A., Kormelink, R., Rottier, P. & Goldbach, R. (2001) Tomato spotted wilt virus glycoproteins exhibit trafficking and localization signals that are functional in mammalian cells. *J. Virol.*, **75**, 1004–1012.

Kikkert, M., van Lent, J., Storms, M., Bodegom, P., Kormelink, R. & Goldbach, R. (1999) Tomato spotted wilt virus particle morphogenesis in plant cells. *J. Virol.*, **73**, 2288–2297.

Kristen, U. & Lockhausen, J. (1983) Estimation of Golgi membrane flow rates in ovary glands of *Aptenia cordifolia* using cytochalasin B. *Eur. J. Cell Biol.*, **29**, 262–267.

Lee, W.M., Ishikawa, M. & Ahlquist, P. (2001) Mutation of host delta9 fatty acid desaturase inhibits brome mosaic virus RNA replication between template recognition and RNA synthesis. *J. Virol.*, **75**, 2097–2106.

Mansour, S., Skaug, J., Zhao, X., Giordano, J., Scherer, S. & Melancon, P. (1999) p200 ARF-GEP1: a Golgi-localized guanine nucleotide exchange protein whose Sec7 domain is targeted by the drug brefeldin A. *Proc. Natl. Acad. Sci. USA*, **96**, 7968–7973.

Matlin, K. & Simons, K. (1983) Reduced temperature prevents transfer of a membrane glycoprotein to the cell surface but does not prevent terminal glycosylation. *Cell*, **34**, 233–243.

Merisko, E.M., Fletcher, M. & Palade, G.E. (1986) The reorganization of the Golgi complex in anoxic pancreatic acinar cells. *Pancreas*, **1**, 95–109.

Miles, S., McManus, H., Forsten, K.E. & Storrie, B. (2001) Evidence that the entire Golgi apparatus cycles in interphase HeLa cells: sensitivity of Golgi matrix proteins to an ER exit block. *J. Cell Biol.*, **155**, 543–555.

Miller, D.J., Schwartz, M.D. & Ahlquist, P. (2001) Flock house virus RNA replicates on outer mitochondrial membranes in Drosophila cells. *J. Virol.*, **75**, 11664–11676.

Mollenhauer, H.H. & Morré, D.J. (1976) Cytochalasin B, but not colchicine, inhibits migration of secretory vesicles in root tips of maize. *Protoplasma*, **87**, 39–48.

Mollenhauer, H.H., Morré, D.J. & Rowe, L.D. (1990) Alteration of intracellular traffic by monensin: mechanism, specificity and relationship to toxicity. *Biochim. Biophys. Acta*, **1031**, 225–246.

Mollenhauer, H.H., Morré, D.J. & Vanderwoude, W.J. (1975) Endoplasmic reticulum–Golgi apparatus associations in maize root tips. *Mikroskopie*, **31**, 257–272.

Morré, D.J. & Mollenhauer, H.H. (1964) Isolation of the Golgi apparatus from plant cells. *J. Cell Biol.*, **23**, 295–305.

Nebenführ, A., Gallager, L., Dunahay, T.G., Frohlick, J.A., Masurkiewicz, A.M., Meehl, J.B. & Staehelin, L.A. (1999) Stop-and-go movements of plant Golgi stacks are mediated by the acto-myosin system. *Plant Physiol.*, **121**, 1127–1141.

Nebenführ, A., Ritzenthaler, C. & Robinson, D.G. (2002) Brefeldin A: deciphering an enigmatic inhibitor of secretion. *Plant Physiol.*, **130**, 1102–1108.

Orci, L., Stamnes, M., Ravazzola, M., Amherdt, M., Perrelet, A., Sollner, T.H. & Rothman, J.E. (1997) Bidirectional transport by distinct populations of COPI-coated vesicles. *Cell*, **90**, 335–349.

Pimpl, P., Movafeghi, A., Coughlan, S., Denecke, J., Hillmer, S. & Robinson, D.G. (2000) *In situ* localization and *in vitro* induction of plant COPI-coated vesicles. *Plant Cell*, **12**, 2219–2236.

Pope, D.G., Thorpe, J.R., Al-Azzawi, M.J. & Hall, J.L. (1979) The effect of cytochalasin B on the rate of growth and ultrastructure of wheat coleoptiles and maize roots. *Planta*, **144**, 373–383.

Prod'homme, D., Le Panse, S., Drugeon, G. & Jupin, I. (2001) Detection and subcellular localization of the turnip yellow mosaic virus 66K replication protein in infected cells. *Virology*, **281**, 88–101.

Quader, H. & Fast, H. (1990) Influence of cytosolic pH changes on the organization of the endoplasmic reticulum in epidermal cells of onion bulb scales: acidification by loading with weak organic acids. *Protoplasma*, **157**, 216–224.

Quader, H., Hofmann, A. & Schnepf, E. (1989) Reorganization of the endoplasmic reticulum in epidermal cells of onion bulb scales after cold stress: involvement of cytoskeletal elements. *Planta*, **177**, 273–280.

Ratajczak, R., Hinz, G. & Robinson, D.G. (1999) Localization of pyrophosphatase in membranes of cauliflower inflorescence cells. *Planta*, **208**, 205–211.

Reichel, C., Mas, P. & Beachy, R.N. (1999) The role of the ER and cytoskeleton in plant viral trafficking. *Trends Plant Sci.*, **4**, 458–462.

Reichel, C. & Beachy, R.N. (1998) Tobacco mosaic virus infection induces severe morphological changes of the endoplasmic reticulum. *Proc. Natl. Acad. Sci. USA*, **95**, 11169–11174.

Restrepo-Hartwig, M.A. & Ahlquist, P. (1996) Brome mosaic virus helicase- and polymerase-like proteins colocalize on the endoplasmic reticulum at sites of viral RNA synthesis. *J. Virol.*, **70**, 8908–8916.

Restrepo-Hartwig, M.A. & Ahlquist, P. (1999) Brome mosaic virus RNA replication proteins 1a and 2a colocalize and 1a independently localize on the yeast endoplasmic reticulum. *J. Virol.*, **73**, 10303–10309.

Ritzenthaler, C., Nebenführ, A., Movafeghi, A., Stussi-Garaud, C., Behnia, L., Pimpl, P., Staehelin, L.A. & Robinson, D.G. (2002a) Reevaluation of the effects of brefeldin A on plant cells using tobacco bright yellow 2 cells expressing Golgi-targeted green flourescent protein and COPI-antisera. *Plant Cell*, **14**, 237–261.

Ritzenthaler, C., Laporte, C., Gaire, F., Dunoyer, P., Schmitt, C., Duval, S., Piéquet, A., Loudes, A.M., Rohfritsch, O., Stussi-Garaud, C. & Pfeiffer, P. (2002b) Grapevine fanleaf virus replication occurs on endoplasmic reticulum-derived membranes. *J. Virol.*, **76**, 8808–8819.

Robinson, D.G. (1981) Membrane flow in relation to secretion in higher plant cells: mew results and concepts, in *Cell Wall '81* (eds D.G. Robinson & H. Quader), pp. 47–56, Wiss. Verlagsges, Stuttgart.

Robinson, D.G., Eisinger, W.R. & Ray, P.M. (1976) Dynamics of the Golgi system in wall matrix polysaccharide synthesis and secretion by pea cells. *Ber. Deutsch. Bot. Ges.*, **89**, 147–161.

Robinson, D.G. & Kristen, U. (1982) Membrane flow via the Golgi apparatus of higher plant cells. *Int. Rev. Cytol.*, **77**, 89–127.

Robinson, D.G. & Ray, P.M. (1977) The reversible cyanide inhibition of Golgi secretion in pea cells. *Cytobiologie*, **15**, 65–77.

Rothman, J.E. & Wieland, F.T. (1996) Protein sorting by transport vesicles. *Science*, **272**, 227–234.

Rubino, L. & Russo, M. (1998) Membrane targeting sequences in tombusvirus infections. *Virology*, **252**, 431–437.

Rubino, L., Weber-Lotfi, F., Dietrich, A., Stussi-Garaud, C. & Russo, M. (2001) The open reading frame 1-encoded (36K) protein of Carnation Italian ringspot virus localizes to mitochondria. *J. Gen. Virol.*, **82**, 29–34.

Russo, M., Martelli, G.P., Rana, G.L. & Kyriakopoulou, P.E. (1978) The ultrastructure of artichoke yellow ringspot virus infections. *Microbiologica*, **1**, 81–99.

Rust, R.C., Landmann, L., Gosert, R., Tang, B.L., Hong, W., Hauri, H.P., Egger, D. & Bienz, K. (2001) Cellular COPII proteins are involved in production of the vesicles that form the poliovirus replication complex. *J. Virol.*, **75**, 9808–9818.

Saraste, J. & Kuismanen, E. (1984) Pre- and post-Golgi vacuoles operate in the transport of Semliki Forest virus membrane glycoproteins to the cell surface. *Cell*, **38**, 535–549.

Saraste, J., Palade, G.E. & Farquhar, M.G. (1986) Temperature-sensitive steps in the transport of secretory proteins through the Golgi complex in exocrine pancreatic cells. *Proc. Natl. Acad. Sci. USA*, **83**, 6425–6429.

Satiat-Jeunemaitre, B., Cole, L., Bourett, T., Howard, R. & Hawes, C. (1996) Brefeldin A effects in plant and fungal cells: something new about vesicle trafficking. *J. Microsc.*, **181**, 162–177.

Schaad, M.C., Jensen, P.E. & Carrington, J.C. (1997) Formation of plant RNA virus replication complexes on membranes: role of an endoplasmic reticulum-targeted viral protein. *EMBO J.*, **16**, 4049–4059.

Schnepf, E. (1963) Zur Cytologie und Physiologie pflanzlicher Drüsen. 2. Über die Wirkung von Sauerstoffentzug und von Atmungsinhibitoren auf die Sekretion des Fangschleimes von *Drosophyllum* und auf die Feinstruktur der Drüsenzellen. *Flora*, **153**, 23–48.

Sevier, C.S., Weisz, O.A., Davis, M. & Machamer, C.E. (2000) Efficient export of the vesicular stomatitis virus G protein from the endoplasmic reticulum requires a signal in the cytoplasmic tail that includes both tyrosine-based and di-acidic motifs. *Mol. Biol. Cell.*, **11**, 13–22.

Shannon, T.M. & Steer, M.W. (1984) The root cap as a test system for the evaluation of Golgi inhibitors. II. Effect of potential inhibitors on slime droplet formation and structure of the secretory system. *J. Exp. Bot.*, **35**, 1708–1714.

Spang, A., Herrmann, J.M., Hamamoto, S. & Schekman, R. (2001) The ADP ribosylation factor-nucleotide exchange factors Gea1p and Gea2p have overlapping, but not redundant functions in retrograde transport from the Golgi to the endoplasmic reticulum. *Mol. Biol. Cell*, **4**, 1035–1045.

Stearns, T., Willingham, M.C., Botstein, D. & Kahn, R.A. (1990) ADP-ribosylation factor is functionally and physically associated with the Golgi complex. *Proc. Natl. Acad. Sci. USA*, **87**, 1238–1242.

Stinissen, H.M., Peumans, W.J. & Chrispeels, M.J. (1985) Post-translational processing of proteins in vacuoles and protein bodies is inhibited by monensin. *Plant Physiol.*, **77**, 495–498.

Tartakoff, A. (1986) Temperature and energy dependence of secretory protein transport in the exocrine pancreas. *EMBO J.*, **5**, 1477–1482.

van Bokhoven, H., van Lent, J.W., Custers, R., Vlak, J.M., Wellink, J. & van Kammen, A. (1992) Synthesis of the complete 200K polyprotein encoded by cowpea mosaic virus B-RNA in insect cells. *J. Gen. Virol.*, **73**, 2775–2784.

Van Der Heijden, M.W., Carette, J.E., Reinhoud, P.J., Haegi, A. & Bol, J.F. (2001) Alfalfa mosaic virus replicase proteins P1 and P2 interact and colocalize at the vacuolar membrane. *J. Virol.*, **75**, 1879–1887.

Volkmann, D. & Czaja, A.W.P. (1981) Reversible inhibition of secretion in root cap cells of cress after treatment with cytochalasin B. *Exp. Cell Res.*, **135**, 229–236.

Ward, T.H., Polishchuk, R.S., Caplan, S., Hirschberg, K. & Lippincott-Schwartz, J. (2001) Maintenance of Golgi structure and function depends on the integrity of ER export. *J. Cell Biol.*, **155**, 557–570.

Xu, A., Bellamy, A.R. & Taylor, J.A. (2000) Immobilization of the early secretory pathway by a virus glycoprotein that binds to microtubules. *EMBO J.*, **19**, 6465–6474.

Zhang, G.F., Driouich, A. & Staehelin, L.A. (1993) Effect of monensin on plant Golgi: re-examination of the monensin-induced changes in cisternal architecture and functional activities of the Golgi apparatus of sycamore suspension-cultured cells. *J. Cell Sci.*, **104**, 819–831.

Zhang, G.F., Driouich, A. & Staehelin, A. (1996) Monensin-induced redistribution of enzymes and products from Golgi stacks to swollen vesicles in plant cells. *Eur. J. Cell Biol.*, **71**, 332–340.

13 SNARE components and mechanisms of exocytosis in plants

M.R. Blatt and G. Thiel

13.1 Introduction

The plant endomembrane system is responsible for the synthesis, delivery and maintenance of some of the most biologically interesting and commercially important products in plants, including the cell wall, various alkaloids, anti-cancer drugs, dyes and enzymes. These products and their processing depend on biosynthetic activities localised to the plant ER and Golgi, with subsequent targeting to the plasma membrane, endomembrane and vacuolar structures. Vesicle trafficking and membrane fusion are integral to these events in eukaryotes and are facilitated by so-called SNARE proteins and associated regulatory proteins, including binding proteins such as Sec1/Munc18 and various tethering factors (Novick & Zerial, 1997; Jahn & Sudhof, 1999; Jahn, 2000; Zerial & McBride, 2001). SNAREs are associated with each membrane and provide a core of inter-action for membrane fusion and secretion throughout the endomembrane system (Jahn & Sudhof, 1999).

The overall pattern of vesicle trafficking in plants shares common features with mammalian and yeast cells (Lauber *et al.*, 1997; Thiel & Battey, 1998; Blatt *et al.*, 1999; Sanderfoot & Raikhel, 1999; Sutter *et al.*, 2000), but many aspects of membrane organisation and its maintenance on which they depend appear unique to plants. For example, in the absence of cell migration it is the control of cell and vacuolar expansion, assymetric growth and cell division that underlies development and morphogenesis in plants. These features are shared by many tissues (epidermal and vascular tissues, root hairs, pollen, etc.) and they imply a selective trafficking of different secretory cargoes to particular membrane domains (see Lauber *et al.*, 1997; Steinmann *et al.*, 1999; Geldner *et al.*, 2001). Finally, it is clear that plant cells contain multiple vacuolar compartments with distinct vesicle-trafficking pathways, some of which may bypass the Golgi (Rogers, 1998; Jauh *et al.*, 1999; Jiang & Rogers, 1999). In each of these, and a number of other examples we have single snap-shots that point to the underlying processes and one or two molecular components that, probably, also contribute to these events. The challenge, to fully define and analyse the dominant mem-brane trafficking pathways of plants in molecular, mechanistic and kinetic terms, remains.

13.2   Integration and regulation of secretion in plants

The majority of the best known examples of secretion in plants follow patterns of constitutive secretion, including the secretion of polysaccharide mucigel by the cells of root caps (Chaboud & Rougier, 1981), the secretion by floral nectaries and by gland cells of carnivorous plants (Findlay, 1988), and amylase release during the germination of barley (Jones & Jacobsen, 1991). These examples are marked by cargo passage to the apoplast in which secretion is prolonged or continuous, is not preceded by a large accumulation of vesicles beneath the target membrane, and does not appear tightly coupled to endocytosis or vesicle recycling that are characteristic of regulated secretion in mammalian tissues. Nonetheless, even these examples entail targeting to specific cellular domains and traffic that is often triggered by external stimuli. Growing evidence also points to vesicle pools that are primed for release on stimulation (Homann & Thiel, 2002; Sutter *et al.*, 2000), thus blurring the distinctions between conceptual models for secretion.

Plant secretory vesicles are commonly 100–150 nm in diameter, they bud off each *trans*-Golgi face at rates of 2–4 per minute and, in hypersecretory cells, including maize root cap cells and pollen, vesicles arrive at the plasma membrane at rates of several thousands per minute (Picton & Steer, 1983; Steer & Steer, 1989). In tip-growing cells such as root hairs and pollen, in some algae and in grass coleoptile mesophyll cells that show substantial cell elongation, fusion occurs in discrete zones, often in the first few micrometers of the tip apex (Ott & Brown, 1974). Secretion in the root cap and gland cells is localised primarily to the outer tangential surface of the cell (Chaboud & Rougier, 1981). In all of these cell types new plasma membrane – and membrane proteins – must be supplied constantly by exocytotic activity in order to provide for expansion of the total cell surface area and to supply materials for cell wall biosynthesis. So it must be assumed that processes which control cell growth are also able to integrate these events with a control of exocytosis. Indeed, the importance of integrating membrane trafficking with cell growth is underscored by the fact that inhibitors such as Brefeldin A that affect different stages of membrane traffic to the plasma membrane in plant cells also rapidly suppress cell growth (Schindler *et al.*, 1994; Baskin & Bivens, 1995; Cho & Hong, 1995; Morris & Robinson, 1998).

13.2.1   *Intracellular movement of vesicles*

Intracellular translocation is a key element in targeting individual vesicles to their final destinations beyond the Golgi apparatus and may also contribute to the sorting process. The actin cytoskeleton is a major contributor to vesicle movements. Electron micrographs of exocytotically active plant cells have shown a block of vesicle delivery to the plasma membrane from the Golgi following treatments with the actin depolymerising agent cytochalasin (Picton & Steer, 1983; see also Chapter 10). These observations are supported further by the finding that washing with fresh medium, results in a burst of secretory cargo from Golgi vesicles accumulated in the presence of cytochalasin (Robinson & Kristen, 1982). Structural data also argue for a role of microfilaments in the transport of exocytotic vesicles to target membranes. For example, in tip-growing pollen tubes the actin cytoskeleton extends into the growing tip and terminates in a ring of microfilaments just short of the tip itself, close to the predominant site of vesicle fusion with the plasma membrane (Tang *et al.*, 1998b).

The association of secretory vesicles with actin-based dynamics is also exemplified by the response of many algae to wounding. Foissner *et al.* (1996) used video cinemicrography to quantify the movements of individual vesicles along cortical actin strands in the giant alga *Chara*. They found that wounding dramatically altered the pattern of vesicle motion and related the effect to a reorganisation of the actin cytoskeleton to a fine network around the wound site, and to the discharge of the vesicle contents to the apoplast at this site.

Wounding commonly leads to local deposition of callose, complex carbohydrates and lignification (Labavitch, 1981; Barber & Mitchell, 1997; Yang *et al.*, 1997; Cassab, 1998). Indeed, in coenocytic algae, including *Nitella* and *Vaucheria*, wounding is followed by extremely rapid mobilisation of the vesicular structures, matrix material, and membrane to the disturbed site (Walker, 1955; Foissner, 1988, 1990; Crooks *et al.*, 1999; M. Blatt, unpublished observations) a response which in *Vaucheria* is fast enough to prevent the lysis of protoplasts during their isolation without osmotic balance (Blatt *et al.*, 1980). The mechanics of this response has yet to be examined in any detail, but probably depends on a reorganisation of the actin cytoskeleton (Blatt *et al.*, 1980) and redirection of secretory vesicles that contain sulphated polysaccharides (Wessels, 1986; Bolwell, 1993) such as is observed during early responses to pathogen invasion (Bolwell, 1993; Brown *et al.*, 1998; McLusky *et al.*, 1999).

The role of microtubules in vesicle trafficking in plants is less clear, although the *Arabidopsis* genome encodes almost a dozen putative v-SNARE-binding proteins that, in mammalian cells, also interact with these cytoskeletal elements (see Section 13.4.5). In dividing cells, microtubules are closely associated with the flow of membrane during cell-plate formation (Baskin & Cande, 1990). Recently Strompen *et al.* (2002) found that the *Hinkel* mutation of *Arabidopsis*, which leads to incomplete cell-plate formation and persistent microtubule structures, encodes a plant-specific kinesin-related protein. In this instance, the association with vesicle traffic is less well defined, as transfer of the cytokinesis-specific syntaxin Syp111 (see Section 13.4.1) and lateral expansion of the phragmoplast appeared unaffected. However, disruption of microtubules is well known to influence cell-wall formation which depends on vesicle-mediated deposition and its localisation (Emons *et al.*, 1992). Because microtubules – like the actin cytoskeleton – are dynamic elements in the cell, their structure may be critical in defining the microdomains for vesicle fusion. Thus, microtubules may be more closely linked to the final events of vesicle targeting rather than their translocation between regions within the cell per se. For example, tip growth of root hairs is severely disorganised when the microtubule cytoskeleton is either depolymerised or stabilised experimentally

(Bibikova *et al.*, 1999). These processes are also intimately connected with cytosolic-free [Ca^{2+}] and Ca^{2+} entry across the plasma membrane (Taylor & Hepler, 1997; Bibikova *et al.*, 1999; Hepler *et al.*, 2001), suggesting parallels to Ca^{2+}-dependent secretion in mammalian neuromuscular tissues.

13.2.2 Coupling secretion to cytosolic-free Ca^{2+} concentration

More detailed information on the coupling exocytosis to cytosolic-free [Ca^{2+}] ([Ca^{2+}]$_i$) has come from experiments with maize coleoptile protoplasts. Sutter *et al.* (2000) used photolysis with caged Ca^{2+} buffers (so-called caged Ca^{2+}) to elevate [Ca^{2+}]$_i$ rapidly while measuring the total membrane capacitance, C_m, at the same. The specific capacitance of biological membranes is remarkably constant (close to $1\,\mu F\,cm^{-2}$), making it an ideal reference for recording real-time changes in total surface area (Gillis, 1995; Angleson & Betz, 1997). Sutter *et al.* found that elevating [Ca^{2+}]$_i$ gave a biphasic rise in C_m, with a rapid increase followed by a second, slower phase of variable nature. This second phase often gave a significant further rise in C_m, but under certain conditions was marked by a decline in C_m.

Analysis of these data uncovered three key elements to the secretion process (Sutter *et al.*, 2000). First, the coleoptile mesophyll cells contain only a limited number of vesicles immediately available for insertion into the plasma membrane following a [Ca^{2+}]$_i$ stimulus. Second, insertion of these vesicles is a direct function of the amplitude of the [Ca^{2+}]$_i$ rise over a dynamic range of approximately 100 nM, that is, within the range of concentrations relevant to the cell. Finally, after depleting this primary pool of vesicles, net traffic at the plasma membrane is governed predominantly by antagonistic processes of exocytosis and endocytosis, each with comparably slow kinetics. Sutter *et al.* (2000) successfully formulated a simple, mass-action kinetic model for [Ca^{2+}]$_i$ coupling to C_m changes based on the assumption of two discrete pools of vesicles with transfer of vesicles between pools, a first pool comprising vesicles close to the plasma membrane, ready to fuse and subject to a [Ca^{2+}]$_i$-dependent step, and the second pool that slowly resupplies the first, also dependent on a [Ca^{2+}]$_i$-sensitive step. Finally, the membrane was assumed to recycle by endocytosis through a process insensitive to [Ca^{2+}]$_i$ and with no direct connection to either of these vesicle pools.

Of course, such a kinetic model is only a first and crude approximation and lacks many of the complexities of plant secretory processes. Homann and Tester (1997) found that the Ca^{2+}-sensitive increases in capacitance in barley aleurone cells were paralleled by a small but steady increase in surface area (capacitance) that was independent of [Ca^{2+}]$_i$. Furthermore, the capacitance increases evoked at micromolar [Ca^{2+}]$_i$ also required the presence of GTP at the cytosolic face of the plasma membrane. Similar results have been obtained from *Zea mays* root cap protoplasts (Carroll *et al.*, 1998). These observations might be accommodated in the model of Sutter *et al.* (2000) with the inclusion of additional kinetic pathways, for example a [Ca^{2+}]$_i$-independent slippage of vesicles from one or the other of the two vesicle pools.

Other vesicle pools must also exist, since much larger excursions in membrane surface area have been associated with osmotic and hydrostatic pressure changes in protoplasts from barley aleurone, guard cell and maize coleoptile mesophyll protoplasts (Zorec & Tester, 1993; Homann, 1998; Thiel *et al.*, 2000; Weise *et al.*, 2000; Bick *et al.*, 2001). In guard cell protoplasts, these changes are known to be mediated through exo- and endocytosis at the plasma membrane (Homann & Thiel, 1999). These additional pools are entirely consistent with our knowledge of the low elastic extensibility of biological membranes and the need for co-ordination of vesicle trafficking with increases in cell surface area during growth. How they integrate with $[Ca^{2+}]_i$-dependent exocytosis is less clear. In maize coleoptile protoplasts, large increases in C_m can be triggered by hypo-osmotic challenge, even when C_m increases associated with elevated $[Ca^{2+}]_i$ have been exhausted (see Fig. 13.1; also Thiel *et al.*, 2000). Thus, it appears that plant cells possess at least two different mechanisms for exocytosis at the plasma membrane, one sensitive to $[Ca^{2+}]_i$ that draws vesicles from a relatively small pool, the other insensitive to $[Ca^{2+}]_i$, drawing on a vesicle pool of at least 10-fold greater surface area, and somehow coupled to tensile stress in the membrane. Whether the coupling mechanism depends on stretch-activation of other ion channels such as Cl^- channels (Rupnik & Zorec, 1995) is an open question.

Some quantitative details are available for pressure-driven exocytosis. With the patch pipette linked directly to a hydrostatic line, Homann (1998), Kubitscheck *et al.* (2000) and Bick *et al.* (2001) challenged guard cell protoplasts to step changes in positive and negative pressure. The results showed that C_m responded

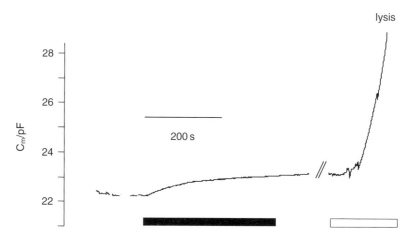

Figure 13.1 Surface area changes (C_m) respond differently to elevated $[Ca^{2+}]_i$ and hypo-osmotic treatment (see also Sutter *et al.*, 2000). Sustained elevation of cytosolic-free $[Ca^{2+}]$ was obtained by cage release on exposure to 380 nm light (solid bar) to give a sustained rise in $[Ca^{2+}]_i$ from about 4 nM to >1 µM (data not shown). Sustained osmotic step was obtained by transfer of the protoplast from 520 mOsm to 290 mOsm media and resulted in lysis of the protoplast.

immediately, changing steadily with time and that the changes were proportional to, and of the same sign, as the pressure step (Bick *et al.*, 2001). It is difficult to see how these large excursions could be accommodated with the same immediacy unless vesicular membrane was already available for fusion within the cell. With large pressure steps, the surface area (C_m) recorded in these studies increased at a constant rate until the protoplasts burst. We speculate that lysis in these cases represents the point at which all vesicular membrane is exhausted.

13.3   Temporal and spatial analysis of vesicle fusion

Membrane capacitance recordings are useful not only to investigate gross control of exo- and endocytosis. The technique can also resolve these events at a much higher resolution to define the biophysical and kinetic properties of individual fusion events and, in this sense, provide a microscopic assay for the final stages of exocytosis. The technique has been employed to investigate conductance and physical nature, in animal cells, of the fusion pore formed between a vesicle as it begins to merge with the target membrane (Curran *et al.*, 1993; Monck & Fernandez, 1994; Lollike *et al.*, 1998; Debus & Lindau, 2000). Combining capacitance recordings with electrochemical measurements of vesicle cargo release can also be used to resolve the biophysical complexities of fusion into a series of discrete steps, each with distinct kinetic characteristics and physical correlates (Detoledo *et al.*, 1993; Albillos *et al.*, 1997).

13.3.1   Kinetics and vesicle cycling in plants

Resolution of single fusion events has been possible in a few cases in plants. Exocytotic vesicles in maize coleoptile protoplasts have proven to be of the same diameters and size distributions as were previously anticipated from electron micrographs of sections through these cells (Thiel *et al.*, 1998). Furthermore, Homann and Thiel (1999) found that single fusion events in protoplasts from *Vicia* guard cells were correlated with the pressure used to drive increases in cell surface area, suggesting that fusion of the same pools of vesicles were responsible for the excursions in cell surface area.

High temporal resolution of single fusion events in these cells has highlighted some similarities as well as differences to vesicular fusion in animals. For example, it seems that the first stages of fusion, leading to the formation of the fusion pore, are complete within a few milliseconds, both in animal and plant cells (Weise *et al.*, 2000). However, in plant cells there appear to be different modes of fusion. Analysis of many elementary events has shown that fusion may lead to complete integration of the vesicle into the plasma membrane, or the fusion pore may form only transiently in a process reminiscent of what is called flickering exocytosis in animal cells (Albillos *et al.*, 1997; Lollike *et al.*, 1998). As in animal cells (Ales *et al.*, 1999), both modes of fusion are observed to draw on a single pool of vesicles (Weise *et al.*, 2000).

Weise *et al.* (2000) examined the relative occurrence of permanent versus transient fusion events and their kinetics, drawing the conclusion that the dwell time of the transient fusion events exhibited a clear maximum about 100 ms. From reaction kinetic considerations (Colquhoun & Hawkes, 1977; Colquhoun & Sigworth, 1995), these results indicate that transient fusion is not a Markov process, but is strongly biased in the forward direction. In other words, there must be an input of energy to drive the system towards fusion. Such a kinetic system can be described with a cyclic three-state model comprising two fused and one non-fused state. Using the information from the lifetime distributions of transient fusion events, the molecular mechanism of fusion can be described with the rate constants shown in Fig. 13.2.

The kinetics of transient and permanent fusion in several animal cells is well complemented by data from total internal reflection (evanescent) fluorescence microscopy (Steyer *et al.*, 1997; Oheim *et al.*, 1998; Steyer & Almers, 2001). The technique enables measurements in living cells of real-time events of vesicle tethering, docking and fusion with the plasma membrane. A major difficulty in its application to plant cells is the lack of a sizeable contact area of plant protoplasts

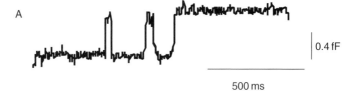

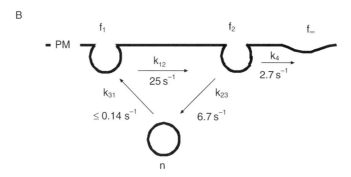

Figure 13.2 Kinetic analysis indicates a series of communicating vesicle pools contribute to fusion at the plasma membrane. (A) Spontaneous capacitance steps recorded in cell attached patch on plasma membrane of maize coleoptile protoplast including two transient and one irreversible step increase in C_m. (B) Kinetic minimal model for vesicle fusion with the plasma membrane (PM). Cyclic reaction model with two kinetically distinct fused states (f_1, f_2) and a non-fused state (n). Rate constants obtained from lifetime distributions of the transient fusion events and from the relative occurrence of transient and irreversible fusion events (Weise *et al.*, 2000).

when settled on coverslips and, still more important, the relatively low rate of exocytosis in plant cells (Loerke & Thiel, unpublished data). Nevertheless, similar information has been forthcoming using less sophisticated microscopic studies of vesicle movements, at least in *Chara* (Foissner *et al.*, 1996), and indicates that secretory vesicles in this cell type also undergo a similar sequence of steps prior to fusing with the plasma membrane. Finally, it is worth remembering that much of this process is reversible. Indeed, capacitance recordings both in animal (Henkel *et al.*, 2000) and plant cells (G. Thiel, unpublished) have revealed transient fusion events that occur in oscillatory fashion. This remarkable observation suggests that a single vesicle can oscillate between fused and non-fused states, possibly dependent on similar mechanisms in both cell types.

13.3.2   Fusion domains and hot-pots

Although indications are plentiful for localised secretion in many plant cell types, notably root hairs and pollen tubes (see Section 13.2), direct evidence that fusion events take place in discrete, microdomains – what we refer to as *hot-pots* (an oblique reference to the localised pattern of boiling in a simmering pan of water) to distinguish them from the hot-spots of $[Ca^{2+}]_i$ elevation often associated with Ca^{2+}-induced Ca^{2+} release (Bootman *et al.*, 1997; Neher, 1998) – is only now becoming available through capacitance measurements of single fusion events. A simple model of fusion events distributed randomly over the cell surface predicts that the number of events averaged over time should be directly proportional to the surface area over which the measurements are taken. The impedance of a patch electrode is proportional to the area of the pipette opening (Sakmann & Neher, 1983). Thus, assuming measurements are taken randomly over the surface of the cell, one might expect that the number of fusion events recorded over many experiments would correlate with the pipette resistance determined before sealing the plasma membrane (Kreft & Zorec, 1997). By contrast, a lack of correlation between patch area and secretory activity must therefore indicate that exo- and endocytosis take place in distinct hot-pots that are distributed over the surface at intervals similar to, or greater than the diameter of the largest patch pipette openings. From measurements of this kind on maize coleoptile protoplasts (Fig. 13.3), it turned out that the distribution of single exocytotic event was not correlated with patch area. It follows that exocytosis does not occur randomly over the surface of the plasma membrane. Much the same, conclusion was drawn from microscopical observations of wound wall formation in *Chara*. Foissner *et al.* (1996) observed individual vesicles following the tracks of other vesicles, attached to a few, discrete sites on the plasma membrane. These observations, together with similar studies in mammalian and yeast cells (Kreft & Zorec, 1997; Schneider *et al.*, 1997; Holthuis *et al.*, 1998b; Lehman *et al.*, 1999; Cho *et al.*, 2002) suggest that all eukaryotic organisms integrate a common subset of mechanisms for targeting exocytosis.

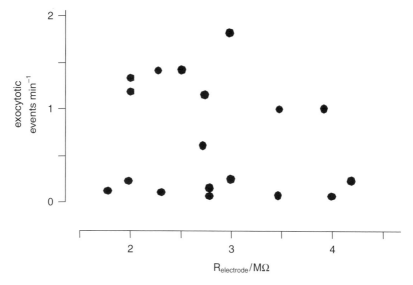

Figure 13.3 Secretory activity does not correlate with patch surface area. Frequency of appearance of exocytotic (C_m) steps in cell attached patches as a function of the patch pipette resistance. Data from maize coleoptile protoplasts (see Thiel *et al.*, 1998).

13.4 SNARE genomics

While major advances have been made recently in developing biophysical and kinetic tools as well as in our understanding of vesicle trafficking at the (sub-) cellular level, in almost no case is it yet possible to relate such mechanistic information to the molecular nuts and bolts of the plant cell. A major challenge now lies in associating specific proteins and their interacting partners with events such as localised secretion, flickering fusion and Ca^{2+}-dependent exocytosis. The groundwork is already well laid through similar advances in model systems such as yeast and neurons for which many of the proteins in question have been identified and their precise function in the exocytotic apparatus well characterised (Kaiser *et al.*, 1997; Jahn & Sudhof, 1999; Lin & Scheller, 2000). Some parallels may be drawn to proteins with quasi-identical functions in plant cells, but only after the key proteins and their interactions are better known will it be possible to precisely identify their role(s) in membrane trafficking.

The availability of the complete genomic sequence for several model eukaryotes, now including *Arabidopsis* (Arabidopsis genome initiative, 2000), has provided considerable information about vesicle trafficking elements, not only in terms of their overall conservation but with respect to the remarkable degree of complexity within each subfamily of these proteins. Comparisons between animal (*H. sapiens*, *C. elegans*), insect (*Drosophila*) and yeast (*S. cerevisiae*) models turn up a number of important distinctions, notably the absence of certain proteins in yeast and, in *Arabidopsis*, a diversity of vesicle trafficking proteins as well as the presence of several new subfamilies of these proteins with wholly unknown function(s). It is likely that new functions underlie the increased diversity among vesicle trafficking proteins in plants. However, the multiplicity among proteins within several sub-classes in *Arabidopsis* makes it impossible to draw any real conclusions at this time. Indeed, because it is now clear that a single SNARE protein may perform different functions at different sites within the cell – to the extent of functional exchange in some instances between cognate interactors (Mcnew *et al.*, 2000; Marash & Gerst, 2001) – assigning biological roles to these trafficking elements on the basis of sequence similarities is difficult, even where close homologies do occur across phylogenetic boundaries.

The *Arabidopsis* genome includes coding sequences for a core of 63 putative vesicle trafficking proteins predicted on the basis of sequence domain homology (Table 13.1). These comprise 24 syntaxin-like t-SNAREs, 3 SNAP25 homologues and 14 VAMP-like v-SNAREs along with several homologues to other SNAREs, including yeast Bet1p, Gos1p, Vti1p and their mammalian counterparts that are thought to play roles in ER and Golgi and in post-Golgi vesicle trafficking (Mcnew *et al.*, 1997; Hay *et al.*, 1998; Chao *et al.*, 1999; vonMollard & Stevens, 1999). The *Arabidopsis* genome also encodes 3 SNAP homologues, one NSF ATPase and 3 large MW (CDC48-like) ATPases (Bednarek *et al.*, 1994), and for 6 of the Sec1 family of peripheral binding proteins that in mammals, yeast and *Drosophila* regulate syntaxin accessibility and SNARE interactions (Wu *et al.*, 1998; Jahn, 2000; Yang *et al.*, 2000).

13.4.1   Arabidopsis *t-SNAREs*

The *Arabidopsis* syntaxin-like t-SNAREs are divided into four broad groups based on sequence homologies (see Fig. 13.4; also Blatt *et al.*, 1999). Sanderfoot *et al.* (2000) have further subdivided these sequences between eight classes (Syntaxins of plants, Syp1 through Syp8) based on genomic analyses. The proteins predicted are of molecular weights between 27 kDa and 40 kDa, and all of the sequences show the hallmarks of syntaxins (Fig. 13.5), including three N-terminal coiled-coil domains (HA, HB, HC), and the H3 and C-terminal transmembrane domains. The H3 domain is highly conserved, even across phylogenetic boundaries, and probably determines the interactions with cognate elements of the SNARE complex as it does in mammals and yeast (see Blatt *et al.*, 1999; Jahn & Sudhof, 1999). Not surprisingly, the greatest sequence divergence between the *Arabidopsis* syntaxin clusters is found towards the N-termini of the syntaxin proteins (Fig. 13.5).

Broadly speaking, these clusters or subgroups almost certainly reflect the division of functions within the cell. The largest of the subgroups corresponds to the Syp1 gene class which comprises nine members with close homology to the plasma membrane syntaxins Sso1/Sso2 of yeast (Aalto *et al.*, 1993) and human Syn1A (Zhang *et al.*, 1995). The Syp1 class includes AtSyr1 (=Syp121), which is localised to the

Table 13.1 *Arabidopsis* SNAREs, related proteins and their interactions

Name[1]	Synonym[2]	AGI designation[3]	Genomic locus[3]	Expressed[4]	Localisation[5]	Partner interactions[6]
(syntaxin-like)						
AtSyp111	Knolle	At1g08350	F22O13.4	Y	phragm	AtSNnp11, AtSNnp12, AtSnp13
AtSyp112		At2g18260	T30D6.23	?		
AtSyp121	AtSyr1	At3g11820	F26K24.11	Y	plasma membr	
AtSyp122	AtSyr4	At3g52400	F22O6.220	Y	plasma membr	AtSnp11
AtSyp123		At4g03330	F4C21.26	Y		
AtSyp124		At1g55410	T1F9.22	?		
AtSyp125		At1g10980	T28P6.10	?		
AtSyp131		At3g03800	F20H23.28	?		
AtSyp132		At5g08080	T22D6.20	Y		
AtSyp21	AtPep12	At5g16830	F5E19.180	Y	pre-vacuole	AtVti11, AtSyp51
AtSyp22	AtVam3	At5g46860	MSD23.4	Y	pre-vacuole	
AtSyp23	AtPlp	At4g17730	dl14901c	Y		
AtSyp31	AtSed5	At5g05760	MJJ3.17	Y		
AtSyp32		At3g24350	K7M2.10	Y		
AtSyp41	AtTlg2a	At5g26980	F2P16.240	Y	*trans*-Golgi	AtSyp61, AtVps45
AtSyp42	AtTlg2b	At4g02195	T10M13.19	Y	*trans*-Golgi	AtSyp61, AtVps45
AtSyp43		At3g05710	F18C1.4	Y		
AtSyp51		At1g15930	F3O9.4	Y	*trans*-Golgi, pre-vacuole	AtSyp61, AtSyp21
AtSyp52		At1g73260	F20B17.2	Y		
AtSyp61		At1g27550	F3M18.7	Y	*trans*-Golgi, pre-vacuole	AtSyp41, AtSyp51
AtSyp71		At3g09740	F11F8.33	Y		
AtSyp72		At3g45280	F18N11.40	Y		
AtSyp73		At3g61450	F2A19.50	Y		
AtSyp81		At1g47920	F19C24.5	Y		

					plasma membr, phragm	
(SNAP25-like)						
AtSnp11	AtSNAP33	At5g61210	MAF19.6	Y	plasma membr, phragm	AtSyp111, AtSyp122
AtSnp12	AtSNAP29	At5g07880	MXM12.4	Y		AtSyp111
AtSnp13	AtSNAP30	At1g13530	F16A14.10	?		AtSyp111
v-SNAREs (VAMP-like)						
AtVAMP711	VAMP7C	At4g32150	F10N7.4	Y		
AtVAMP712		At2g25340	T22F11.7	Y		
AtVAMP713		At5g11150	F2I11.40	?		
AtVAMP714		At5g22360	MWD9.16	?		
AtVAMP721	Sar1	At1g04630	T1G11.1	Y		
AtVAMP722	VAMP7B	At2g33120	F25I18.14	Y		
AtVAMP723		At2g33110	F25I18.15	Y		
AtVAMP724		At4g15780	dl3930c	Y		
AtVAMP725		At2g32670	F24L7.19	?		
AtVAMP726		At1g04650	F13M7.25	?		
AtVAMP727		At3g54300	F24B22.260	?		
AtSec22		At1g11610	F12F1.27	Y		
AtYkt61		At5g58060	K21L19.5	Y		
AtYkt62		At5g58180	MCK7.5	?		
(Vti1-like)						
AtVti11	AtVTI1a	At5g38510	MUL8.190	Y	*trans*-Golgi, pre-vacuole	AtSyp21, AtSyp51
AtVti12	AtVTI1b	At1g25740	T24P13.5	Y	*trans*-Golgi, pre-vacuole	AtSyp61, AtSyp51
AtVti13		At3g29100	MXE2.6	Y		
(Bet1-like)						
AtBet11		At3g58170	F9D24.80	Y		
AtBet12		At4g14450	dl3265w	?		
Other SNAREs						
(Gos1-like)						
AtGos11		At1g15590	F7H2.21	Y		
AtGos12		Atg245200	F4L23.29	Y		

Contd

Table 13.1 (continued)

	Name[1]	Synonym[2]	AGI designation[3]	Genomic locus[3]	Expressed[4]	Localisation[5]	Partner interactions[6]
(Membrin-like)							
	AtMemb11		At2g36900	T1J8.8	Y		
	AtMemb12		At5g50440	MXI22.9	?		
(Plant specific)							
	AtNpsn11		At2g35190	T4C15.14	Y		
	AtNpsn12		At1g44640	F11A17.20	Y		
	AtNpsn13		At3g17440	MTO12.3	Y		
SNARE-associated elements (SNAP-like)							
	AtAsnp11	α-SNAP1	At3g56450	T5P219.100	Y		
	AtAsnp12	α-SNAP2	At3g56190	F18O21.150	Y		
	AtAsnp21	β-SNAP	At4g20410	F9F13.60	Y		
(ATPases)							
	AtNSF		At4g04910	T1J1.4	Y		
	AtCDC48a		At3g09840	F8A24.11	Y		
	AtCDC48b		At2g03670	F19B11.22	Y		
	AtCDC48c		At3g01610	F4P13.15	?		
(Sec1-like)							
	AtSec11	Keule	At1g12080	F5O11.8	Y		
	AtSec21	AtSEC1a	At1g01980	F22M8.14	Y	phragm	
	AtSec22	AtSEC1b	At4g12120	F16I13.190	Y		
(Vps- and Sly-like)							
	AtVps45		At1g70890	T14N5.4	Y		
	AtVps33		At3g54860	F28P10.160	Y	*trans*-Golgi	
	AtSly1		At2g17980	T27K22.15	Y		AtSyp41, AtSyp42

(VAP33-like)

AtPva11	At3g60600	T4L20.10	Y
AtPva12	At2g45140	T14P1.27	Y
AtPva13	At4g00170	F6N15.21	Y
AtPva14	At1g51270	F11M15.13	?
AtPva21	At5g47180	MQL5.3	Y
AtPva22	At1g08820	F22O13.31	?
AtPva31	At2g23830	T29E15.3	Y
AtPva41	At5g54110	MIP23.9	Y
AtPva42	At4g21450	F18E5.70	Y
AtPva43	At4g05060	T32N4.12	Y

(Tomosyn-like)

AtTyn11	At5g05570	MOP10.11	Y
AtTyn12	At4g35560	F8D20.70	Y

[1] Naming conventions according to Sanderfoot *et al.* (2000) or as indicated in the text.
[2] Synonyms, original or previously used names (see text for references).
[3] *Arabidopsis* Genome Initiative designations and BAC-annotated gene loci.
[4] Indicates whether gene is known to be expressed, e.g. through the presence of annotated EST(s).
[5] Localisation (see text for details).
[6] Known partner interactions either *in vitro* or *in vivo* (see text for details).

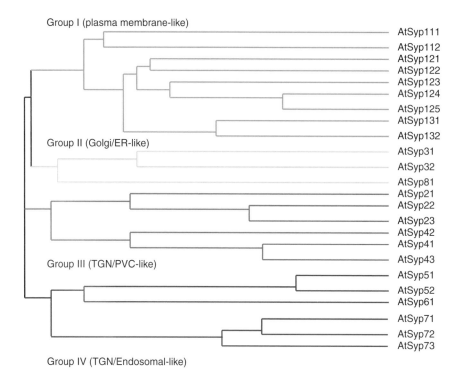

Group I (plasma membrane-like)
— AtSyp111
— AtSyp112
— AtSyp121
— AtSyp122
— AtSyp123
— AtSyp124
— AtSyp125
— AtSyp131
— AtSyp132
Group II (Golgi/ER-like)
— AtSyp31
— AtSyp32
— AtSyp81
— AtSyp21
— AtSyp22
— AtSyp23
— AtSyp42
— AtSyp41
— AtSyp43
Group III (TGN/PVC-like)
— AtSyp51
— AtSyp52
— AtSyp61
— AtSyp71
— AtSyp72
— AtSyp73
Group IV (TGN/Endosomal-like)

Figure 13.4 Phylogenetic relationships of *Arabidopsis* syntaxins based on sequence (CLUSTALW; www.ebi.ac.uk) and exon splicing (TAIR; www.arabidopsis.org). Broad groupings indicated in grey and further subdivided into eight classes as indicated on right. See text for details.

plasma membrane (Leyman *et al.*, 1999), but also includes Knolle (=Syp111), which is localised exclusively to the phragmoplast of dividing cells (Lukowitz *et al.*, 1996; Lauber *et al.*, 1997). The three members of the Syp3/Syp8 form a group (see Fig. 13.4) that includes proteins with homologies to Golgi and ER syntaxins Sed5p, Ufe1p of yeast and human Syn18 (Banfield *et al.*, 1994; Patel *et al.*, 1998). The six Syp2/Syp4 sequences form a third group that includes AtVam3 (=Syp22), AtPep12 (=Syp21) and AtTlg2 (=Syp41) with closest orthologues found in *trans*-Golgi network (TGN) and pre-vacuolar compartments (PVC) of yeast (Gotte & Gallwitz, 1997; Nichols & Pelham, 1998; Seron *et al.*, 1998). Of these, the *Arabidopsis* Syp2 proteins appear to be identified with distinct subpopulations of pre-vacuolar and vacuolar compartments (Conceicao *et al.*, 1997; Sato *et al.*, 1997; Sanderfoot *et al.*, 1999; Zheng *et al.*, 1999) and Syp41 whereas Syp42 localises to the TGN (Bassham *et al.*, 2000). The remaining six genes of the Syp5/Syp6/Syp7 group show homologies to Tlg1p and Tlg2p of yeast and to human Syn6 and Syn8, which contribute to endosomal trafficking (Bock *et al.*, 1996; Advani *et al.*, 1998; Holthuis *et al.*, 1998a,b; Tang *et al.*, 1998b; Prekeris *et al.*, 1999). In fact, Syp 51 and Syp61 have been found associated with a PVC

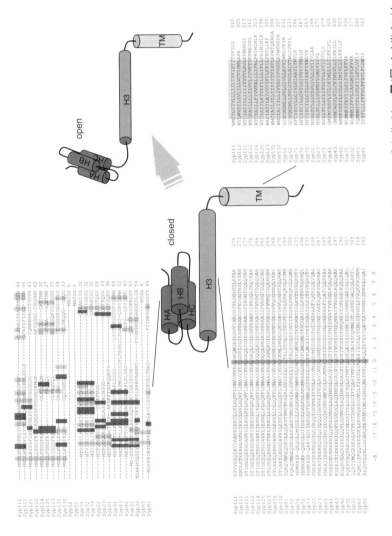

Figure 13.5 *Arabidopsis* syntaxin sequences diverge at the N-terminus (alignment *top left*) where the pattern of similarities (■/■ / ■/■ : dark/light, identity/homology) is most notable between Syp classes. All 24 sequences share a common signature sequence pattern in the H3 coil (alignment *bottom left*) of helically repeated hydrophobic residues (□ bars) centred about a core glutamine (■ bar), and a C-terminal membrane anchor (alignment *bottom right*) of 18–23 hydrophobic residues (■ bar).

and with the Golgi and TGN, respectively, following immunogold labelling (Sanderfoot *et al.*, 2001a).

It is worth noting that the multiplicity of proteins within these groups does not necessarily reflect genetic redundancy. Indeed, all evidence to date has indicated otherwise. Gene disruptions of single members in three of the Syp classes lead to lethal phenotypes although at least one additional gene occurs within the (sub-) classes. Thus, disruption of members of the Syp2 and Syp4 gene classes shows gametophytic-lethal phenotypes despite the presence of close homologues (Sanderfoot *et al.*, 2001b). A similar degree of specialisation may also be reflected in the embryo-lethal *knolle* mutation of Syp111 (Lukowitz *et al.*, 1996) for which a homologue also occurs in the genome.

Some specialisation may also relate to differences in developmental- and/or tissue-specific expression. AtSyp111 (=Knolle) is essential for cell plate formation (Lauber *et al.*, 1997) and its localisation is restricted to the phragmoplast of dividing cells. Temporal and spatial expression of AtSyp111 is tightly controlled and appears critical to its functioning. Volker *et al.* (2001) have reported that ectopic expression of the wild-type gene under control of the 35S promoter leads to Syp111 accumulation at the plasma membrane, but does not rescue the *knolle* mutant *Arabidopsis*. By contrast, AtSyp121 (=AtSyr1), like its tobacco ortho-logue NtSyp121 (=NtSyr1) (Leyman *et al.*, 2000; Geelen *et al.*, 2002), appears to be expressed in a cell-cycle independent manner (Leyman *et al.*, 1999; and unpublished data) and probably has a more general role in vesicle trafficking at the plasma membrane. Geelen *et al.* (2002) found that expression of a dominant-negative mutant of the tobacco orthologue of Syp121 did not increase the number of dividing initials or give rise to multinucleate cells within the root meristem that might indicate cells trapped late in division. The absence of such trapping, a characteristic of the *knolle* mutant, also argues against a functional overlap between these two syntaxins. In the light of these studies, it will be interesting to know if a third member of this class, AtSyp122 (=AtSyr4) shows a functional redundancy with AtSyp121. AtSyp122 is also thought to reside at the plasma membrane (L. Sticher, Fribourg, personal communication), but nothing is yet known of its role in the cell.

Of the three *Arabidopsis* SNAP25 orthologues, only one has been characterised to date. In mammalian cells, these t-SNAREs contribute two coiled-coil domains to the hetero-tetrameric supercoil (the so-called SNAREpin) of the SNARE complex (Weber *et al.*, 1998; Scales *et al.*, 2001). Heese *et al.* (2001) have reported that AtSNAP33 is a ubiquitously expressed membrane-associated protein. It interacts with AtSyp111 in two-hybrid screens (see Table 13.1), accumulates at the plasma membrane and, during cell division, colocalises with AtSyp111 at the phragmo-plast. Heese *et al.* (2001) found that a knockout of AtSNAP33 resulted in a dwarf phenotype that was eventually lethal. AtSyp111 also interacted with the two other SNAP25-like proteins in yeast two-hybrid assays (Table 13.1), suggesting that AtSNAP33 function in cytokinesis may be replaced partially by the other SNAP25 homologues. A general role in secretion may be assumed if AtSNAP33 interacts with other members of the Syp1 cluster as it does with the tobacco orthologue NtSyp121 (Kargul *et al.*, 2001).

13.4.2   Arabidopsis *brevins and other v-SNAREs*

The vesicle-associate membrane proteins (VAMPs) – also known as synaptobrevins or v-SNAREs – like the syntaxins are commonly anchored to the membrane by a C-terminal transmembrane domain and harbour an adjacent coiled-coil domain centred on an arginine residue. The canonical mammalian VAMPs are found on vesicle membranes. These are lower molecular weight (18–24 kDa) proteins that conjoin with the cognate t-SNAREs in the SNARE complex to bring the two membrane bilayers together leading to fusion. Mammals appear to contain a large variety of VAMPs (Chen & Scheller, 2001) whereas the yeast genome contains only five (Kaiser *et al.*, 1997; Jahn & Sudhof, 1999; Sanderfoot *et al.*, 2000). Both mammals and yeast also contain additional groups of proteins that encompass v-SNARE (and in some cases also t-SNARE) functions in endomembrane trafficking (Mcnew *et al.*, 1997; Tang *et al.*, 1998a; Conchon *et al.*, 1999; vonMollard & Stevens, 1999; Ossig *et al.*, 2000; Parlati *et al.*, 2000).

Arabidopsis appears to fall between these extremes, but as yet very little is known about the localisation, function and partner interactions of any of these proteins (Table 13.1). The *Arabidopsis* genome harbours 14 VAMP isoforms, the majority of which show greatest homology to the mammalian VAMP7 proteins which are associated with trafficking to endosomal and lysosomal compartments (Advani *et al.*, 1998, 1999; Prekeris *et al.*, 1999), as well as one homologue of the yeast and mammalian Sec22 v-SNAREs (Tang *et al.*, 1998a; Parlati *et al.*, 2000) and two homologues of the lipid-anchored, yeast v-SNARE Ykt6p (Sogaard *et al.*, 1994; McNew *et al.*, 1997). Counterparts to other v-SNAREs are missing, notably the VAMP1/syntaptobrevin 1 class of proteins associated with secretory traffic to the plasma membrane (Calakos & Scheller, 1994; Rossetto *et al.*, 1996; Taubenblatt *et al.*, 1999).

Two of the three homologues of the Vti1p v-SNARE, AtVTI11 (=AtVti1a) and AtVti12 (=AtVTI1b), have been examined and appear to parallel elements of the yeast TGN-to-vacuole pathways (Zheng *et al.*, 1999). The *Arabidopsis* orthologues are expressed throughout the plant and colocalise with the putative vacuolar cargo receptor AtELP, an orthologue of the BP80 group of vacuolar sorting receptors (Paris *et al.*, 1997; Cao *et al.*, 2000). Zheng *et al.* (1999) found that AtVti11 interacted with AtSyp21 in coimmunoprecipitation assays and, when expressed in Vti1p-mutant yeast, both AtVti11 and AtVti12 partially suppressed temperature-sensitive Vti1p-mutant phenotypes. Subsequent studies have suggested additional interactions of the v-SNAREs with the putative endosomal syntaxins Syp51 and Syp61 (Table 13.1) (Sanderfoot *et al.*, 2001a). Intriguingly, a mutation in AtVti11, isolated in screens for altered shoot gravitropism, gives rise to a partial agravitropic phenotype (Kato *et al.*, 2002), possibly as a consequence of disfunctional vesicle traffic to the vacuole.

Arabidopsis also harbours two genes with sequence similarity to the Bet1p/Sft1p subfamilies that contribute to intra-Golgi trafficking (Banfield *et al.*, 1995; McNew *et al.*, 1997; Wooding & Pelham, 1998). Tai and Banfield (2001) have reported that AtBS14a (=AtBet11) and AtBS14b (=AtBet12) suppress the temperature-sensitive growth defect in *sft1–1* mutant and Δ*sft1*, but not Δ*bet1* yeast. Relating these data to the situation in planta is difficult, but it is possible that the *Arabidopsis* proteins take part in retrograde traffic within the Golgi cisternae. The remaining groups of putative v-SNAREs have homologues thought to be involved in ER-to-Golgi, intra-Golgi and/or TGN trafficking (Nagahama *et al.*, 1996; Paek *et al.*, 1997; Hay *et al.*, 1998; Tang *et al.*, 1998a; Chao *et al.*, 1999) but no further information is available as yet.

13.4.3   NSF ATPase and attachment proteins (SNAPs)

Once formed, the supercoil of SNARE helices is remarkably stable and requires both energy input (Fasshauer *et al.*, 1998) and a specific juxtaposition of so-called zero layer residues (Scales *et al.*, 2001) for disassembly. The process of disassembly for re-priming of the SNARE proteins is driven by ATP hydrolysis, mediated by the NEM-sensitive factor (NSF) ATPase and its associated soluble NSF-attachment protein (SNAP) which recruits the ATPase to the SNARE complex (Jahn & Sudhof, 1999; Chen & Scheller, 2001). In *Arabidopsis*, as in other eukaryotes, the NSF ATPase is a single-copy gene (Feiler *et al.*, 1995) while two SNAP orthologues of the mammalian α-SNAP and one of γ-SNAP also occur in the plant (here designated AtSNP11, AtSNP12 and AtSNP21, respectively; see Table 13.1). The *Arabidopsis* genome also encodes three isoforms of a second ATPase homologous to the yeast CDC48 and mammalian p97 ATPases that play similar roles in so-called homotypic fusion between vesicles derived from the same source (Patel *et al.*, 1998; Rabouille *et al.*, 1998).

13.4.4   Sec1 homologues

Although many of the plant SNAREs are likely to function in distal steps, including traffic to and from the plasma membrane, the proteins are synthesised on the ER and probably pass through the Golgi and TGN. Thus, a large number of SNARE partners are potentially available, often on the same membrane, and control of the interactions between cognate SNARE elements is of paramount importance to the cell. The *Arabidopsis* genome encodes 6 proteins with homologies to the Sec1 family of proteins (for review, see Jahn, 2000). The Sec1 family comprises large molecular weight peripheral binding proteins that, in mammals, appear to maintain a so-called closed conformation of the syntaxin – with the HA, HB and HC domains bound to the H3 coil (see Fig. 13.5) – as well as facilitating the conversion to an open conformation, preparing the syntaxin for binding with its SNARE partners (Misura *et al.*, 2000; Yang *et al.*, 2000; Bracher & Weissenhorn, 2001).

Three of the proteins in *Arabidopsis* (here designated AtSec11, AtSec21 and AtSec22, respectively; see Table 13.1), show closest homology to the neuronal protein nSec1, its relative Unc18 and the yeast Sec1p (Carr *et al.*, 1999). AtSec11 (=Keule) is required for cytokinesis in *Arabidopsis* and was isolated in mutant screens along with AtSyp111 (Table 13.1) (Assaad *et al.*, 2001; Heese *et al.*, 2001). The protein is expressed throughout the plant, is enriched in dividing tissues, and appears to be required at all stages of the plant's life cycle. Like its mammalian orthologues, AtSec11 exists in two forms, one soluble and the other peripherally associated with membranes, and it binds AtSyp111 (Assaad *et al.*, 2001). The remaining Sec1-like proteins are more closely related to each other, but their situation(s) and function(s) within the cell have yet to be determined.

A fourth member of the peripheral binding proteins, AtVps45, is present on the TGN but, unlike its orthologue Vps45p in yeast (Peterson *et al.*, 1999), AtVps45 does not interact with, or colocalise with, the prevacuolar t-SNARE AtSyp21. Instead, Bassham *et al.* (2000) found that AtVps45 interacts with two t-SNAREs, AtSyp41 and AtSyp42, that show similarity to the yeast t-SNARE Tlg2p and colocalise with AtVps45 at the TGN (Table 13.1). Again, differences to the yeast counterparts, as with the patterns of localisation for the t-SNAREs, imply functional complexities that are unique to the plant.

Although information is still lacking for the remaining two Sec1-family proteins, homologies to their counterparts in yeast and mammals, Sly1p and Vps33p (Dascher & Balch, 1996; Lupashin *et al.*, 1996; Darsow *et al.*, 1997), suggest that these may be involved in ER-to-Golgi and vacuolar trafficking. These issues apart, it is clear that the number of Sec1-family proteins in *Arabidopsis* falls well short of the count of syntaxins. Possibly, the interactions of the peripheral binding proteins *in vivo* are pathway-, rather than syntaxin-specific, but these characteristics have yet to be explored in any detail.

13.4.5   Actin- and other SNARE-associated proteins

Given the importance of the cytoskeleton in translocating membrane vesicles to their target(s) within the cell (see Section 13.2.1), associations with microfilaments or microtubules are to be expected. Several SNARE-binding proteins are known to that either interact with microtubules, actin microfilaments or affect actin-related functions in the cell. These proteins include the polyphosphoinositide phosphatase synaptojanin that associates with syndapins and dynamin in early stages of vesicle endocytosis in nerve terminals (Cremona *et al.*, 1999; Ringstad *et al.*, 1999; Qualmann & Kelly, 2000) and affect cell polarity in yeast (SingerKruger *et al.*, 1998), the VAMP-binding protein VAP33 (Skehel *et al.*, 2000), and *Drosophila* tomosyn and its homologues (Fujita *et al.*, 1998; Masuda *et al.*, 1998; Lehman *et al.*, 1999).

Our surveys of the *Arabidopsis* genome failed to uncover any obvious homologues of mammalian synaptojanins, but unearthed the coding sequences for 10 VAP33-like proteins. In *Aplesia*, VAP33 is thought to regulate neurotransmission through its binding with VAMP to modulate excitatory post-synaptic potentials, but the mouse homologue mVAP33 co-localises with VAMP in the neuronal cell body rather than the dendrites. Skehel *et al.* (2000) have reported that mVAP33 associates with microtubules and intracellular vesicles of heterogeneous size, but not with synaptic vesicles at the plasma membrane. They suggest that association of mVAP33 and VAMP/synaptobrevin may be important for delivery of components to synaptic terminals rather than directly participating in synaptic vesicle exocytosis.

The *Arabidopsis* VAP33-like proteins show moderate to high homology (56–73% identical and 88–93% homologous over the core helix of 49 amino acid residues) to mVAP33 and its human orthologue (see Fig. 13.6). At least 8 of these proteins are expressed and, bar one (=At2g23830), show a similar structure to the mammalian counterparts with variable domains of putative α-helix and/or, β-sheet toward the N-terminus plus two extended, putative central and C-terminal α-helices (PELE; biowb.sdsc.edu). For the present, we have designated the proteins as *p*lant *V*AMP-*a*ssociated proteins (AtPva), though clearly any role in vesicle trafficking and VAMP binding remains to be ascertained. Gene structure and sequence analysis indicate that the proteins are divided into four classes (Table 13.1) which also reflect the predicted localisation of the proteins within the cell as based on consensus sorting and targeting signal motifs (PSORT; biowb.sdsc.edu). The first two groups include proteins with a high probability of ER or cytosolic residence and, with one exception at a little N-terminal extension to the core helix, like the mammalian VAP33 proteins. One of these (AtPva22) includes a 221-residue extension to the C-terminus. The third group comprises a single protein, AtPva31, that shows the least homology to either the mammalian or plant sequences and uncertain localisation. The last group comprises three, very similar proteins with the highest homology to mammalian VAP33 proteins but significant probability of nuclear or cytosolic localisation.

Finally, the *Arabidopsis* genome encodes two proteins of approximately 130 kDa and 105 kDa, designated here AtTyn11 and AtTyn12 (*Arabidopsis thaliana Tomosyn*-like proteins), with homologies to mammalian and *Drosophila* tomosyns (Table 13.1). The functioning of tomosyn is particularly intriguing, as the mammalian homologue forms extremely stable complexes with syntaxin 1 and is capable of dissociating the Sec1 homologue Munc18 to form a novel 10S complex with syntaxin 1, SNAP25 and synaptotagmin (Fujita *et al.*, 1998). Thus, apart from any link to the actin cytoskeleton, tomosyn may influence t-SNARE activity by destabilising interactions that retain the closed conformation of syntaxin. However, no details are available for the *Arabidopsis* proteins at present.

13.5   Conclusions and future directions

Between mammals and yeast more than half a dozen pathways are known to exist for vesicle traffic from the ER and Golgi apparatus to the cell surface and to the vacuole (Kaiser *et al.*, 1997; Keller & Simons, 1997; Nichols & Pelham, 1998).

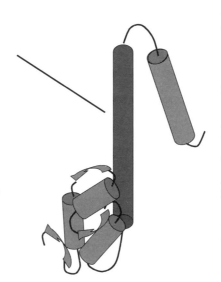

Figure 13.6 *Arabidopsis* Pva sequences show a common structure with homology to human (also mouse and rat, not shown) VAP33 sequences centred on a core of highly conserved proline, positively charged and hydrophobic residues (alignment *above*; ▮ / ▯ shading: dark/light, identity/homology). Of the *Arabidopsis* sequences, AtPva31 (=At2g23830) diverges significantly and does not share extended homology with the mammalian counterparts (outlined boxes).

For plants, kinetic analyses and biochemical studies outlined above suggest that at least three different pools of vesicles are available for delivery to the plasma membrane alone. These data and the evidence for specialised vacuoles in plant cells (Beevers & Raikhel, 1998; Neuhaus & Rogers, 1998) imply that the situation in plants is likely to prove significantly more complex.

Developments in applying capacitance methods to secretion at the plasma membrane are yielding information about intracellular vesicle pools and their mobilisation. In combination with styryl markers (Carroll *et al.*, 1998; Homann, 1998) and other fluorescent dyes these techniques should help to establish the dynamics of vesicle cycling between pools. It may be possible to obtain further detail about the molecular kinetics of vesicle docking, fusion and secretion using related technologies, including voltammetry (Angleson & Betz, 1997).

These technical advances aside, the complexity of vesicle trafficking in the plant cell still leaves major technical hurdles to be overcome (i) in selecting and identifying secretory pathways for different groups of proteins, (ii) in determining the targets for the various SNARE proteins already known in plants as well as isolating and characterising their partner proteins, and (iii) in demonstrating their functioning in secretion within the intact plant cell. For these purposes, a selection of protein probes will be invaluable to mark the sequence(s) of secretory processing events. A number of targeted proteins are available (Paris *et al.*, 1997; Neuhaus & Rogers, 1998; Boevink *et al.*, 1999; Batoko *et al.*, 2000; Hawes *et al.*, 2001; Satiat-Jeunemaitre & Hawes, 2001; Geelen *et al.*, 2002) and should facilitate progress in understanding secretion. Determining the functional targets of the various plant SNAREs and their interacting partners poses further difficulties, not least because *in vitro* they may interact promiscuously with other related partner proteins. Ultimately, we will need evidence *in vivo* for the roles of each SNARE element to confirm their specific functions, the nature of interactions between SNARE elements, the mechanism(s) by which such interactions are controlled, and whether additional elements and/or patterns of interaction occur that are unique to plant cells.

Acknowledgements

Preparation of this review and related work in the authors' laboratories is currently supported by grants from the Deutsche Forschungsgemeinschaft (GT) and from the Biotechnology and Biological Sciences Research Council (MRB).

References

Aalto, M.K., Ronne, H. & Keranen, S. (1993) Yeast syntaxins sso1p and sso2p belong to a family of related membrane-proteins that function in vesicular transport. *EMBO J.*, **12**, 4095–4104.

Advani, R.J., Bae, H.R., Bock, J.B., Chao, D.S., Doung, Y.C., Prekeris, R., Yoo, J.S. & Scheller, R.H. (1998) Seven novel mammalian SNARE proteins localize to distinct membrane compartments. *J. Biol. Chem.*, **273**, 10317–10324.

Advani, R.J., Yang, B., Prekeris, R., Lee, K.C., Klumperman, J. & Scheller, R.H. (1999) VAMP-7 mediates vesicular transport from endosomes to lysosomes. *J. Cell Biol.*, **146**, 765–775.

Albillos, A., Dernick, G., Horstmann, H., Almers, W., Detoledo, G.A. & Lindau, M. (1997) The exocytotic event in chromaffin cells revealed by patch amperometry. *Nature*, **389**, 509–512.

Ales, E., Tabares, L., Poyato, J.M., Valero, V., Lindau, M. & de Toledo, G.A. (1999) High calcium concentrations shift the mode of exocytosis to the kiss-and-run mechanism. *Nat. Cell Biol.*, **1**, 40–44.

Angleson, J.K. & Betz, W.J. (1997) Monitoring secretion in real time: capacitance, amperometry and fluorescence compared. *Trends Neurosci.*, **20**, 281–287.

Arabidopsis genome initiative (2000) Analysis of the genome sequence of the flowering plant *Arabidopsis thaliana*. *Nature*, **408**, 796–815.

Assaad, F.F., Huet, Y., Mayer, U. & Jurgens, G. (2001) The cytokinesis gene KEULE encodes a Sec1 protein that binds the syntaxin KNOLLE. *J. Cell Biol.*, **152**, 531–543.

Banfield, D.K., Lewis, M.J. & Pelham, H.R.B. (1995) A SNARE-like protein required for traffic through the Golgi complex. *Nature*, **375**, 806–809.

Banfield, D.K., Lewis, M.J., Rabouille, C., Warren, G. & Pelham, H.R.B. (1994) Localization of sed5, a putative vesicle targeting molecule, to the *cis*-golgi network involves both its transmembrane and cytoplasmic domains. *J. Cell Biol.*, **127**, 357–371.

Barber, M.S. & Mitchell, H.J. (1997) Regulation of phenylpropanoid metabolism in relation to lignin biosynthesis in plants. *Int. Rev. Cytol. – Surv. Cell Biol.*, **172**, 243–293.

Baskin, T.I. & Bivens, N.J. (1995) Stimulation of radial expansion in *Arabidopsis* roots by inhibitors of actomyosin and vesicle secretion but not by various inhibitors of metabolism. *Planta*, **197**, 514–521.

Baskin, T. & Cande, W.Z. (1990) The structure and function of the mitotic spindle in flowering plants. *Annu. Rev. Plant Phys. Mol. Biol.*, **41**, 277–315.

Bassham, D.C., Sanderfoot, A.A., Kovaleva, V., Zheng, H.Y. & Raikhel, N.V. (2000) AtVPS45 complex formation at the *trans*-Golgi network. *Mol. Biol. Cell*, **11**, 2251–2265.

Batoko, H., Zheng, H.Q., Hawes, C. & Moore, I. (2000) A Rab1 GTPase is required for transport between the endoplasmic reticulum and Golgi apparatus and for normal Golgi movement in plants. *Plant Cell*, **12**, 2201–2217.

Bednarek, S.Y., Reynolds, T.L., Schroeder, M., Grabowski, R., Hengst, L., Gallwitz, D. & Raikhel, N.V. (1994) A small GTP-binding protein from *Arabidopsis thaliana* functionally complements the yeast YPT6 null mutant. *Plant Physiol.*, **104**, 591–596.

Beevers, L. & Raikhel, N.V. (1998) Transport to the vacuole: receptors and trans elements. *J. Exp. Bot.*, **49**, 1271–1279.

Bibikova, T.N., Blancaflor, E.B. & Gilroy, S. (1999) Microtubules regulate tip growth and orientation in root hairs of *Arabidopsis thaliana*. *Plant J.*, **17**, 657–665.

Bick, I., Thiel, G. & Homann, U. (2001) Cytochalasin D attenuates the desensitisation of pressure-stimulated vesicle fusion in guard cell protoplasts. *Eur. J. Cell Biol.*, **80**, 521–526.

Blatt, M.R., Leyman, B. & Geelen, D. (1999) Molecular events of vesicle trafficking and control by SNARE proteins in plants. *New Phytol.*, **144**, 389–418.

Blatt, M.R., Wessells, N. & Briggs, W.R. (1980) Actin and cortical fiber reticulation in the siphonaceous alga *Vaucheria sessilis*. *Planta*, **147**, 363–375.

Bock, J.B., Lin, R.C. & Scheller, R.H. (1996) A new syntaxin family member implicated in targeting of intracellular transport vesicles. *J. Biol. Chem.*, **271**, 17961–17965.

Boevink, P., Martin, B., Oparka, K., Cruz, S.S. & Hawes, C. (1999) Transport of virally expressed green fluorescent protein through the secretory pathway in tobacco leaves is inhibited by cold shock and brefeldin A. *Planta*, **208**, 392–400.

Bolwell, G.P. (1993) Dynamic aspects of the plant extracellular-matrix. *Int. Rev. Cytol. – Surv. Cell Biol.*, **146**, 261–324.

Bootman, M.D., Berridge, M.J. & Lipp, P. (1997) Cooking with calcium: the recipes for composing global signals from elementary events. *Cell*, **91**, 367–373.

Bracher, A. & Weissenhorn, W. (2001) Crystal structures of neuronal squid Sec1 implicate inter-domain hinge movement in the release of t-SNAREs. *J. Mol. Biol.*, **306**, 7–13.

Brown, I., Trethowan, J., Kerry, M., Mansfield, J. & Bolwell, G.P. (1998) Localization of components of the oxidative cross-linking of glycoproteins and of callose synthesis in papillae formed during the interaction between non-pathogenic strains of *Xanthomonas campestris* and French bean mesophyll cells. *Plant J.*, **15**, 333–343.

Calakos, N. & Scheller, R.H. (1994) Vesicle-associated membrane protein and synaptophysin are associated on the synaptic vesicle. *J. Biol. Chem.*, **269**, 24534–24537.

Cao, X.F., Rogers, S.W., Butler, J., Beevers, L. & Rogers, J.C. (2000) Structural requirements for ligand binding by a probable plant vacuolar sorting receptor. *Plant Cell*, **12**, 493–506.

Carr, C.M., Grote, E., Munson, M., Hughson, F.M. & Novick, P.J. (1999) Sec1p binds to SNARE complexes and concentrates at sites of secretion. *J. Cell Biol.*, **146**, 333–344.

Carroll, A.D., Moyen, C., VanKesteren, P., Tooke, F., Battey, N.H. & Brownlee, C. (1998) Ca^{2+}, annexins, and GTP modulate exocytosis from maize root cap protoplasts. *Plant Cell*, **10**, 1267–1276.

Cassab, G.I. (1998) Plant cell wall proteins. *Annu. Rev. Plant Physiol. Plant Mol. Biol.*, **49**, 281–309.

Chaboud, A. & Rougier, M. (1981) Mucilaginous secretion by plant-roots and their function within the rhizosphere. *Annee Biologique*, **20**, 313–326.

Chao, D.S., Hay, J.C., Winnick, S., Prekeris, R., Klumperman, J. & Scheller, R.H. (1999) SNARE membrane trafficking dynamics *in vivo*. *J. Cell Biol.*, **144**, 869–881.

Chen, Y.A. & Scheller, R.H. (2001) SNARE-mediated membrane fusion. *Nat. Rev. Mol. Cell Biol.*, **2**, 98–106.

Cho, H.T. & Hong, Y.N. (1995) Effect of IAA on synthesis and activity of the plasma membrane H^+-ATPase of sunflower hypocotyls in relation to IAA-induced cell elongation and H^+ secretion. *J. Plant Physiol.*, **145**, 717–725.

Cho, S.J., Quinn, A.S., Stromer, M.H., Dash, S., Cho, J., Taatjes, D.J. & Jena, B.P. (2002) Structure and dynamics of the fusion pore in live cells. *Cell Biol. Int.*, **26**, 35–42.

Colquhoun, D. & Hawkes, A.G. (1977) Relaxation and fluctuations of membrane currents that flow through drug-operated channels. *Proc. Roy. Soc. Lond. B. Biol. Sci.*, **199**, 231–262.

Colquhoun, D. & Sigworth, F.J. (1995) Practical analysis of single channel records, in *Single Channel Recording* (eds B. Sakmann & E. Neher), pp. 397–482, Plenum: New York.

Conceicao, A.S., MartyMazars, D., Bassham, D.C., Sanderfoot, A.A., Marty, F. & Raikhel, N.V. (1997) The syntaxin homologue AtPEP12p resides on a late post-Golgi compartment in plants. *Plant Cell*, **9**, 571–582.

Conchon, S., Cao, X.C., Barlow, C. & Pelham, H.R.B. (1999) Got1p and Sft2p: membrane proteins involved in traffic to the Golgi complex. *EMBO J.*, **18**, 3934–3946.

Cremona, O., DiPaolo, G., Wenk, M.R., Luthi, A., Kim, W.T., Takei, K., Daniell, L., Nemoto, Y., Shears, S.B., Flavell, R.A., McCormick, D.A. & Decamilli, P. (1999) Essential role of phospho-inositide metabolism in synaptic vesicle recycling. *Cell*, **99**, 179–188.

Crooks, K., Coleman, J. & Hawes, C. (1999) The turnover of cell surface proteins of carrot protoplasts. *Planta*, **208**, 46–58.

Curran, M.J., Cohen, F.S., Chandler, D.E., Munson, P.J. & Zimmerberg, J. (1993) Exocytotic fusion pores exhibit semi-stable states. *J. Membr. Biol.*, **133**, 61–75.

Darsow, T., Rieder, S.E. & Emr, S.D. (1997) A multispecificity syntaxin homologue, Vam3p, essential for autophagic and biosynthetic protein transport to the vacuole. *J. Cell Biol.*, **138**, 517–529.

Dascher, C. & Balch, W.E. (1996) Mammalian sly1 regulates syntaxin-5 function in endoplasmic reticulum to Golgi transport. *J. Biol. Chem.*, **271**, 15866–15869.

Debus, K. & Lindau, M. (2000) Resolution of patch capacitance recordings and of fusion pore conductances in small vesicles. *Biophys. J.*, **78**, 2983–2997.

Detoledo, G.A., Fernandezchacon, R. & Fernandez, J.M. (1993) Release of secretory products during transient vesicle fusion. *Nature*, **363**, 554–558.

Emons, A.M.C., Derksen, J. & Sassen, M.M.A. (1992) Do microtubules orient plant cell wall microfibrils. *Physiologia Plantarum*, **84**, 486–493.

Fasshauer, D., Eliason, W.K., Brunger, A.T. & Jahn, R. (1998) Identification of a minimal core of the synaptic SNARE complex sufficient for reversible assembly and disassembly. *Biochemistry*, **37**, 10354–10362.

Feiler, H.S., Desprez, T., Santoni, V., Kronenberger, J., Caboche, M. & Traas, J. (1995) The higher-plant *Arabidopsis thaliana* encodes a functional CDC48 homologue which is highly expressed in dividing and expanding cells. *EMBO J.*, **14**, 5626–5637.

Findlay, G.P. (1988) Nectaries and other glands, in *Solute Transport in Plant Cells and Tissues*, pp. 538–560, Longman Press: Harlow.

Foissner, I. (1988) Chlortetracycline-induced formation of wall appositions (callose plugs) in internodal cells of *Nitella flexilis* (Characeae). *J. Phycol.* **24**, 458–467.

Foissner, I. (1990) Wall appositions induced by ionophore a23187, cacl2, lacl3, and nifedipine in Characean cells. *Protoplasma*, **154**, 80–90.

Foissner, I., Lichtscheidl, I.K. & Wasteneys, G.O. (1996) Actin-based vesicle dynamics and exocytosis during wound wall formation in characean internodal cells. *Cell Motil. Cytoskeleton*, **35**, 35–48.

Fujita, Y., Shirataki, H., Sakisaka, T., Asakura, T., Ohya, T., Kotani, H., Yokoyama, S., Nishioka, H., Matsuura, Y., Mizoguchi, A., Scheller, R.H. & Takai, Y. (1998) Tomosyn: a syntaxin-1-binding protein that forms a novel complex in the neurotransmitter release process. *Neuron*, **20**, 905–915.

Geelen, D., Leyman, B., Batoko, H., Sansebastiano, G.-P., Moore, I. & Blatt, M.R. (2002) The ABA-related SNARE homolog NtSyr1 contributes to secretion and growth: evidence from competition with its cytosolic domain. *Plant Cell*, **14**, 387–406.

Geldner, N., Friml, J., Stierhof, Y.D., Jurgens, G. & Palme, K. (2001) Auxin transport inhibitors block PIN1 cycling and vesicle traficking. *Nature*, **413**, 425–428.

Gillis, K.D. (1995) Techniques for membrane capacitance measurements, in *Single-Channel Recording* (eds B. Sakmann & E. Neher), pp. 155–198, Plenum: New York.

Gotte, M. & Gallwitz, D. (1997) High expression of the yeast syntaxin-related Vam3 protein suppresses the protein transport defects of a pep12 null mutant. *FEBS Lett.*, **411**, 48–52.

Hawes, C., Saint-Jore, C.M., Brandizzi, F., Zheng, H.Q., Andreeva, A.V. & Boevink, P. (2001) Cytoplasmic illuminations: *in planta* targeting of fluorescent proteins to cellular organelles. *Protoplasma*, **215**, 77–88.

Hay, J.C., Klumperman, J., Oorschot, V., Steegmaier, M., Kuo, C.S. & Scheller, R.H. (1998) Localization, dynamics, and protein interactions reveal distinct roles for ER and Golgi SNAREs. *J. Cell Biol.*, **141**, 1489–1502.

Heese, M., Gansel, X., Sticher, L., Wick, P., Grebe, M., Granier, F. & Jurgens, G. (2001) Functional characterization of the KNOLLE-interacting t-SNARE AtSNAP33 and its role in plant cytokinesis. *J. Cell Biol.*, **155**, 239–249.

Henkel, A.W., Meiri, H., Horstmann, H., Lindau, M. & Almers, W. (2000) Rhythmic opening and closing of vesicles during constitutive exo- and endocytosis in chromaffin cells. *EMBO J.*, **19**, 84–93.

Hepler, P.K., Vidali, L. & Cheung, A.Y. (2001) Polarized cell growth in higher plants. *Ann. Rev. Cell Dev. Biol.*, **17**, 159–187.

Holthuis, J.C.M., Nichols, B.J., Dhruvakumar, S. & Pelham, H.R.B. (1998a) Two syntaxin homologues in the TGN/endosomal system of yeast. *EMBO J.*, **17**, 113–126.

Holthuis, J.C.M., Nichols, B.J. & Pelham, H.R.B. (1998b) The syntaxin Tlg1p mediates trafficking of chitin synthase III to polarized growth sites in yeast. *Mol. Biol. Cell*, **9**, 3383–3397.

Homann, U. (1998) Fusion and fission of plasma membrane material accommodates for osmotically induced changes in the surface area of guard cell protoplasts. *Planta*, **206**, 329–333.

Homann, U. & Tester, M. (1997) Ca^{2+}-independent and Ca^{2+}/GTP-binding protein-controlled exocytosis in a plant cell. *Proc. Natl. Acad. Sci. USA*, **94**, 6565–6570.

Homann, U. & Thiel, G. (1999) Unitary exocytotic and endocytotic events in guard cell protoplasts during osmotically driven volume changes. *FEBS Lett.*, **460**, 495–499.

Homann, U. & Thiel, G. (2002) The number of K^+ channels in the plasma membrane of guard cell protoplasts changes in parallel with the surface area. *Proc. Natl. Acad. Sci. USA*, **99**, 10215–10220.

Jahn, R. (2000) Sec1/Munc18 proteins: mediators of membrane fusion moving to center stage. *Neuron*, **27**, 201–204.

Jahn, R. & Sudhof, T.C. (1999) Membrane fusion and exocytosis. *Ann. Rev. Biochem.*, **68**, 863–911.

Jauh, G.Y., Phillips, T.E. & Rogers, J.C. (1999) Tonoplast intrinsic protein isoforms as markers for vacuolar functions. *Plant Cell*, **11**, 1867–1882.

Jiang, L.W. & Rogers, J.C. (1999) The role of BP-80 and homologues in sorting proteins to vacuoles. *Plant Cell*, **11**, 2069–2071.

Jones, R.L. & Jacobsen, J.V. (1991) Regulation of synthesis and transport of secreted proteins in cereal aleurone. *Int. Rev. Cytol. – Surv. Cell Biol.*, **126**, 49–88.

Kaiser, C.A., Gimeno, R.E. & Shaywitz, D.A. (1997) Protein secretion, membrane biogenesis and endocytosis, in *The Molecular and Cellular Biology of the Yeast Saccharomyces* (eds J.R. Pringle, J.R. Broach & E.W. Jones), pp. 91–227, Cold Spring Harbor Laboratory Press: Boston.

Kargul, J., Gansel, X., Tyrrell, M., Sticher, L. & Blatt, M.R. (2001) Protein-binding partners of the tobacco syntaxin NtSyr1. *FEBS Lett.*, **508**, 253–258.

Kato, T., Morita, M.T., Fukaki, H., Yamauchi, Y., Uehara, M., Niihama, M. & Tasaka, M. (2002) SGR2, a phospholipase-like protein, and ZIG/SGR4, a SNARE, are involved in the shoot gravitropism of *Arabidopsis*. *Plant Cell*, **14**, 33–46.

Keller, P. & Simons, K. (1997) Post-Golgi biosynthetic trafficking. *J. Cell Sci.*, **110**, 3001–3009.

Kreft, M. & Zorec, R. (1997) Cell-attached measurements of attofarad capacitance steps in rat melanotrophs. *Pflugers Arch. – Eur. J. Physiol.*, **434**, 212–214.

Kubitscheck, U., Homann, U. & Thiel, G. (2000) Osmotically evoked shrinking of guard-cell protoplasts causes vesicular retrieval of plasma membrane into the cytoplasm. *Planta*, **210**, 423–431.

Labavitch, J.M. (1981) Cell-wall turnover in plant development. *Annu. Rev. Plant Physiol. Plant Mol. Biol.*, **32**, 385–406.

Lauber, M.H., Waizenegger, I., Steinmann, T., Schwarz, H., Mayer, U., Hwang, I., Lukowitz, W. & Jurgens, G. (1997) The *Arabidopsis* KNOLLE protein is a cytokinesis-specific syntaxin. *J. Cell Biol.*, **139**, 1485–1493.

Lehman, K., Rossi, G., Adamo, J.E. & Brennwald, P. (1999) Yeast homologues of tomosyn and lethal giant larvae function in exocytosis and are associated with the plasma membrane SNARE, Sec 9. *J. Cell Biol.*, **146**, 125–140.

Leyman, B., Geelen, D. & Blatt, M.R. (2000) Localization and control of expression of Nt-Syr1, a tobacco SNARE protein. *Plant J.*, **24**, 369–381.

Leyman, B., Geelen, D., Quintero, F.J. & Blatt, M.R. (1999) A tobacco syntaxin with a role in hormonal control of guard cell ion channels. *Science*, **283**, 537–540.

Lin, R.C. & Scheller, R.H. (2000) Mechanisms of synaptic vesicle exocytosis. *Ann. Rev. Cell Dev. Biol.*, **16**, 19–49.

Lollike, K., Borregaard, N. & Lindau, M. (1998) Capacitance flickers and pseudoflickers of small granules, measured in the cell-attached configuration. *Biophys. J.*, **75**, 53–59.

Lukowitz, W., Mayer, U. & Jurgens, G. (1996) Cytokinesis in the *Arabidopsis* embryo involves the syntaxin related *knolle* gene product. *Cell*, **84**, 61–71.

Lupashin, V.V., Hamamoto, S. & Schekman, R.W. (1996) Biochemical requirements for the targeting and fusion of ER-derived transport vesicles with purified yeast Golgi membranes. *J. Cell Biol.*, **132**, 277–289.

Marash, M. & Gerst, J.E. (2001) t-SNARE dephosphorylation promotes SNARE assembly and exocytosis in yeast. *EMBO J.*, **20**, 411–421.

Masuda, E.S., Huang, B.C.B., Fisher, J.M., Luo, Y. & Scheller, R.H. (1998) Tomosyn binds t-SNARE proteins via a VAMP-like coiled coil. *Neuron*, **21**, 479–480.

McLusky, S.R., Bennett, M.H., Beale, M.H., Lewis, M.J., Gaskin, P. & Mansfield, J.W. (1999) Cell wall alterations and localized accumulation of feruloyl-3′-methoxytyramine in onion epidermis at sites of attempted penetration by *Botrytis allii* are associated with actin polarisation, peroxidase activity and suppression of flavonoid biosynthesis. *Plant J.*, **17**, 523–534.

McNew, J.A., Parlati, F., Fukuda, R., Johnston, R.J., Paz, K., Paumet, F., Sollner, T.H. & Rothman, J.E. (2000) Compartmental specificity of cellular membrane fusion encoded in SNARE proteins. *Nature*, **407**, 153–159.

McNew, J.A., Sogaard, M., Lampen, N.M., Machida, S., Ye, R.R., Lacomis, L., Tempst, P., Rothman, J.E. & Sollner, T.H. (1997) Ykt6p, a prenylated SNARE essential for endoplasmic reticulum Golgi transport. *J. Biol. Chem.*, **272**, 17776–17783.

Misura, K.M.S., Scheller, R.H. & Weis, W.I. (2000) Three-dimensional structure of the neuronal-Sec1-syntaxin 1a complex. *Nature*, **404**, 355–362.

Monck, J.R. & Fernandez, J.M. (1994) The exocytotic fusion pore and neurotransmitter release. *Neuron*, **12**, 707–716.

Morris, D.A. & Robinson, J.S. (1998) Targeting of auxin carriers to the plasma membrane: differential effects of brefeldin A on the traffic of auxin uptake and efflux carriers. *Planta*, **205**, 606–612.

Nagahama, M., Orci, L., Ravazzola, M., Amherdt, M., Lacomis, L., Tempst, P., Rothman, J.E. & Sollner, T.H. (1996) A v-SNARE implicated in intra-Golgi transport. *J. Cell Biol.*, **133**, 507–516.

Neher, E. (1998) Vesicle pools and Ca^{2+} microdomains: new tools for understanding their roles in neurotransmitter release. *Neuron*, **20**, 389–399.

Neuhaus, J.M. & Rogers, J.C. (1998) Sorting of proteins to vacuoles in plant cells. *Plant Mol. Biol.*, **38**, 127–144.

Nichols, B.J. & Pelham, H.R.B. (1998) SNAREs and membrane fusion in the Golgi apparatus. *Biochim. Biophys. Acta Mol. Cell Res.*, **1404**, 9–31.

Novick, P. & Zerial, M. (1997) The diversity of Rab proteins in vesicle transport. *Curr. Opin. Cell Biol.*, **9**, 496–504.

Oheim, M., Loerke, D., Stuhmer, W. & Chow, R.H. (1998) The last few milliseconds in the life of a secretory granule – docking, dynamics and fusion visualized by total internal reflection fluorescence microscopy (TIRFM). *Eur. Biophys. J. Biophys. Lett.*, **27**, 83–98.

Ossig, R., Schmitt, H.D., de Groot, B., Riedel, D., Keranen, S., Ronne, H., Grubmuller, H. & Jahn, R. (2000) Exocytosis requires asymmetry in the central layer of the SNARE complex. *EMBO J.*, **19**, 6000–6010.

Ott, D. & Brown, M. Jr. (1974) Developmental cytology of the genus *Vaucheria*. I. Organization of the vegetative filament. *Br. Phycol. J.*, **9**, 111–126.

Paek, I., Orci, L., Ravazzola, M., Erdjumentbromage, H., Amherdt, M., Tempst, P., Sollner, T.H. & Rothman, J.E. (1997) ERS-24, a mammalian v-SNARE implicated in vesicle traffic between the ER and the Golgi. *J. Cell Biol.*, **137**, 1017–1028.

Paris, N., Rogers, S.W., Jiang, L.W., Kirsch, T., Beevers, L., Phillips, T.E. & Rogers, J.C. (1997) Molecular cloning and further characterization of a probable plant vacuolar sorting receptor. *Plant Physiol.*, **115**, 29–39.

Parlati, F., McNew, J.A., Fukuda, R., Miller, R., Sollner, T.H. & Rothman, J.E. (2000) Topological restriction of SNARE-dependent membrane fusion. *Nature*, **407**, 194–198.

Patel, S.K., Indig, F.E., Olivieri, N., Levine, N.D. & Latterich, M. (1998) Organelle membrane fusion: a novel function for the syntaxin homologue Ufe1p in ER membrane fusion. *Cell*, **92**, 611–620.

Peterson, M.R., Burd, C.G. & Emr, S.D. (1999) Vac1p coordinates Rab and phosphatidylinositol 3-kinase signaling in Vps45p-dependent vesicle docking/fusion at the endosome. *Curr. Biol.*, **9**, 159–162.

Picton, J.M. & Steer, M.W. (1983) Membrane recycling and the control of secretory activity in pollen tubes. *J. Cell Sci.*, **63**, 303–310.

Prekeris, R., Yang, B., Oorschot, V., Klumperman, J. & Scheller, R.H. (1999) Differential roles of syntaxin 7 and syntaxin 8 in endosomal trafficking. *Mol. Biol. Cell*, **10**, 3891–3908.

Qualmann, B. & Kelly, R.B. (2000) Syndapin isoforms participate in receptor-mediated endocytosis and actin organization. *J. Cell Biol.*, **148**, 1047–1061.

Rabouille, C., Kondo, H., Newman, R., Hui, N., Freemont, P. & Warren, G. (1998) Syntaxin 5 is a common component of the NSF- and p97-mediated reassembly pathways of Golgi cisternae from mitotic Golgi fragments *in vitro*. *Cell*, **92**, 603–610.

Ringstad, N., Gad, H., Low, P., DiPaolo, G., Brodin, L., Shupliakov, O. & Decamilli, P. (1999) Endophilin/SH3p4 is required for the transition from early to late stages in clathrin-mediated synaptic vesicle endocytosis. *Neuron*, **24**, 143–154.

Robinson, D.G. & Kristen, U. (1982) Membrane flow via the Golgi apparatus in higher-plant cells. *Int. Rev. Cytol.*, **77**, 89–127.

Rogers, J.C. (1998) Compartmentation of plant cell proteins in separate lytic and protein storage vacuoles. *J. Plant Physiol.*, **152**, 653–658.

Rossetto, O., Gorza, L., Schiavo, G., Schiavo, N., Scheller, R.H. & Montecucco, C. (1996) VAMP synaptobrevin isoforms 1 and 2 are widely and differentially expressed in nonneuronal tissues. *J. Cell Biol.*, **132**, 167–179.

Rupnik, M. & Zorec, R. (1995) Intracellular Cl⁻ modulates Ca²⁺-induced exocytosis from rat melano-trophs through GTP-binding proteins. *Pflugers Arch. – Eur. J. Physiol.*, **431**, 76–83.

Sakmann, B. & Neher, E. (1983) Geometric parameters of pipettes and membrane patches, in *Single Channel Recording* (eds B. Sakmann & E. Neher), pp. 37–51, Plenum: New York.

Sanderfoot, A.A., Assaad, F.F. & Raikhel, N.V. (2000) The *Arabidopsis* genome. An abundance of sol-uble *N*-ethylmaleimide-sensitive factor adaptor protein receptors. *Plant Physiol.*, **124**, 1558–1569.

Sanderfoot, A.A., Kovaleva, V., Bassham, D.C. & Raikhel, N.V. (2001a) Interactions between syntaxins identify at least five SNARE complexes within the Golgi/prevacuolar system of the *Arabidopsis* cell. *Mol. Biol. Cell*, **12**, 3733–3743.

Sanderfoot, A.A., Kovaleva, V., Zheng, H.Y. & Raikhel, N.V. (1999) The t-SNARE AtVAM3p resides on the prevacuolar compartment in *Arabidopsis* root cells. *Plant Physiol.*, **121**, 929–938.

Sanderfoot, A.A., Pilgrim, M., Adam, L. & Raikhel, N.V. (2001b) Disruption of individual members of *Arabidopsis* syntaxin gene families indicates each has essential functions. *Plant Cell*, **13**, 659–666.

Sanderfoot, A.A. & Raikhel, N.V. (1999) The specificity of vesicle trafficking: coat proteins and SNAREs. *Plant Cell*, **11**, 629–641.

Satiat-Jeunemaitre, B. & Hawes, C. (2001) Immunocytochemistry for light microscopy, in *Plant Cell Biology: A Practical Approach* (eds C. Hawes & B. Satiat-Jeunemaitre), pp. 207–233, Academic: London.

Sato, M.H., Nakamura, N., Ohsumi, Y., Kouchi, H., Kondo, M., HaraNishimura, I., Nishimura, M. & Wada, Y. (1997) The AtVAM3 encodes a syntaxin-related molecule implicated in the vacuolar assembly in *Arabidopsis thaliana*. *J. Biol. Chem.*, **272**, 24530–24535.

Scales, S.J., Yoo, B.Y. & Scheller, R.H. (2001) The ionic layer is required for efficient dissociation of the SNARE complex by alpha-SNAP and NSF. *Proc. Natl. Acad. Sci. USA*, **98**, 14262–14267.

Schindler, T., Bergfeld, R., Hohl, M. & Schopfer, P. (1994) Inhibition of Golgi apparatus function by brefeldin A in maize coleoptiles and its consequences on auxin-mediated growth, cell-wall extensibility and secretion of cell wall proteins. *Planta*, **192**, 404–413.

Schneider, K., Wells, B., Dolan, L. & Roberts, K. (1997) Structural and genetic analysis of epidermal cell differentiation in *Arabidopsis* primary roots. *Development*, **124**, 1789–1798.

Seron, K., Tieaho, V., PrescianottoBaschong, C., Aust, T., Blondel, M.O., Guillaud, P., Devilliers, G., Rossanese, O.W., Glick, B.S., Riezman, H., Keranen, S. & HaguenauerTsapis, R. (1998) A yeast t-SNARE involved in endocytosis. *Mol. Biol. Cell*, **9**, 2873–2889.

SingerKruger, B., Nemoto, Y., Daniell, L., Ferronovick, S. & Decamilli, P. (1998) Synaptojanin family members are implicated in endocytic membrane traffic in yeast. *J. Cell Sci.*, **111**, 3347–3356.

Skehel, P.A., FabianFine, R. & Kandel, E.R. (2000) Mouse VAP33 is associated with the endoplasmic reticulum and microtubules. *Proc. Natl. Acad. Sci. USA*, **97**, 1101–1106.

Sogaard, M., Tani, K., Ye, R.R., Geromanos, S., Tempst, P., Kirchhausen, T., Rothman, J.E. & Sollner, T. (1994) A rab protein is required for the assembly of SNARE complexes in the docking of transport vesicles. *Cell*, **78**, 937–948.

Steer, M.W. & Steer, J.M. (1989) Pollen tube tip growth. *New Phytologist*, **111**, 323–358.

Steinmann, T., Geldner, N., Grebe, M., Mangold, S., Jackson, C.L., Paris, S., Galweiler, L., Palme, K. & Jurgens, G. (1999) Coordinated polar localization of auxin efflux carrier PIN1 by GNOM ARF GEF. *Science*, **286**, 316–318.

Steyer, J.A. & Almers, W. (2001) A real-time view of life within 100 nm of the plasma membrane. *Nat. Rev. Mol. Cell Biol.*, **2**, 268–275.

Steyer, J.A., Horstmann, H. & Almers, W. (1997) Transport, docking and exocytosis of single secretory granules in live chromaffin cells. *Nature*, **388**, 474–478.

Strompen, G., El Kasmi, F., Richter, S., Lukowitz, W., Assaad, F.F., Jurgens, G. & Mayer, U. (2002) The *Arabidopsis HINKEL* gene encodes a kinesin-related protein involved in cytokinesis and is expressed in a cell cycle-dependent manner. *Current Biol.*, **12**, 153–158.

Sutter, J.U., Homann, U. & Thiel, G. (2000) Ca^{2+}-stimulated exocytosis in maize coleoptile cells. *Plant Cell*, **12**, 1127–1136.

Tai, W.C.S. & Banfield, D.K. (2001) *AtBS14a* and *AtBS14b*, two Bet1/Sft1-like SNAREs from *Arabidopsis thaliana* that complement mutations in the yeast *SFT1* gene. *FEBS Lett.*, **500**, 177–182.

Tang, B.L., Low, D.Y.H. & Hong, W.J. (1998a) Hsec22c: a homologue of yeast Sec22p and mammalian rsec22a and msec22b/ERS-24. *Biochem. Biophys. Res. Commun.*, **243**, 885–891.

Tang, B.L., Low, D.Y.H., Tan, A.E.H. & Hong, W.J. (1998b) Syntaxin 10: a member of the syntaxin family localized to the trans-Golgi network. *Biochem. Biophys. Res. Commun.*, **242**, 345–350.

Taubenblatt, P., Dedieu, J.C., GulikKrzywicki, T. & Morel, N. (1999) VAMP (synaptobrevin) is present in the plasma membrane of nerve terminals. *J. Cell Sci.*, **112**, 3559–3567.

Taylor, L.P. & Hepler, P.K. (1997) Pollen germination and tube growth. *Annu. Rev. Plant Physiol. Plant Mol. Biol.*, **48**, 461–491.

Thiel, G. & Battey, N. (1998) Exocytosis in plants. *Plant Mol. Biol.*, **38**, 111–125.

Thiel, G., Kreft, M. & Zorec, R. (1998) Unitary exocytotic and endocytotic events in *Zea mays* L. coleoptile protoplasts. *Plant J.*, **13**, 117–120.

Thiel, G., Sutter, J.U. & Homann, U. (2000) Ca^{2+}-sensitive and Ca^{2+}-insensitive exocytosis in maize coleoptile protoplasts. *Pflugers Arch. – Eur. J. Physiol.*, **439**, R152–R153.

Volker, A., Stierhof, Y.D. & Jurgens, G. (2001) Cell cycle-independent expression of the *Arabidopsis* cytokinesis-specific syntaxin KNOLLE results in mistargeting to the plasma membrane and is not sufficient for cytokinesis. *J. Cell Sci.*, **114**, 3001–3012.

vonMollard, G.F. & Stevens, T.H. (1999) The *Saccharomyces cerevisiae* v-SNARE Vti1p is required for multiple membrane transport pathways to the vacuole. *Mol. Biol. Cell*, **10**, 1719–1732.

Walker, N.A. (1955) Microelectrode experiments on *Nitella*. *Aust. J. Biol. Sci.*, **8**, 476–489.

Weber, T., Zemelman, B.V., McNew, J.A., Westermann, B., Gmachl, M., Parlati, F., Sollner, T.H. & Rothman, J.E. (1998) SNARE pins: minimal machinery for membrane fusion. *Cell*, **92**, 759–772.

Weise, R., Kreft, M., Zorec, R., Homann, U. & Thiel, G. (2000) Transient and permanent fusion of vesicles in *Zea mays* coleoptile protoplasts measured in the cell-attached configuration. *J. Membr. Biol.*, **174**, 15–20.

Wessels, J.G.H. (1986) Cell-wall synthesis in apical hyphal growth. *Int. Rev. Cytol. – Surv. Cell Biol.*, **104**, 37–79.

Wooding, S. & Pelham, H.R.B. (1998) The dynamics of Golgi protein traffic visualized in living yeast cells. *Mol. Biol. Cell*, **9**, 2667–2680.

Wu, M.N., Littleton, J.T., Bhat, M.A., Prokop, A. & Bellen, H.J. (1998) *ROP*, the *Drosophila Sec1* homologue, interacts with syntaxin and regulates neurotransmitter release in a dosage-dependent manner. *EMBO J.*, **17**, 127–139.

Yang, B., Steegmaier, M., Gonzalez, L.C. & Scheller, R.H. (2000) nSec1 binds a closed conformation of syntaxin1A. *J. Cell Biol.*, **148**, 247–252.

Yang, Y.O., Shah, J. & Klessig, D.F. (1997) Signal perception and transduction in defense responses. *Genes Dev.*, **11**, 1621–1639.

Zerial, M. & McBride, H. (2001) Rab proteins as membrane organizers (pg. 107). *Nat. Rev. Mol. Cell Biol.*, **2**, 216.

Zhang, R.E., Maksymowych, A.B. & Simpson, L.L. (1995) Cloning and sequence analysis of a cDNA encoding human syntaxin 1A, a polypeptide essential for exocytosis. *Gene*, **159**, 293–294.

Zheng, H.Y., vonMollard, G.F., Kovaleva, V., Stevens, T.H. & Raikhel, N.V. (1999) The plant vesicle-associated SNARE AtVTI1a likely mediates vesicle transport from the trans-Golgi network to the prevacuolar compartment. *Mol. Biol. Cell*, **10**, 2251–2264.

Zorec, R. & Tester, M. (1993) Rapid pressure-driven exocytosis endocytosis cycle in a single plant cell – capacitance measurements in aleurone protoplasts. *FEBS Lett.*, **333**, 283–286.

14 Cytokinesis: membrane trafficking by default?

Gerd Jürgens and Tobias Pacher

14.1 Introduction

Cytokinesis partitions a dividing cell after the two sets of daughter chromosomes have moved to opposite poles during anaphase. Unlike yeast or animal cells, somatic cells of higher plants form the partitioning membrane compartment *de novo*, starting at the centre of the division plane. This new compartment, called the cell plate, then expands laterally towards the periphery of the cell. The cell-plate margin fuses with the plasma membrane in cortical division sites that were marked at the entry into mitosis by the transient formation of the preprophase band (reviewed by Staehelin & Hepler, 1996; Heese *et al.*, 1998; Otegui & Staehelin, 2000). Recently, polarised cytokinesis has been described in vacuolate cells of *Arabidopsis* (Cutler & Ehrhardt, 2002). Although the initial stage of cell-plate formation by vesicle fusion is the same as in non-vacuolate cells, the cell plate then appears to attach locally to the plasma membrane before expanding along the circumference to the opposite end of the dividing cell.

Plant cytokinesis is a distinct membrane trafficking process in at least three ways. First, this process is confined to the mitotic phase of the cell cycle whereas other trafficking processes occur during the interphase when the cell is growing in size. Second, unlike vesicles destined to fuse with target membranes in the non-dividing cell, cytokinetic vesicles initially fuse with one another to form a new membrane compartment, the cell plate, which becomes the target membrane for fusion of later arriving vesicles. Third, vesicle trafficking during cytokinesis is assisted by a plant-specific cytoskeletal array, the phragmoplast, that forms only in late anaphase and disassembles upon the completion of cytokinesis. To fully appreciate the specific conditions of cytokinetic vesicle trafficking, one has to bear in mind that the completion of cytokinesis is not a checkpoint for M/G1 progression (Waizenegger *et al.*, 2000). One-third of the original cell surface has to be made during about 4 % of the cell cycle (McClinton & Sung, 1997). This figure translates into less than 30 min in early embryogenesis of *Arabidopsis* or in synchronised tobacco BY-2 cells. Thus, membrane trafficking to the division plane has to be organised very efficiently.

This review specifically focuses on vesicle trafficking during phragmoplast-assisted cytokinesis. Other aspects of cytokinesis have been reviewed recently and are not addressed here (Heese *et al.*, 1998; Otegui & Staehelin, 2000; Smith, 2001; Brown & Lemmon, 2001). Starting with a brief description of cytokinesis in somatic cells, we review recent findings related to the cytokinetic machinery of vesicle trafficking and fusion. We then discuss the possible mechanisms that may underlie the apparent specificity of vesicle trafficking and the high efficiency of cell-plate formation.

14.2 A dynamic view of cell plate formation

Our current view of phragmoplast-assisted cytokinesis originated from detailed electron-microscopic studies performed on high-pressure frozen/freeze-substituted tobacco BY-2 cells (Samuels *et al.*, 1995) and from studies on the dynamics of phragmoplast microtubules and actin filaments (Staehelin & Hepler, 1996; Granger & Cyr, 2000). Following the segregation of two sets of daughter chromosomes to opposite poles, a new cytoskeletal structure is organised in the interzone from remnants of the mitotic spindle. This phragmoplast consists of two bundles, each of microtubules and actin filaments with opposite polarity. Whereas the plus ends of microtubules overlap at the plane of division, those of the actin filaments merely abut each other. Membrane vesicles traffic along the phragmoplast to the centre of the division plane where they fuse with one another to form the cell plate. Upon fusion, the vesicles form a mesh connected by tubules, and this incipient cell plate undergoes a series of morphologically distinct intermediates leading to a solid disk-shaped membrane compartment that secretes callose into its lumen (Staehelin & Hepler, 1996). Superimposed with this process of maturation is the lateral expansion of the cell plate, which is achieved by a reorganisation of the phragmoplast. Initially a solid bundle that spans the width of the daughter nuclei, the phragmoplast is transformed into a widening hollow cylinder. This transformation is achieved by a translocation process that depolymerises existing microtubules and actin filaments in the centre and polymerises new ones along the lateral margin. As a consequence, additional membrane vesicles are now delivered to the growing margin of the cell plate. The lateral translocation of the phragmoplast and the centrifugal expansion of the cell plate continue in synchrony until the cortical division site is reached. Recent studies suggest that actomyosin provides the necessary tension for the lateral translocation of the phragmoplast (Molchan *et al.*, 2001). Finally, the cell plate fuses with the parental plasma membrane of the dividing cell, maturing into the new plasma membrane and cell wall, whereas the phragmoplast is dismantled. Table 14.1 gives a list of proteins that have been localised to the phragmoplast or the cell plate, and indicates their putative functions during cytokinesis.

14.3 Formation of cytokinetic vesicles

Cytokinetic vesicles measure 60–80 nm in diameter and are thus similar to transport vesicles in other trafficking pathways (Lauber *et al.*, 1997; Waizenegger *et al.*, 2000). It is commonly thought that the cell plate-forming vesicles originate from

Table 14.1 Proteins located at the cell plate and/or phragmoplast

Protein	Function	Proposed role during cytokinesis	Localisation during cytokinesis	Other localisation beside cytokinesis	Reference
KNOLLE (*Arabidopsis*)	t-SNARE	Vesicle docking	Cell plate	–	Lauber et al., 1997
SNAP33 (*Arabidopsis*)	t-SNARE	Vesicle docking	Cell plate	Plasma membrane	Heese et al., 2001
NPSN11 (*Arabidopsis*)	v-SNARE? (novel plant SNARE)	Vesicle docking?	Cell plate, unknown subcellular organelles	Plasma membrane	Zheng et al., 2002
TKRP125 (tobacco)	Kinesin-related protein	Maintenance of phragmoplast microtubule organisation?	Phragmoplast	Not analysed	Asada & Shibaoka, 1994
KatAp (*Arabidopsis*)	Kinesin-like protein	?	Phragmoplast	Cytoplasm, nucleus, midzone during mitosis	Liu et al., 1996
KCBP (*Arabidopsis*, tobacco)	Kinesin-like calmodulin-binding protein	Minus end-directed motor protein possibly involved in microtubule organisation and/or function	Phragmoplast	Preprophase band, mitotic spindle	Bowser & Reddy, 1997
DcKRP120-2 (carrot)	BimC class kinesin-related proteins	Maintenance of phragmoplast microtubule organisation?	Phragmoplast mid-line	Spindle	Barroso et al., 2000
AtPAKRP1 (*Arabidopsis*)	Phragmoplast-associated kinesin-related protein	Establishment and/or maintenance of phragmoplast MT array	Microtubule plus-ends	Microtubules towards the spindle midzone during late anaphase	Lee & Liu, 2000
AtPAKRP2 (*Arabidopsis*)	Phragmoplast-associated ungrouped N-terminal motor kinesin	Transport of Golgi-derived vesicles in the phragmoplast?	Concentrated near the division site, also in the phragmoplast	Interzonal MTs during late anaphase	Lee et al., 2001
HINKEL (*Arabidopsis*)	Kinesin-like protein with N-terminal motor domain	Regulation of phragmoplast microtubule stability during cell-plate expansion?	Phragmoplast?	Not analysed	Strompen et al., 2002

Protein (organism)	Protein type	Function	Localisation	Subcellular localisation / timing	Reference
NACK1 (tobacco)	Kinesin-related protein with N-terminal motor domain	Localisation of NPK1 for expansion of the cell plate	Equatorial zone of the phragmoplast	Dispersed (prometaphase – early anaphase) spindle midzone (late anaphase)	Nishihama et al., 2002
Phragmoplastin (soy bean)	Dynamin-like GTPase	Formation of fusion tubules of the TVN?	Growing margins of the cell plate	?	Gu & Verma, 1996
ADL1A (Arabidopsis)	Dynamin-like protein (related to phragmoplastin)	Vesicle trafficking from the Golgi apparatus? Membrane recycling from the forming cell plate? Formation of fusion tubules of the TVN?	Phragmoplast	Unknown subcellular compartments (Golgi stacks?) in interphase	Kang et al., 2001
KORRIGAN (Arabidopsis)	Endo-1,4-beta-glucanase (cellulase)	Maturation of the new cell wall	Cell plate	Golgi apparatus, cell wall-associated (interphase)	Nicol et al., 1998 Zuo et al., 2000
CalS1 (Arabidopsis)	Cell plate-specific callose synthase	Callose deposition during cell-plate formation	Cell plate	–	Hong et al., 2001a
UGT1 (Arabidopsis)	UDP-glucose transferase	Delivery of UDP-glucose to the cell plate-specific callose synthase	Cell plate	–	Hong et al., 2001b
EXGT (tobacco)	Cell wall-associated endoxyloglucan transferase	Processing of xyloglucans?	Phragmoplast, cell plate	Secretory pathway and apoplast (interphase), equatorial plate (metaphase)	Yokoyama & Nishitani, 2001
AtCDC48 (Arabidopsis)	AAA-ATPase (related to yeast Cdc48p)	?	Phragmoplast	Nucleus	Feiler et al., 1995
CDC2aAt (Arabidopsis)	Ser/Thr-protein kinase	Cell-cycle regulator	Cytoplasm, phragmoplast	Nucleus and cytoplasm (interphase), condensed chromosomes (onset of	Stals et al., 1997

Contd

Table 14.1 (continued)

Protein	Function	Proposed role during cytokinesis	Localisation during cytokinesis	Other localisation beside cytokinesis	Reference
p43(Ntf6) (tobacco)	Cell cycle-regulated MAP kinase	?	Phragmoplast	anaphase), preprophase band, spindle	Calderini et al., 1998
MMK3 (Medicago)	MAP kinase	?	Phragmoplast, cell plate (late telophase)	Cell plate (transiently in anaphase)	Bögre et al., 1999
NPK1 (tobacco)	MAPKKK	Expansion of the cell plate	Equatorial zone of the phragmoplast	Between the segregating chromosomes (anaphase)	Nishihama et al., 2001
WLIM1 (sunflower)	LIM-domain protein	?	Phragmoplast	Nucleus, cytoplasm, spindle midzone	Mundel et al., 2000
Vinculin-like antigen	?	Membrane/F-actin anchorage of protein complexes? (cell wall/extracellular matrix/plasma membrane/actin cytoskeleton continuum)	Phragmoplast, cell plate	Cytoplasm, nucleus, often associated with plastids and smaller subcellular structures	Endle et al., 1998
PIN1 (Arabidopsis)	Putative auxin efflux carrier	–	Cell plate	Not analysed	Geldner et al., 2001

Golgi stacks that gather near the plane of division during M phase (Nebenführ *et al.*, 2000). The cytokinesis-specific syntaxin KNOLLE (see below) is first localised to large patches, presumably Golgi stacks, in mitotic cells before being relocated to the forming cell plate (Lauber *et al.*, 1997). More recently, KNOLLE has been detected at the *trans*-Golgi network by immuno-gold labelling (Völker *et al.*, 2001). Whether all Golgi stacks of the dividing cell produce cytokinetic vesicles remains to be determined. Furthermore, it is unclear how the formation of cytokinetic vesicles is regulated. In yeast and mammals, vesicle budding has been shown to require a small GTPase as well as its cognate guanine-nucleotide exchange factor (GEF) and activating protein (GAP) for recruitment of coat proteins and for cargo selection (Springer *et al.*, 1999; Kirchhausen, 2000). For plant cytokinetic vesicle budding, neither a GTPase and its effectors nor coat proteins have been identified.

Cytokinetic vesicles deliver both membrane and soluble proteins as well as other cargo to the cell plate. A major component of the membrane protein cargo is callose synthase that spins out the extracellular callose into the lumen of the cell plate (Hong *et al.*, 2001a,b; Lane *et al.*, 2001). Some of the transported proteins are enzymes involved in cell wall formation, such as the membrane-bound endo-glucanase KORRIGAN (Nicol *et al.*, 1998; Zuo *et al.*, 2000) and a soluble endoxylo-glucan transferase (Yokoyama & Nishitani, 2001). In addition, precursors of cell wall material, such as pectin and xyloglucans, need to be transported within the lumen of the cytokinetic vesicles (reviewed by Dupree & Sherrier, 1998; Verma, 2001). Proper localisation of KORRIGAN to the cell plate has been proposed to involve tyrosine-based (YppΦ) and di-leucine (LL) sorting motifs, which in non-plant cells mediate sorting during endocytosis and polarised protein secretion (Zuo *et al.*, 2000; see Kirchhausen, 2000, for review). However, mutation of the sorting motifs did not affect localisation to the cell plate although the protein was also observed at the plasma membrane (Zuo *et al.*, 2000). At present, there is no convincing evidence for active selection of cargo proteins during cytokinetic vesicle budding, which is thus similar to the apparent lack of cargo selection during the budding of secretory vesicles from the TGN in interphase cells.

14.4 Delivery of cytokinetic vesicles

Membrane vesicles are cargo of specific motor molecules that move along cytoskeletal tracks in a directional manner. The phragmoplast consists of both microtubules and actin filaments, and either cytoskeletal system could provide the tracks for the delivery of cytokinetic vesicles to the plane of cell division. Although vesicles have been detected on phragmoplast microtubules (Samuels *et al.*, 1995), a role of actin filaments in their delivery had not been ruled out until recently. Genetic evidence from *Arabidopsis* now favours a prominent role of microtubules in delivering cytokinetic vesicles to the plane of division. Mutations eliminating tubulin-folding cofactors block cell division and trafficking of the cytokinesis-specific syntaxin KNOLLE but do not affect actin-dependent trafficking of PIN1 to the plasma membrane (Mayer *et al.*, 1999; Geldner *et al.*, 2001; Steinborn *et al.*, 2002). In addition, electron tomography of cell plate formation in cellularising endosperm detected noncoated vesicles linked to microtubules via proteins that resemble kinesin molecules (Otegui *et al.*, 2001).

Cytokinetic vesicle transport along phragmoplast microtubules would require a plus end-directed kinesin motor protein similar to conventional kinesin heavy chain (KHC) in fungal and animal cells (Woehlke & Schliwa, 2000). Although 61 putative kinesin-related proteins are encoded in the *Arabidopsis* genome, only a truncated putative conventional KHC has been identified (Reddy & Day, 2001). Whether this protein is functional and plays a role in cytokinetic vesicle trafficking is not clear. It is also conceivable that a plant-specific motor delivers the cytokinetic vesicles, as there are several plant-specific kinesins that appear to be involved in the dynamic stability or reorganisation of the phragmoplast microtubules during cytokinesis (Table 14.1; Asada *et al.*, 1997; Lee & Liu, 2000; Nishihama *et al.*, 2002; Strompen *et al.*, 2002). Another phragmoplast-associated kinesin-related protein, AtPAKRP2, accumulates towards the equator and appears to be associated with membrane compartments (Lee *et al.*, 2001). However, AtPAKRP2 does not co-localise with the cytokinesis-specific syntaxin KNOLLE. Its involvement in cytokinetic vesicle transport remains to be determined.

14.5   Vesicle fusion in the plane of cell division

Upon arrival at the plane of cell division, the membrane vesicles fuse with one another, forming a wide tubular network (Samuels *et al.*, 1995). The tubules of this network are locally constricted by spiral-shaped structures consisting of *Arabidopsis* Dynamin-Like 1 (ADL1) protein (Otegui *et al.*, 2001), which is the homologue of the dynamin-related GTPase phragmoplastin (Gu & Verma, 1996, 1997; Zhang *et al.*, 2000; Verma, 2001). Phragmoplastin and ADL1 are peripheral membrane proteins that accumulate to high levels at the growing cell plate (Gu & Verma, 1996; Lauber *et al.*, 1997). In addition to its putative role in membrane constriction, phragmoplastin also interacts with the catalytic subunit of callose synthase CalS1 and the UDP-glucose transferase UGT1, both of which are localised at the cell plate (Hong *et al.*, 2001a,b). Thus, phragmoplastin could also assist in callose deposition, which transforms the tubular network into a fenestrated sheet (Staehelin & Hepler, 1996).

The fusion of cytokinetic vesicles with one another bears similarities to the homotypic fusion process of yeast vacuoles, which has been studied in detail (review: Wickner & Haas, 2000). Homotypic fusion involves distinct steps of priming, tethering, docking and fusion. In the initial step of priming, *cis*-SNARE (soluble *N*-ethylmaleimide-sensitive factor adaptor protein receptor) complexes residing on the same membrane are broken up by the action of Sec18p, an AAA-type ATPase homologous to the mammalian *N*-ethylmaleimide-sensitive factor (NSF). A homologue of yeast AAA-type ATPase Cdc48p has been localised to the cell plate in dividing cells of *Arabidopsis* but its role in cytokinesis is not clear (Feiler *et al.*, 1995). Tethering of vesicles occurs in both homotypic and hetero-typic fusion events and is mediated by Rab-type GTPases and their effector protein complexes (Zerial & McBride, 2001). Subsequently, *trans*-SNARE complexes dock the vesicles to one another or a target membrane before actual fusion of the interacting membranes occurs (Chen & Scheller, 2001; Mayer, 2001). How vesicle priming, tethering and fusion occur during plant cytokinesis is still unclear whereas some components involved in the docking step have been analysed.

In yeast and mammals, vesicle docking is brought about by the formation of *trans*-SNARE complexes mediated by alpha-helical bundles of four coiled-coil domains from v-SNAREs on the vesicular membrane and t-SNAREs on the target membrane (Chen & Scheller, 2001). Regardless of their composition, the alpha-helical bundles of *trans*-SNARE complexes are virtually identical (Antonin *et al.*, 2002). Endomembrane SNARE complexes are formed by a v-SNARE (synaptobrevin) interacting with three t-SNAREs (a syntaxin and two t-SNARE light chains) which each contribute one alpha-helix. By contrast, plasma-membrane SNARE com-plexes are trimeric, consisting of a synaptobrevin, a syntaxin and a SNAP25 homologue, which is a t-SNARE that contributes two alpha-helices (Chen & Scheller, 2001). Being members of large protein families, synaptobrevins and syntaxins form specific pairs involved in a given trafficking pathway. By contrast, SNAP25 homologues and t-SNARE light chains appear to be more promiscuous, participating in several trafficking pathways (Fukuda *et al.*, 2000; McNew *et al.*, 2000; Parlati *et al.*, 2000; review: Chen & Scheller, 2001).

That the general SNARE complex model of vesicle docking may also apply in plant cytokinesis was suggested by the identification of the cytokinesis-specific syntaxin KNOLLE on the basis of its mutant phenotype (Lukowitz *et al.*, 1996). In the absence of KNOLLE protein, cytokinetic vesicle fusion is impaired and unfused vesicles persist well into interphase (Lauber *et al.*, 1997). KNOLLE orthologues exist in plants but not in yeast or animals. KNOLLE mRNA accu-mulates in a cell cycle-dependent manner (Lukowitz *et al.*, 1996) and KNOLLE protein is only present during M phase. Initially accumulating in large patches, which presumably are Golgi stacks, KNOLLE protein relocates to the cell divi-sion plane upon the formation of the phragmoplast and disappears at the comple-tion of cytokinesis (Lauber *et al.*, 1997). In contrast to their rapid turnover in dividing cells, KNOLLE mRNA and protein are stable when misexpressed in non-proliferating cells, and KNOLLE protein is trafficked to the plasma membrane (Völker *et al.*, 2001). The biological significance of the tight regulation of KNOLLE expression during the cell cycle is not clear. It is conceivable, however, that the strong transcriptional activation at G2/M phase is to ensure that sufficient amounts of KNOLLE protein are available for the efficient fusion of cytokinetic vesicles during cell-plate formation and expansion. This may be crucial because the completion of cytokinesis is not a checkpoint for cell-cycle progression (Mayer *et al.*, 1999; Waizenegger *et al.*, 2000). Thus, large number of vesicles have to be delivered and fused during a brief period of time before the cell enters the next stage of the cell cycle.

Another member of the cytokinetic SNARE complex is SNAP33, a SNAP25 homologue, which was identified as a KNOLLE-interacting protein in a yeast two-hybrid screen (Heese *et al.*, 2001). Mutational elimination of SNAP33 results in a mild cytokinesis defect, presumably because two closely related proteins, SNAP29 and SNAP30, that also interact with KNOLLE in the yeast two-hybrid assay are able to partially compensate for the lack of SNAP33. The *snap33* mutants die before flowering and display symptoms of necrosis in leaves and stem. Consistent with the mutant phenotype, SNAP33 protein not only co-localises with KNOLLE during cytokinesis but also accumulates at the plasma membrane in non-dividing cells (Heese *et al.*, 2001). Thus, SNAP33 appears to be a promiscuous t-SNARE, being involved in several trafficking pathways. However, KNOLLE–SNAP33 interaction suggests that the cytokinetic SNARE complex resembles in composition the trimeric SNARE complex found at the plasma membrane in non-plant organisms (see above). Thus, cytokinesis may be a special case of protein secretion that involves components of the exocytic machinery in combination with cytokinesis-specific components such as KNOLLE. Very recently, a novel plant SNARE, NPSN11, has been identified as another *in vitro* interactor of KNOLLE and shown to localise to the cell plate as well as to the plasma membrane (Zheng *et al.*, 2002). Since NPSN11 is not a canonical v-SNARE it remains to be determined whether the cytokinetic SNARE complex consists of NPSN11, SNAP33 and KNOLLE. Alternatively, KNOLLE may form multiple SNARE complexes with changing partners, including as yet unknown v-SNAREs. At present, the syntaxin KNOLLE is the only cytokinesis-specific component of the SNARE complex(es) involved in cell-plate formation.

In yeast and mammals, the formation of SNARE complexes may be regulated by interaction of syntaxins with Sec1 family members. For example, binding of neuronal Sec1 stabilises a *closed* conformation of syntaxin 1A that inhibits SNARE complex formation (Misura *et al.*, 2000; Yang *et al.*, 2000). In yeast, the Sec1 family member Sly1p interacts with the N-terminus of the *cis*-Golgi SNARE Sed5 (Yamaguchi *et al.*, 2002), and this interaction may contribute to the specificity of SNARE complex formation (Peng & Gallwitz, 2002). In *Arabidopsis*, the KEULE gene encodes a Sec1 homologue that interacts with KNOLLE (Assaad *et al.*, 2001). Moreover, *keule* mutants accumulate unfused cytokinetic vesicles, very much like *knolle* mutants (Waizenegger *et al.*, 2000). It is thus conceivable that KEULE activates KNOLLE before cytokinetic vesicle docking. Alternatively, KEULE may contribute to the specificity of SNARE complex formation, by analogy to the Sly1p action in yeast. However, *knolle keule* double mutant embryos are lethal, consisting of multinucleate single cells whereas each single mutant completes embryogenesis but fails to undergo cytokinesis at the seedling stage (Waizenegger *et al.*, 2000). This observation suggests some functional redundancy of cytokinetic vesicle fusion during embryogenesis but not at later stages of development.

14.6   What determines the specificity of cytokinetic vesicle trafficking?

Vesicle trafficking in interphase or post-mitotic cells requires sorting and targeting to one of the several possible destinations. For example, anterograde transport from the *trans*-Golgi network can deliver vesicles to the plasma membrane, the endosome or the pre-vacuolar compartment (Sanderfoot & Raikhel, 1999). In order to distinguish between these destinations, the vesicles carrying specific cargo have to be *labelled* appropriately. In yeast, vesicles are actively sorted for trafficking to the pre-vacuolar compartment whereas vesicles destined to the plasma membrane appear to traffic by default. However, the actual situation may be more complex. As shown recently in yeast, some vesicles do not traffic directly to the plasma membrane but first pass through the endosome, and this trafficking route seems to share targeting components with vesicles destined for the vacuole (Harsay & Schekman, 2002). In non-proliferating cells of higher plants, trafficking to the vacuole(s) requires specific sorting signals whereas protein secretion to the apoplast seems to be the default pathway (Sanderfoot & Raikhel, 1999; Bassham & Raikhel, 2000). For example, GFP fused to an N-terminal signal peptide for uptake into the ER lumen is secreted to the apoplast (Batako *et al.*, 2000; Geelen *et al.*, 2002).

In dividing cells of higher plants, vesicle traffic appears to be predominantly routed to the plane of cell division, as the following examples illustrate. The putative auxin efflux carrier PIN1 is located at the basal plasma membrane in non-dividing root cells but accumulates in the cell plate during cytokinesis (Geldner *et al.*, 2001). Similarly, two cell wall-modifying enzymes, the beta-glucanase KORRIGAN and an endoxyloglucan transferase, as well as GFP fused to the N-terminal 24 amino acids of endoxyloglucan transferase are targeted to the apoplast in interphase but accumulate in the cell plate during cytokinesis of tobacco BY-2 cells (Nicol *et al.*, 1998; Zuo *et al.*, 2000; Yokoyama & Nishitani, 2001). Conversely, the cytokinesis-specific syntaxin KNOLLE is targeted to the plasma membrane when misexpressed in non-proliferating cells (Völker *et al.*, 2001). Although the number of case studies is very limited, these observations suggest that trafficking to the cell plate is the predominant pathway during cytokinesis, with cargo selection comparable to trafficking to the plasma membrane in interphase or non-proliferating cells.

The apparent specificity of trafficking during cytokinesis may be due to several factors. The phragmoplast microtubules provide tracks along which a specific plus end-directed kinesin motor delivers vesicles to the plane of division. The kinesin motor would associate with vesicles budding from the TGN during M phase and attach them to the phragmoplast microtubules. If this scenario holds true then proteins and other cargo present in the TGN at the right time will be transported to the plane of cell division. Of course, there would be exceptions. For example, resident Golgi proteins would be excluded because of their retention signals. In addition, proteins with sorting/targeting signals to other post-Golgi destinations, such as PVC or vacuole, might also be excluded. However, there is no experimental evidence for counter-selection of proteins with vacuolar targeting signals during

M phase. Surprisingly, the plasma membrane H^+-ATPase does not accumulate in the cell plate (Lauber *et al.*, 1997). Rather than being actively excluded from cytokinetic vesicles, the plasma membrane H^+-ATPase may simply not be present in the Golgi apparatus during cytokinesis. Similar considerations could apply to other proteins.

In conclusion, cytokinetic trafficking may be a default pathway during M phase, transporting proteins that are present in the TGN to the plane of cell division. If the cell plate is not their proper destination, proteins might subsequently be retrieved by endocytosis, as shown for PIN1 (Geldner *et al.*, 2001). The biological significance of the default mechanism could rather be to ensure that all necessary materials for building the cell plate are actually delivered in sufficient quantities.

14.7   How can cytokinesis be carried out so fast?

Considering the speed with which the cell plate is formed and laterally expanded, it seems questionable that the synthetic capacity of the Golgi stacks is sufficient to produce the necessary membrane material in time. Based on quantitative data from electron tomography, Otegui *et al.* (2001) estimate that approximately 500 000 membrane vesicles are required to form a cell plate with a surface area of $400\,\mu m^2$ during endosperm cellularisation. Assuming that a surface area of about one-third that size is made by vesicle fusion during somatic cytokinesis, each of the approximately 20 Golgi stacks in the vicinity of the forming cell plate (*Golgi belt*, Nebenführ *et al.*, 2000) has to produce about 8000 vesicles in a period of less than 30 min. One way to solve the problem of delivering the necessary membrane material in time would be pre-fabrication at the Golgi stacks before the onset of cytokinesis. This idea is supported by the observation that the cytokinesis-specific syntaxin KNOLLE starts to accumulate in large patches, probably Golgi stacks, from the onset of mitosis, and the signal intensity increases up to the formation of the phragmoplast at late anaphase/early telophase. As soon as the phragmoplast has formed, KNOLLE-positive material stretches across the mid-plane of the phragmoplast, which suggests very rapid delivery of vesicles (Lauber *et al.*, 1997). In addition, treatment of BY-2 cells at metaphase with the fungal toxin brefeldin A (BFA) only blocks cell-plate expansion but not the early stage of cell-plate formation (Yasuhara *et al.*, 1995; Yasuhara & Shibaoka, 2000). BFA specifically inhibits large ARF-GEFs involved in vesicle budding in yeast (Donaldson & Jackson, 2000) and was also shown to inhibit the *Arabidopsis* ARF-GEF GNOM *in vitro* (Steinmann *et al.*, 1999). At what membrane compartments BFA acts in dividing cells remains to be determined (see below).

Another way to cope with the tight schedule of cell-plate formation would be recycling of the excess membrane material that was delivered to form the initial cell plate. Again, there is some evidence supporting this idea. Electron tomography data suggest that approximately 75% of the membrane material initially delivered to the plane of division is subsequently removed during maturation of the cell plate (Otegui *et al.*, 2001). Clathrin-coated buds suggestive of endocytosis have been observed on the convoluted fenestrated sheets before the cell plate shrinks in thickness and with numerous clathrin-coated vesicles near the cell plate at that stage. A possible role in endocytosis from the cell plate via clathrin-coated vesicles has been discussed for a recently identified *Arabidopsis* dynamin homologue that resembles animal dynamin II which is involved in endocytosis from the plasma membrane (unpublished observation cited in Verma, 2001). Further evidence for membrane recycling to the cell plate has been obtained by BFA-treatment of dividing *Arabidopsis* root cells. Treatment with the drug resulted in the formation of *BFA compartments* that accumulated both the cytokinesis-specific syntaxin KNOLLE and the recycling plasma membrane protein PIN1 (Geldner *et al.*, 2001). The BFA effect on KNOLLE localisation was completely reversible as KNOLLE was relocated to the cell plate after removal of BFA. The *BFA compartments* were shown to consist of perinuclear aggregates of large membrane vesicles that were distinct from and adjacent to intact Golgi stacks, suggesting that BFA treatment has trapped recycling KNOLLE protein in a post-Golgi endomembrane compartment.

Based on these observations, we propose the following tentative model of cell-plate formation by two-phase cytokinetic vesicle trafficking (Fig. 14.1). In phase I, cytokinetic vesicles traffic from the TGN to the centre of the division plane where

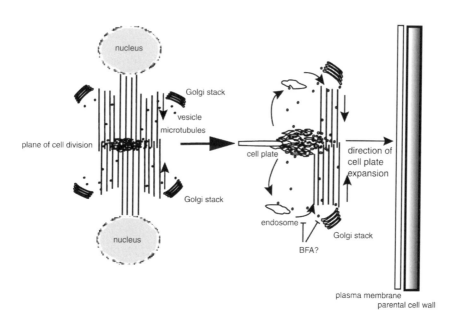

Figure 14.1 Model of vesicle trafficking during cytokinesis – Phase I (left): Golgi-derived vesicles traffic along phragmoplast microtubules to the plane of cell division where they fuse with one another to form the transient membrane compartment of the cell plate. Phase II (right): Recycling vesicles transport excess membrane material via endosomes from the centre to the growing margin of the cell plate. Golgi-derived vesicles also deliver new membrane material to the cell-plate margin. BFA treatment may block vesicle budding from recycling endosomes and Golgi stacks.

they fuse with one another to form the membrane compartment of the initial cell plate. This early phase is very rapid and makes use of pre-fabricated membrane material accumulated in the TGN. As soon as the phragmoplast forms, vesicles can be delivered along the microtubule tracks. The distance from the TGN to the plane of division is so short that an average kinesin motor would deliver its cargo within several seconds (Woehlke & Schliwa, 2000). The rate-limiting step may rather be vesicle budding from the TGN, which probably involves recruitment of clathrin-adaptor coat proteins and SNARE complex components, including KNOLLE and possibly SNAP33 and an unknown v-SNARE, or NPSN11. In addition, a substantial amount of cargo needs to be taken up, especially membrane-bound callose synthase, soluble enzymes for cell wall biosynthesis and modification, as well as pectin and xyloglucan.

Phase II of cytokinesis coincides with the lateral expansion of the cell plate. Vesicles are targeted to its growing margin via phragmoplast microtubules that have been translocated laterally (Staehelin & Hepler, 1996). During this phase, excess membrane material, and possibly cargo as well, is locally recycled: KNOLLE-containing vesicles budding from the maturing cell plate are trafficked to endosomal compartments from which they are re-delivered to the growing margin (Fig. 14.1). At the same time, additional membrane material may be transported from the TGN to the cell-plate margin. Both populations of vesicles would have to carry essentially the same cargo in order to take full advantage of the extended period of pre-fabrication.

Although recycling through endosomes would account for the fast delivery of cell-plate material across the plane of cell division, this model raises a number of questions that cannot be adequately answered at present. For example, is KNOLLE involved in vesicle docking to the endosomes and, if this is the case, what are its SNARE partners? Do the recycling vesicles traffic along cytoskeletal tracks to the endosomes? What determines recycling of KNOLLE to the cell-plate margin as opposed to trafficking to the vacuole for degradation?

14.8   Concluding remarks

In this review, we have summarised recent findings on vesicle trafficking to and fusion at the plane of cell division. We have focused on two striking features of this process, its apparent specificity and high efficiency. The apparent specificity is in part due to the cytokinesis-specific microtubule tracks for vesicle delivery, although no specific plus end-directed motor protein has been identified up to now. However, there is no experimental evidence for cargo sorting during vesicle budding from the Golgi stacks. Thus, cytokinetic vesicle trafficking looks like a cell cycle-regulated default pathway. The other striking feature is the high efficiency of vesicle trafficking during cytokinesis, which is necessitated by the limited period of time for making the cell plate. This high efficiency could be achieved by pre-fabrication of membrane material before the onset of cytokinesis and by local vesicle recycling via endosomes during cell-plate expansion. Our tentative two-phase model may serve as a conceptual framework for further analysis of mechanisms that underlie the plant-specific execution of cytokinesis.

References

Antonin, W., Fasshauer, D., Becker, S., Jahn, R. & Schneider, T.R. (2002) Crystal structure of the endosomal SNARE complex reveals common structural principle of all SNAREs. *Nat. Struct. Biol.*, **9**, 107–111.

Asada, T. & Shibaoka, H. (1994) Isolation of polypeptides with microtubule-translocating activity from phragmoplasts of tobacco BY-2 cells. *J. Cell Sci.*, **107**, 2249–2257.

Asada, T., Kuriyama, R. & Shibaoka, H. (1997) TKRP125, a kinesin-related protein involved in the centrosome-independent organization of the cytokinetic apparatus in tobacco BY-2 cells. *J. Cell Sci.*, **110**, 179–189.

Assaad, F.F., Huet, Y., Mayer, U. & Jürgens, G. (2001) The cytokinesis gene *KEULE* encodes a Sec1 protein that binds the syntaxin KNOLLE. *J. Cell Biol.*, **152**, 531–543.

Barroso, C., Chan, J., Allan, V., Doonan, J., Hussey, P. & Lloyd, C. (2000) Two kinesin-related proteins associated with the cold-stable cytoskeleton of carrot cells: characterization of a novel kinesin, DcKRP120-2. *Plant J.*, **24**, 859–868.

Bassham, D.C. & Raikhel, N.V. (2000) Unique features of the plant vacuolar sorting machinery. *Curr. Opin. Cell Biol.*, **12**, 491–495.

Batako, H., Zheng, H.Q., Hawes, C. & Moore, I. (2000) A Rab1 GTPase is required for transport between the endoplasmic reticulum and Golgi apparatus and for normal Golgi movement in plants. *Plant Cell*, **12**, 2201–2217.

Bögre, L., Calderini, O., Binarova, P., Mattauch, M., Till, S., Kiegerl, S., Jonak, C., Pollaschek, C., Barker, P., Huskisson, N.S., Hirt, H. & Heberle-Bors, E. (1999) A MAP kinase is activated late in plant mitosis and becomes localized to the plane of cell division. *Plant Cell*, **11**, 101–113.

Bowser, J. & Reddy, A.S. (1997) Localization of a kinesin-like calmodulin-binding protein in dividing cells of *Arabidopsis* and tobacco. *Plant J.*, **12**, 1429–1437.

Brown, R.C. & Lemmon, B.E. (2001) The cytoskeleton and spatial control of cytokinesis in the plant life cycle. *Protoplasma*, **215**, 35–49.

Calderini, O., Bögre, L., Vicente, O., Binarova, P., Heberle-Bors, E. & Wilson, C. (1998) A cell cycle regulated MAP kinase with a possible role in cytokinesis in tobacco cells. *J. Cell Sci.*, **111**, 3091–3100.

Chen, Y.A. & Scheller, R.H. (2001) SNARE-mediated membrane fusion. *Nat. Rev. Mol. Cell Biol.*, **2**, 98–106.

Cutler, S.R. & Ehrhardt, D.W. (2002) Polarized cytokinesis in vacuolate cells of *Arabidopsis*. *Proc. Natl. Acad. Sci. USA*, **99**, 2812–2817.

Donaldson, J.G. & Jackson, C.L. (2000) Regulators and effectors of the ARF GTPases. *Curr. Opin. Cell Biol.*, **12**, 475–482.

Dupree, P. & Sherrier, D.J. (1998) The plant Golgi apparatus. *Biochim. Biophys. Acta*, **1404**, 259–270.

Endle, M.C., Stoppin, V., Lambert, A.M. & Schmit, A.C. (1998) The growing cell plate of higher plants is a site of both actin assembly and vinculin-like antigen recruitment. *Eur. J. Cell Biol.*, **77**, 10–18.

Feiler, H.S., Desprez, T., Santoni, V., Kronenberger, J., Caboche, M. & Traas, J. (1995) The higher plant *Arabidopsis thaliana* encodes a functional CDC48 homologue which is highly expressed in dividing and expanding cells. *EMBO J.*, **14**, 5626–5637.

Fukuda, R., McNew, J.A., Weber, T., Parlati, F., Engel, T., Nickel, W., Rothman, J.E. & Söllner, T.H. (2000) Functional architecture of an intracellular membrane t-SNARE. *Nature*, **407**, 198–202.

Geelen, D., Leyman, B., Batoko, H., Di Sansabastiano, G.P., Moore, I. & Blatt, M.R. (2002) The abscisic acid-related SNARE homolog NtSyr1 contributes to secretion and growth: evidence from competition with its cytosolic domain. *Plant Cell*, **14**, 387–406.

Geldner, N., Friml, J., Stierhof, Y.-D., Jürgens, G. & Palme, K. (2001) Auxin transport inhibitors block PIN1 cycling and vesicle trafficking. *Nature*, **413**, 425–428.

Granger, C.L. & Cyr, R.J. (2000) Microtubule reorganization in tobacco BY-2 cells stably expressing GFP-MBD. *Planta*, **210**, 502–509.

Gu, X. & Verma, D.P. (1996) Phragmoplastin, a dynamin-like protein associated with cell plate formation in plants. *EMBO J.*, **15**, 695–704.

Gu, X. & Verma, D.P. (1997) Dynamics of phragmoplastin in living cells during cell plate formation and uncoupling of cell elongation from the plane of cell division. *Plant Cell*, **9**, 157–169.

Harsay, E. & Schekman, R. (2002) A subset of yeast vacuolar protein sorting mutants is blocked in one branch of the exocytic pathway. *J. Cell Biol.*, **156**, 271–285.

Heese, M., Gansel, X., Sticher, L., Wick, P., Grebe, M., Granier, F. & Jürgens, G. (2001) Functional characterization of the KNOLLE-interacting t-SNARE AtSNAP33 and its role in plant cytokinesis. *J. Cell Biol.*, **155**, 239–250.

Heese, M., Mayer, U. & Jürgens, G. (1998) Cytokinesis in flowering plants: cellular process and developmental integration. *Curr. Opin. Plant Biol.*, **1**, 486–491.

Hong, Z., Delauney, A.J. & Verma, D.P. (2001a) A cell plate-specific callose synthase and its interaction with phragmoplastin. *Plant Cell*, **13**, 755–768.

Hong, Z., Zhang, Z., Olson, J.M. & Verma, D.P. (2001b) A novel UDP-glucose transferase is part of the callose synthase complex and interacts with phragmoplastin at the forming cell plate. *Plant Cell*, **13**, 769–779.

Kang, B.H., Busse, J.S., Dickey, C., Rancour, D.M. & Bednarek, S.Y. (2001) The *Arabidopsis* cell plate-associated dynamin-like protein, ADL1Ap, is required for multiple stages of plant growth and development. *Plant Physiol.*, **126**, 47–68.

Kirchhausen, T. (2000) Three ways to make a vesicle. *Nat. Rev. Mol. Cell Biol.*, **1**, 187–198.

Lane, D.R., Wiedemeier, A., Peng, L., Hofte, H., Vernhettes, S., Desprez, T., Hocart, C.H., Birch, R.J., Baskin, T.I., Burn, J.E., Arioli, T., Betzner, A.S. & Williamson, R.E. (2001) Temperature-sensitive alleles of RSW2 link the KORRIGAN endo-1,4-beta-glucanase to cellulose synthesis and cytokinesis in *Arabidopsis*. *Plant Physiol.*, **126**, 278–288.

Lauber, M.H., Waizenegger, I., Steinmann, T., Schwarz, H., Mayer, U., Hwang, I., Lukowitz, W. & Jürgens, G. (1997) The *Arabidopsis* KNOLLE protein is a cytokinesis-specific syntaxin. *J. Cell Biol.*, **139**, 1485–1493.

Lee, Y.R., Giang, H.M. & Liu, B. (2001) A novel plant kinesin-related protein specifically associates with the phragmoplast organelles. *Plant Cell*, **13**, 2427–2439.

Lee, Y.R. & Liu, B. (2000) Identification of a phragmoplast-associated kinesin-related protein in higher plants. *Curr. Biol.*, **10**, 797–800.

Liu, B., Cyr, R.J. & Palevitz, B.A. (1996) A kinesin-like protein, KatAp, in the cells of *Arabidopsis* and other plants. *Plant Cell*, **8**, 119–132.

Lukowitz, W., Mayer, U. & Jürgens, G. (1996) Cytokinesis in the *Arabidopsis* embryo involves the syntaxin-related *KNOLLE* gene product. *Cell*, **84**, 61–71.

Mayer, A. (2001) What drives membrane fusion in eukaryotes? *Trends Biochem. Sci.*, **26**, 717–723.

Mayer, U., Herzog, U., Berger, F., Inzé, D. and Jürgens, G. (1999) Mutations in the *PILZ* group genes disrupt the microtubule cytoskeleton and uncouple cell cycle progression from cell division in *Arabidopsis* embryo and endosperm. *Eur. J. Cell Biol.*, **78**, 100–108.

McClinton, R.S. & Sung, Z.R. (1997) Organization of cortical microtubules at the plasma membrane in *Arabidopsis*. *Planta*, **201**, 252–260.

McNew, J.A., Parlati, F., Fukuda, R., Johnston, R.J., Paz, K., Paumet, F., Söllner, T.H. & Rothman, J.E. (2000) Compartmental specificity of cellular membrane fusion encoded in SNARE proteins. *Nature*, **407**, 153–159.

Misura, K.M., Scheller, R.H. & Weis, W.I. (2000) Three-dimensional structure of the neuronal-Sec1-syntaxin 1a complex. *Nature*, **404**, 355–362.

Molchan, T.M., Valster, A.H. & Hepler, P.K. (2001) Actomyosin promotes cell plate alignment and late lateral expansion in *Tradescantia* stamen hair cells. *Planta*, **214**, 683–693.

Mundel, C., Baltz, R., Eliasson, A., Bronner, R., Grass, N., Krauter, R., Evrard, J.L. & Steinmetz, A. (2000) A LIM-domain protein from sunflower is localized to the cytoplasm and/or nucleus in a wide variety of tissues and is associated with the phragmoplast in dividing cells. *Plant Mol. Biol.*, **42**, 291–302.

Nebenführ, A., Frohlick, J.A. & Staehelin, L.A. (2000) Redistribution of Golgi stacks and other organelles during mitosis and cytokinesis in plant cells. *Plant Physiol.*, **124**, 135–151.

Nicol, F., His, I., Jauneau, A., Vernhettes, S., Canut, H. & Hofte, H. (1998) A plasma membrane-bound putative endo-1,4-beta-D-glucanase is required for normal wall assembly and cell elongation in *Arabidopsis. EMBO J.*, **17**, 5563–5576.

Nishihama, R., Ishikawa, M., Araki, S., Soyano, T., Asada, T. & Machida, Y. (2001) The NPK1 mitogen-activated protein kinase is a regulator of cell-plate formation in plant cytokinesis. *Genes Dev.*, **15**, 352–363.

Nishihama, R., Soyano, T., Ishikawa, M., Araki, S., Tanaka, H., Asada, T., Irie, K., Ito, M., Terada, M., Banno, H., Yamazaki, Y. & Machida, Y. (2002) Expansion of the cell plate in plant cytokinesis requires a kinesin-like protein/MAPKKK complex. *Cell*, **109**, 87–99.

Otegui, M. & Staehelin, L.A. (2000) Cytokinesis in flowering plants: more than one way to divide a cell. *Curr. Opin. Plant Biol.*, **3**, 493–502.

Otegui, M.S., Mastronarde, D.N., Kang, B.H., Bednarek, S.Y. & Staehelin, L.A. (2001) Three-dimensional analysis of syncytial-type cell plates during endosperm cellularization visualized by high resolution electron tomography. *Plant Cell*, **13**, 2033–2051.

Parlati, F., McNew, J.A., Fukuda, R., Miller, R., Söllner, T.H. & Rothman, J.E. (2000) Topological restriction of SNARE-dependent membrane fusion. *Nature*, **407**, 194–198.

Peng, R. & Gallwitz, D. (2002) Sly1 protein bound to Golgi syntaxin Sed5p allows assembly and contributes to specificity of SNARE fusion complexes. *J. Cell Biol.*, **157**, 645–655.

Reddy, A.S.N. & Day, I.S. (2001) Kinesins in the *Arabidopsis* genome: a comparative analysis among eukaryotes. *BMC Genomics*, **2**, 2.

Samuels, A.L., Giddings, T.H. Jr. & Staehelin, L.A. (1995) Cytokinesis in tobacco BY-2 and root tip cells: a new model of cell plate formation in higher plants. *J. Cell Biol.*, **130**, 1345–1357.

Sanderfoot, A.A. & Raikhel, N.V. (1999) The specificity of vesicle trafficking: coat proteins and SNAREs. *Plant Cell*, **11**, 629–641.

Smith, L.G. (2001) Plant cell division: building walls in the right places. *Nat. Rev. Mol. Cell Biol.*, **2**, 33–39.

Springer, S., Spang, A. & Schekman, R. (1999) A primer on vesicle budding. *Cell*, **97**, 145–148.

Staehelin, L.A. & Hepler, P.K. (1996) Cytokinesis in higher plants. *Cell*, **84**, 821–824.

Stals, H., Bauwens, S., Traas, J., Van Montagu, M., Engler, G. & Inzé, D. (1997) Plant CDC2 is not only targeted to the pre-prophase band, but also co-localizes with the spindle, phragmoplast, and chromosomes. *FEBS Lett.*, **418**, 229–234.

Steinborn, K., Maulbetsch, C., Priester, B., Trautmann, S., Pacher, T., Geiges, B., Küttner, F., Lepiniec, L., Stierhof, Y.-D., Schwarz, H., Jürgens, G. & Mayer, U. (2002) The *Arabidopsis PILZ* group genes encode tubulin-folding cofactor orthologs required for cell division but not cell growth. *Genes Dev.*, **16**, 959–971.

Steinmann, T., Geldner, N., Grebe, M., Mangold, S., Jackson, C.L., Paris, S., Gälweiler, L., Palme, K. & Jürgens, G. (1999) Coordinated polar localization of auxin efflux carrier PIN1 by GNOM ARF GEF. *Science*, **286**, 316–318.

Strompen, G., El Kasmi, F., Richter, S., Lukowitz, W., Assaad, F.F., Jürgens, G. & Mayer, U. (2002) The *Arabidopsis HINKEL* gene encodes a kinesin-related protein involved in cytokinesis and is expressed in a cell cycle-dependent manner. *Curr. Biol.*, **12**, 153–158.

Verma, D.P.S. (2001) Cytokinesis and building of the cell plate in plants. *Annu. Rev. Plant Physiol. Plant Mol. Biol.*, **52**, 751–784.

Völker, A., Stierhof, Y.-D. & Jürgens, G. (2001) Cell cycle-independent expression of the *Arabidopsis* cytokinesis-specific syntaxin KNOLLE results in mistargeting to the plasma membrane and is not sufficient for cytokinesis. *J. Cell Sci.*, **114**, 3001–3012.

Waizenegger, I., Lukowitz, W., Assaad, F., Schwarz, H., Jürgens, G. & Mayer, U. (2000) The *Arabidopsis KNOLLE* and *KEULE* genes interact to promote vesicle fusion during cytokinesis. *Curr. Biol.*, **10**, 1371–1374.

Wickner, W. & Haas, A. (2000) Yeast homotypic vacuole fusion: a window on organelle trafficking mechanisms. *Annu. Rev. Biochem.*, **69**, 247–275.

Woehlke, G. & Schliwa, M. (2000) Walking on two heads: the many talents of kinesin. *Nat. Rev. Mol. Cell Biol.*, **1**, 50–58.

Yamaguchi, T., Dulubova, I., Min, S.W., Chen, X., Rizo, J. & Südhof, T.C. (2002) Sly1 binds to Golgi and ER syntaxins via a conserved N-terminal peptide motif. *Dev. Cell*, **2**, 295–305.

Yang, B., Steegmaier, M., Gonzalez, L.C. & Scheller, R.H. (2000) nSec1 binds a closed conformation of syntaxin 1A. *J. Cell Biol.*, **148**, 247–252.

Yasuhara, H. & Shibaoka, H. (2000) Inhibition of cell-plate formation by brefeldin A inhibited the depolymerization of microtubules in the central region of the phragmoplast. *Plant Cell Physiol.*, **41**, 300–310.

Yasuhara, H., Sonobe, S. & Shibaoka, H. (1995) Effects of brefeldin A on the formation of the cell plate in tobacco BY-2 cells. *Eur. J. Cell Biol.*, **66**, 274–281.

Yokoyama, R. & Nishitani, K. (2001) Endoxyloglucan transferase is localized both in the cell plate and in the secretory pathway destined for the apoplast in tobacco cells. *Plant Cell Physiol.*, **42**, 292–300.

Zerial, M. & McBride, H. (2001) Rab proteins as membrane organizers. *Nat. Rev. Mol. Cell Biol.*, **2**, 107–119.

Zhang, Z., Hong, Z. & Verma, D.P.S. (2000) Phragmoplastin polymerizes into spiral coiled structures via intermolecular interaction of two self-assembly domains. *J. Biol. Chem.*, **275**, 8779–8784.

Zheng, H., Bednarek, S.Y., Sanderfoot, A.A., Alonso, J., Ecker, J.R. & Raikhel, N.V. (2002) NPSN11 is a cell plate-associated SNARE protein that interacts with the syntaxin KNOLLE. *Plant Physiol.*, **129**.

Zuo, J., Niu, Q.W., Nishizawa, N., Wu, Y., Kost, B. & Chua, N.-H. (2000) KORRIGAN, an *Arabidopsis* endo-1,4-beta-glucanase, localizes to the cell plate by polarized targeting and is essential for cytokinesis. *Plant Cell*, **12**, 1137–1152.

Index

DATE DUE

JAN 1 2 2015	